Boiler Plant and Distribution System Optimization Manual

Third Edition

Boiler Plant and Distribution System Optimization Manual

Third Edition

By Harry Taplin, P.E.

River Publishers

Routledge
Taylor & Francis Group
LONDON AND NEW YORK

Published 2020 by River Publishers
River Publishers
Alsbjergvej 10, 9260 Gistrup, Denmark
www.riverpublishers.com

Distributed exclusively by Routledge
4 Park Square, Milton Park, Abingdon, Oxon OX14 4RN
605 Third Avenue, New York, NY 10017, USA

First issued in paperback 2023

Library of Congress Cataloging-in-Publication Data

Taplin, Harry, 1935-
    Boiler plant and distribution system optimization manual / by Harry Taplin, P.E. -- Third edition.
    pages cm
    ISBN 0-88173-740-2 (alk. paper) -- ISBN 978-8-7702-2320-1 (electronic) -- ISBN 978-1-4822-6078-6 (taylor & francis distribution : alk. paper)
1. Steam-boilers--Efficiency. 2. Steam power plants--Design and construction. I. Title.
TJ288.T36 2014
621.1'83--dc23

                                                                2014013184

Boiler Plant and Distribution System Optimization Manual / by Harry Taplin
First published by Fairmont Press in 2014.

Routledge is an imprint of the Taylor & Francis Group, an informa business

Publisher's Note
The publisher has gone to great lengths to ensure the quality of this reprint but points out that some imperfections in the original copies may be apparent.

ISBN 13: 978-0-88173-740-0 (The Fairmont Press, Inc.)
ISBN 13: 978-87-7022-927-2 (pbk)
ISBN 13: 978-1-4822-6078-6 (hbk)
ISBN 13: 978-8-7702-2320-1 (online)
ISBN 13: 978-1-0031-5189-0 (ebook master)

While every effort is made to provide dependable information, the publisher, authors, and editors cannot be held responsible for any errors or omissions.

The views expressed herein do not necessarily reflect those of the publisher.

Table of Contents

Acknowledgments

This book evolved from the Association of Energy Engineer's Boiler Optimization course and similar courses which I have been presenting for more than 25 years. It was compiled over the years in response to the need to manage energy as efficiently as possible and to reduce pollution from combustion sources. This planet's delicate balance is slowly being affected by new products being released into the atmosphere by combustion. Some large plants put out more than 32 different pollutants. Also, every ton of fuel burned each day requires about fifteen tons of air to support combustion so each one of us is competing with combustion processes for the very air we breathe. If energy can be managed more efficiently then there will be less pollution, more profit for companies and fewer undesirable side effects in general.

The sources of information for this book were many and varied: The Association of Energy Engineers, The American Society of Mechanical Engineers, The American Boiler Manufacturers Association, The U.S. Department of Energy and research institutions such as the Navy Civil Engineering Laboratory and Brookhaven National Laboratory all contributed to the information in this book. Many boiler, instrument, control system, and burner manufacturers also contributed in one way or another.

I would like to thank Mr. L. Ron Hubbard for his valuable discoveries and information about education and communication which were vely useful in organizing, understanding and communicating this vast and complex subject.

There are many talented, resourceful and conscientious people involved in the boiler and related industries who have done a great deal to advance our civilization to its modern level. Many of them have attended my courses and shared their knowledge and discoveries with me. I would like to thank each of them for their insights which has contributed in a major way to this book.

Harry R. Taplin, Jr. P.E., CEM

Introduction

This book has been designed to make the job of boiler plant optimization at your facility simple and understandable. This book was written for the plant manager, boiler design engineer, energy engineer, plant operators, troubleshooters and anyone else interested in improving the efficiency of combustion processes.

One of the most productive ways to improve efficiency and profits is to expertly manage the energy costs of utilities, especially in the cost of boiler fuel. Managing an efficient plant involves a different viewpoint from normal day to day plant operations. Unfortunately, operational challenges seem more urgent than managing energy efficiently. One old timer put it well when he said, "improving plant efficiency is not a necessity but keeping it running is a necessity."

Most plant personnel focus on the important job of running a dependable plant. They might need help seeing the challenge from the viewpoint of saving energy and reducing fuel expenses. That's what this book is about.

Why is this so important? The answer lies in the fact that your company has to earn the money it pays for fuel from the sale of some product or service. There are large expenses involved in production of these products or services and it makes no sense to waste these efforts in the boiler plant with huge hidden costs from wasted energy.

Let's examine the situation from another viewpoint. Suppose the net profit of your operation is 5% and that when you have followed some of the suggestions in this book, you discover you can save your company more than $250,000 a year in reduced fuel costs. How much is this really worth? Well, it's about five million dollars. There are a lot of people who would be handsomely rewarded if they could boost annual sales by five million dollars. When you produce "cost avoidance" by increasing the productivity and efficiency of your plant, you are doing the same thing when the bottom line is examined. Avoided costs means profits and profits keep your company in business, assuring you a job into the future. It just may be that your management of boilers and energy systems may have more to do with the success of your company than you have been lead to believe.

How do you get this job done? It seems like it could be a highly technical matter which demands a tremendous amount of time, knowledge and experience with boiler plants and distribution systems. Actually, if this challenge is approached properly, it does not need to be complex or difficult. Read on and discover for yourself.

Harry R. Taplin, Jr., P.E., CEM

Chapter 1

Boiler Plant Orientation
[Fast Track]

MODERN BOILERS

Modern boiler designs go back well over 100 years. It has been a century of changes in design, construction and operation of boiler plants. Pressure vessel design has advanced from riveted to welded construction. Natural draft firing using tall brick stacks has given way to forced draft and balanced draft operations. Controls have taken over most of the tedious watchstander functions and now computerized systems can almost think for plant personnel.

GENERATING STEAM

Boilers must deliver dry clean steam at a specified pressure and temperature. Figure 1.1 shows the basic steam generation cycle. The British thermal unit (Btu) is a term used to identify the heat value of fuel and steam. The scale begins at the freezing point for water or 32°F. *Adding one Btu to a pound of water will raise its temperature 1°F.* It takes 180 Btus to heat water from the freezing point to the boiling point, but we don't have steam yet. At atmospheric pressure it takes an additional 970 Btus to form one pound of steam. This steam is the energy that goes to work for us. When the pound of steam, with its 970 Btus, gives up its heat in the steam distribution system and condenses back into water, its volume shrinks tremendously to 1/700 its steam volume. High pressure replacement steam rushes in to fill the void caused by the shrinkage in volume caused by the steam to water phase change. Because this process is continuous and the replacement steam moves in very quickly, pressure changes are seldom detected.

When water is heated to form steam we are restricted to certain temperatures for the steam. This pressure-temperature relationship is called the saturation temperature. Simply stated, all of the heat transferred to the water just generates more steam. If we want to raise the temperature of the steam above the saturation temperature we take the water out of the process and apply heat to steam only. The steam now becomes superheated as shown on the right-hand side of Figure 1.1.

When steam is generated in a pressure vessel, the same process occurs but conditions change. Figure 1.2 illustrates when steam is generated at 3,000 psi (200 Bar). The saturation [boiling] temperature is now 695°F. It takes a lot more Btus to heat the water to the boiling point, but it takes fewer Btus to form a pound of steam. At atmospheric pressure it takes 970 Btus and at 3,000 psi it takes only 213 Btus. The difference is the 802 Btus needed to bring the water up to the boiling point at 695°F for this high pressure.

Table 1.1 illustrates the relationship of pressure and energy values. It shows the information available from standard *steam tables*, a source of essential information about the properties of steam. Engineers use steam table values for the design and operation of steam plants.

TYPES OF BOILERS

Package boilers are used for most commercial and smaller industrial operations. They are built at a factory and then shipped to the plant via truck, barge and railroad. They are usually complete units which need only be connected up

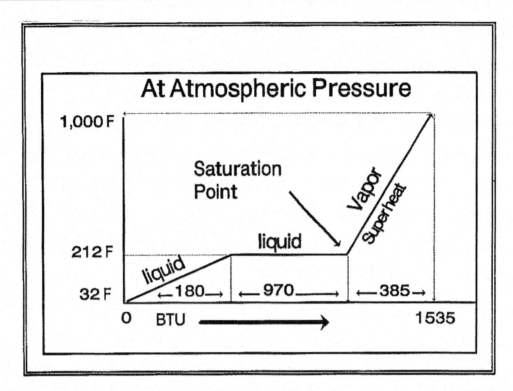

Figure 1.1—This graphic shows the energy [Btu] value of water at atmospheric pressure as it is heated to superheated steam at 1,000°F from the freeze point 32°F.

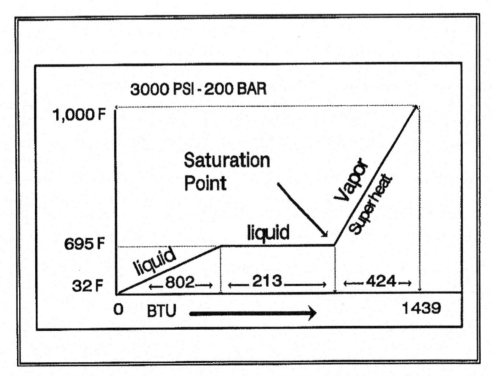

Figure 1.2—This graphic shows the energy [Btu] value of water at 3,000 psig [200 bar] as it is heated to superheated steam at 1,000°F from the freeze point 32°F.

Table 1.1—This is an example of energy values [Btu/lb] of steam in various phases as it is heated to superheated steam at 1,000°F from the freeze point. Note that as pressure goes up the Btu content of the water rises significantly and the Btus necessary to form a pound of steam decrease significantly.

	Heat In Liquid	Heat of Vaporization	Heat @ 1000 F	Total Heat @ 1000 F
Atmospheric Pressure	180	970	385	1535
1500 psi 100 Bar	612	556	322	1490
3000 psi 200 Bar	802	213	424	1439

BTU/lb   VAPOR/4.CH3

to electrical power, fuel piping, steam piping and exhaust systems.

Field erected boilers, because of their large size, have to be built on-site using components either shipped by the manufacturer or fabricated locally.

FIRETUBE BOILERS

An old and very useful boiler type is the fire-tube boiler as shown in Figure 1.3. It consists of a large diameter cylinder filled with water. Tubes extending from end to end serve as the combustion chamber and path for hot gasses inside the pressure vessel. It gets its name, fire tube, from this fact. The firetube boiler design is very useful because of its large amount of stored energy in the large cylindrical design. When pressure drops, the stored energy converts to steam, meeting system demands. This characteristic makes burner and combustion control system design simpler and less sophisticated. Unfortunately, because of the large drum diameter typical of fire tube boilers, it has pressure and horsepower limitations.

A long-standing practical limit on their size has been about 250 psig and 1,000 HP.

Figure 1.4 illustrates a very important fact of boiler pressure vessel design. The force the pressure parts must be designed to withstand is determined by the product of pressure and area. The large diameter drum of the fire tube boiler produces large forces requiring a thick steel shell to retain the pressure limiting their size.

Higher pressure construction requires the use of the water tube design where smaller diameter tubes and drums are used. Because they have smaller cross sectional areas which can withstand much higher pressures.

WATER TUBE BOILERS

By using smaller pressure parts, the water tube design can withstand much higher pressures than the firetube boiler. The water tube boiler design contains much less water than the firetube design and need much more responsive control systems to safely control water level and firing conditions.

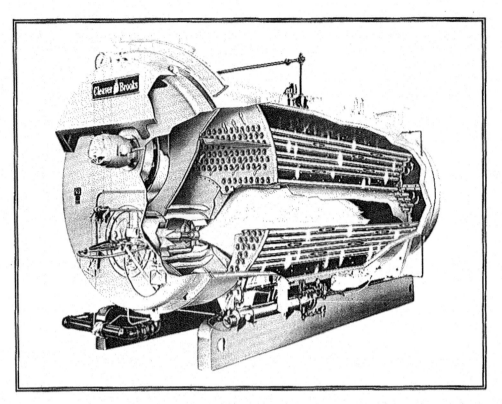

Figure 1.3—Firetube boiler (Courtesy Cleaver Brooks Co.)

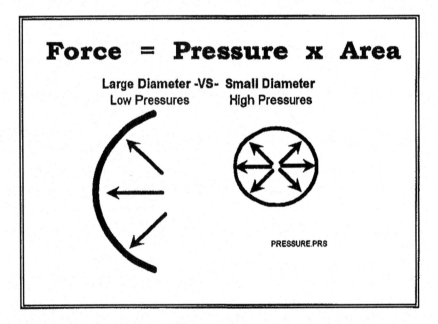

Figure 1.4—This graphic illustrates the difference in pressure vessel construction for fire tube and water tube boilers. The large diameter drum on the left shows how large forces are developed in the large diameter fire tube boiler drum. The advantage of the water tube design, shown on the right, is smaller forces being developed by smaller diameter pressure parts. Fire tube boilers are generally limited to 250 psig and under 1,000 horsepower.

Because the smaller boiler tubes in the water tube design can be bent rather easily, numerous boiler designs have evolved. These designs have been named according to the letter of the alphabet they resemble. For example there is the "A" type [Figure 1.5] with one larger upper steam drum and two lower drums or headers resembling the letter A. The "D" [Figure 1.6] and "0" [Figure 1.7] types are other examples this convention for classifying water tube designs.

Early boiler designs utilized a large percentage of brickwork in the combustion zone to aid stable combustion by reflecting energy back into the wood, coal and oil fires from glowing brickwork. Modern boiler designs have very little brickwork by comparison, usually only around the burner throat in wall fired utility and large industrial units.

The nomenclature for water tube boiler design is quite simple. Like a house they have *floor* tubes, *wall* tubes (front, side and back) and *roof* tubes.

Figure 1.8 illustrates factors concerning boiling in a typical boiler tube. At low heat flux levels, small bubbles are formed referred to as nucleate boiling. As the heat flux Increases more of the tube is occupied by steam bubbles until film boiling be-

gins. As this occurs there is a condition where tube metal temperatures begin to peak, approaching their highest operating temperature. Although designers and manufacturers have compensated for this phenomena in their designs, it is useful to understand this condition with regard to boiler tube

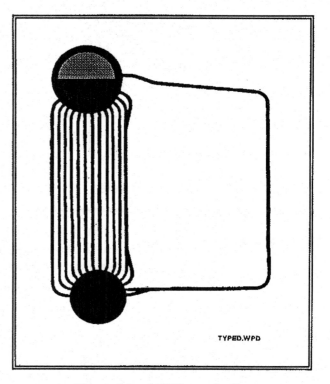

Figure 1.6—A "D" type boiler.

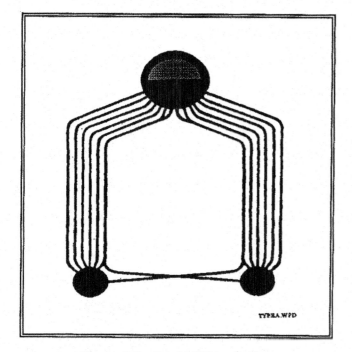

Figure 1.5—An "A" Type boiler.

Figure 1.7—An "O" type boiler.

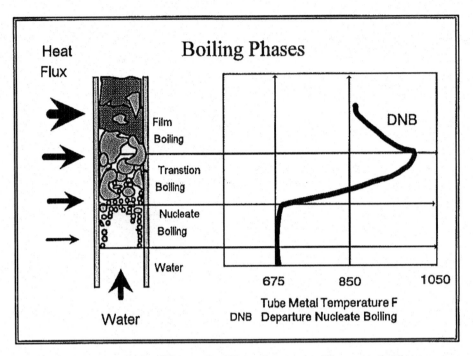

Figure 1.8—On the left, different boiling phases are illustrated. The arrows on the left represent increasing levels of heat flux. The chart on the right shows how tube metal temperature rises as film boiling begins and then falls back from its peak temperature as heat exchange improves with increased velocity due to steam expansion. This peak temperature represents a danger zone for tube failure due to high tube metal temperatures.

failures. This requires an examination of superheated steam as it is not a problem with saturated steam, the product of boiling.

SUPERHEATING OF STEAM

Dry steam is piped from the boiler drum to the superheaters, where its temperature is increased to improve efficiency. Figure 1.9 shows the basic superheater design. For example, temperature of the steam leaving the drum varies between approximately 488 °F [253°C] for a boiler operating pressure of 600 psig to 670°F [355°C] tor a boiler operating pressure of 2500 psig. In a modern boiler, the final temperature of the superheated steam leaving the boiler is usually 1000°F [538°C].

The superheater adds several hundred degrees to the steam temperature. This superheating increases the energy of the steam and allows it to be used more efficiently in the turbine systems.

The higher the temperature of steam, the more efficient the generating unit so it is desirable to maintain high superheat temperatures.

The flow path for the production of superheated steam is from the top of the boiler drum to the entrance of the superheaters. There are two basic types of superheaters, one with tubes facing the flame zone called the "radiant" type and the "convection" type. In the convection type, the superheater is located in the back "pass" of the boiler where the combustion gases are several hundred degrees cooler than they are in the radiant furnace section. The gases passing over the outside of these superheater tubes cause an increase in temperature superheating of the steam inside the tubes.

SUPERHEATER TEMPERATURE CONTROL

Figure 1.10 shows that when tube temperatures go over 1,000 of [538°C], allowable stress for superheater tubes falls off rapidly in various

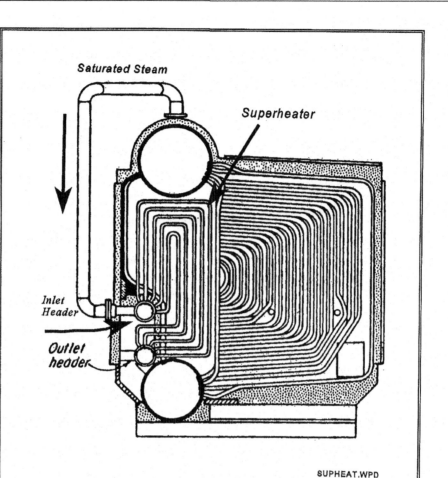

Saturated Steam

Superheater

Inlet
Header

Outlet
header

SUPHEAT.WPD

Figure 1.9 Applying more heat to the mixture of steam and water in the boiler will only produce more saturated steam which is steam at the temperature at which evaporation occurs at a particular pressure. To superheat steam, saturated steam from the boiler is piped into the superheater where its temperature can be raised above its evaporation temperature.

metals used for high temperature high pressure superheater tube design. This imposes a general safety limit on steam temperatures near 1,000°F [538°C]. Designers and operators have the challenge of maintaining safe and stable superheat temperatures below this limit.

Figure 1.11 shows how the boiler designer uses the steam temperature characteristics of convection and radiant superheaters to achieve a constant final steam temperature across the load range of the boiler. The radiant type superheaters are exposed to radiant energy from combustion flames which transfer an intense level of energy compared to convection superheaters that receive their energy from hot combustion gasses. As the firing rate increases a larger percentage of heat is transferred to the convection section compared to the heat absorbed by radiant section. The steam temperature leaving a radiant superheater decreases as the boiler load increases. In a convection superheater the steam temperature increases as boiler load increases. Thus, the two temperature characteristics tend to cancel each other causing a fairly constant average final outlet steam temperature.

The major reason for using several types or designs of superheaters in one boiler is for control of final superheat or reheat steam temperature.

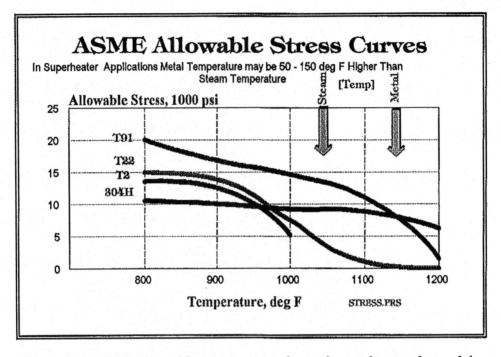

Figure 1.10 ASME Allowable stress curves for various tube metals used for superheater tubes. Note how allowable stress falls off over 1100°F.

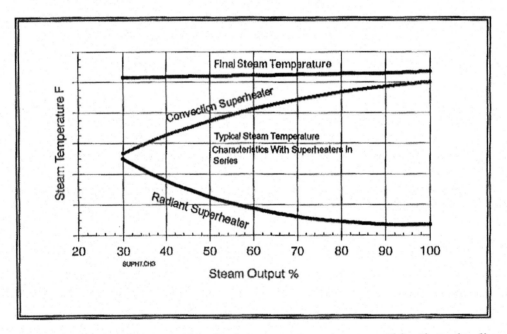

Figure 1.11—To achieve a stable final steam temperature a combination of radiant and convection superheater designs are used.

As the steam leaves the primary superheater it passes through an attemperator, into which relatively cold feedwater is sprayed into the steam to regulate final superheated steam temperature. Figure 1.12 illustrates an attemperator design.

The function of the spray attemperator is to control superheated steam. temperature. The attemperator is installed in the steam header that connects the primary superheater to the secondary superheater. It consists of a thermal sleeve inside the steam header designed to absorb thermal shock and an atomizing water spray nozzle ex-

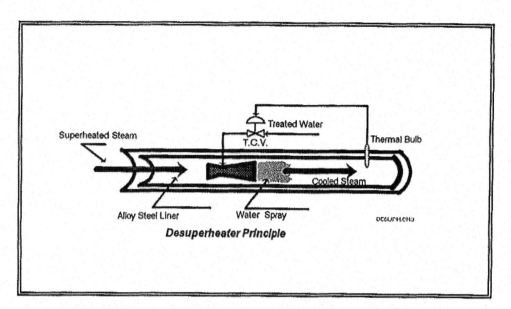

Figure 1.12—Many operating variables affect steam temperature such as load, excess air, feedwater temperature, heating surface cleanliness, fuel and burner settings etc. Attemperator control is necessary to regulate steam temperature within required limits to correct fluctuations caused by these operating variables. This is an attemperator which sprays high purity water into steam to control its temperature. It can be located at the superheater outlet or between superheater sections. There is the possibility that water could carry over into the turbine if control problems develop with the attemperator, however.

tending into the center of the header.

Attemperator water, often supplied by the boiler feed pumps, is sprayed into the hot steam by a nozzle. Since the water is evaporated by the steam, the water must be free of solids if steam purity is to be maintained. The rate of water injection is regulated by an automatic control system to maintain the final steam temperature.

Water enters the attemperator and passes through a spray nozzle located at the entrance to a restricted venturi section. The steam passing the nozzle picks up the water spray, and flows through the venturi. This action evaporates the water and thoroughly mixes it with the steam.

The temperature control system measures the final superheated steam temperature going to the turbine system, and adjusts the attemperator spray flow in order to keep this temperature constant. Suppose that the final temperature is to be kept at 1,000F but that it gradually increases above this value. When the control senses this increase, it causes the attemperator spray control

valve to open further, injecting more cool water into the steam causing a reduction in the final steam temperature leaving the boiler. Similarly, if the final temperature falls below the design value, the control will reduce the attemperator spray flow to allow the final temperature to increase.

From the attemperator, the steam flows to the inlet of the secondary superheater which, like the primary superheater, is a bundle of tubes located in a higher gas temperature zone of the furnace. The primary superheater is located in a zone of lower temperature than the secondary superheater. As in the primary superheater, the steam flowing through the inside of the tubes is heated by the hot combustion gases flowing around the outside of the tubes leaving the secondary superheater at the design final temperature. At this point, the steam in the individual tubes is piped to large manifolds or headers, which tie into one or more large diameter pipes which then run to the turbine system.

NATURAL CIRCULATION IN BOILER TUBES

There is a very large density difference between water and steam. This difference is used to create natural circulation. As the boiler water is heated forming steam in the generating tubes the average density in the generating tubes becomes less than the water filled downcomer tubes which are maintained at a cooler temperature. In some cases the downcomer tubes are much larger than the generating tubes and are installed outside of the hotter steam generation zone. Figure 1.13 illustrates how natural circulation works. However, above about 3206 psig, the critical pressure of steam, the density of water and steam are the same, supercritical boilers operate above this pressure and need special pumps to insure proper boiler circulation.

THE STEAM DRUM

The boiler *steam drum* as shown in Figure 1.14 has five basic functions:

1. Receive feedwater from the economizer and distribute it to the downcomers.

2. Receive the mixture. of steam bubbles and water from the riser tubes.

3. Separate the steam from the water and dry it.

4. Provide a connection for the dried steam to the superheater.

5. Provide a method for distribution of chemicals for treatment of the boiler feedwater (chemical feed)

6. Removal of undesirable materials (blow-down) from the feedwater.

The *feedwater inlet line*, which comes from the boiler economizer, is connected to a distribution manifold that extends almost the entire length of the drum through which the incoming feedwater discharges into the water space of the drum. Here it mixes with the steam and water coming from the riser tubes. The water in the drum also feeds the downcomers to the lower water wall headers and mud drum.

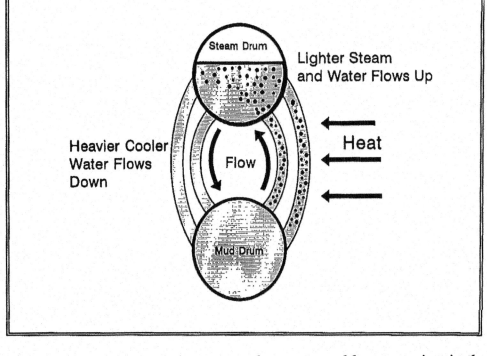

Figure 1.13—Natural circulation occurs when steam and hot water rises in the generating tubes and is replaced with heavier cold water, without steam bubbles, in the downcomer tubes.

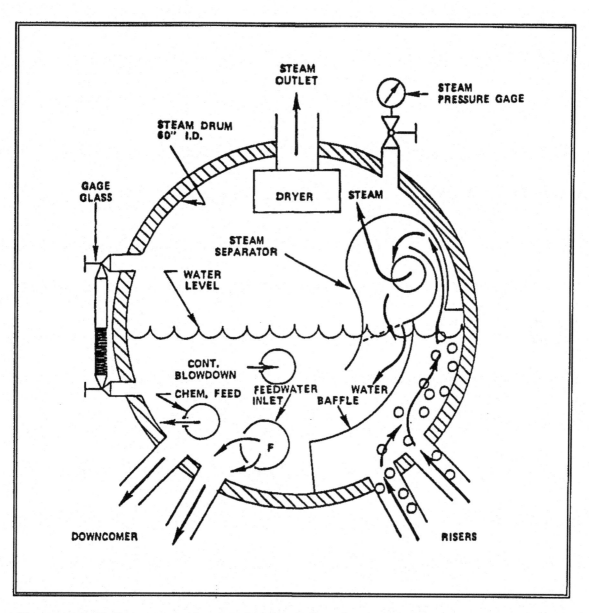

Figure 1.14—Boiler steam and water drum is designed to perform a number of functions. It must insure the outlet steam is dry and free from carryover. It also must provide for chemical injection, blowdown and for feedwater inlet piping. Maintaining the water level in the drum within a safe range is critical to safe boiler operation.

A *continuous blowdown pipe* is also shown. This pipe runs the length of the drum. It contains a series of holes through which contaminated boiler water flows to the blowdown system. By regulating the blowdown rate the concentration of dissolved materials can be controlled.

To ensure that there is at least one reliable indication of water level available to the operator, drums are normally equipped with three (3) water level indicating devices.

HIGH WATER LEVEL CONCERNS

Maximum drum levels should not be exceeded. Higher than permitted drum levels may cause carry over of water droplets with the steam. Water carried over in the steam may reduce final steam temperature and be harmful to downstream equipment (turbines, heaters, etc.). Any slugs of water put in motion by the steam can cause water-hammer and rupture the piping.

For boilers with superheaters, this carryover of water droplets presents an additional problem. Each water droplet contains solids at the same concentration as the boiler water in the steam drum. When the steam passes through the superheater, these water droplets will evaporate, leaving the solids to form deposits' on the inside of the superheater tubes. These solids reduce the transfer of heat from the combustion gases through the tube wall and into the steam. This layer of insulation between the hot tube wall and the cooler steam. With this layer of *"insulation,"* can result in overheating and tube failures.

Solids from carry-over may possibly pass completely through the superheater arid into downstream equipment. Solids entering a steam turbine can damage the turbine blades. Turbine repairs are not only expensive, but can result in long outages.

High drum level conditions can usually be reduced by decreasing the firing rate. Steam bubbles below the water level will tend to reduce causing the water level to fall.

CARRYOVER

There are two types of carryover; mechanical and vaporous.

Mechanical carryover includes:

CARRYOVER

Pure steam becomes contaminated with boiler water

- Spray & mist
- Foaming
- Priming
- Poor steam separation

The water evaporated to produce steam should not contain any contaminating materials, however there may be water droplets carried into steam due to mechanical carryover of boiler water. Conditions in the drum provides a path for introducing solid material into the steam and is referred to as mechanical carryover which may be classified under four headings above. This can result in troublesome deposits in the superheater or on turbine blades.

MIST CARRYOVER

A fine mist is developed as water boils. A bubble of water vapor (steam) reaches the water surface and bursts, leaving a dent in the water. Liquid collapses in on the dent, with the center rising at a faster rate than the edges. This results in a small droplets being tossed free of the boiler water surface. These droplets form a fine mist. This mist is removed to a great extent in various types of separators in the steam drum. However, any mist that remains entrained in the steam will have the same level of contamination as the boiler water.

FOAMING CARRYOVER

The alkalinity, TDS and suspended solids, can interact to create a foam in the boiler. A heavy foam layer is another source of liquid carryover into the steam. The level of foaming can normally be controlled to a reasonable level by maintaining the total alkalinity at a level less than 20 percent of the TDS and maintaining the total suspended

---

FOAMING

Surface layer obstructs steam

Caused by

- Alkalinity
- Organic material
- Oil
- Grease
- Suspended Solids

TDS TOO HIGH

- Corrosive to boiler metal
- Causes foaming and carryover
- Alters boiling patterns in tubes leading to deposits.

THE CAUSES OF PRIMING
Water Surging

- Sudden fluctuations in steam pressure [demand]
- Water level too high
- Operating above boiler rating
- Damaged steam separating equipment

---

solids at a level less than 8 percent of the TDS. Antifoam agents are added to boiler water to help control foaming.

PRIMING CARRYOVER

Priming carryover is caused by liquid surges in the steam drum that throw water into the steam space where it is carried into the steam header. Priming is caused by a mechanical problems or mechanical properties such as oversensitive feed-

water controls or incorrect blowdown procedures. Figure 1.15 illustrates how priming occurs.

START-UP

As steam pressure is increased, the water level should be carefully controlled within normal limits. Prior to picking up load, it is desirable to keep the water level near the lowest safe level to allow for thermal expansion of the water as the steam generation rate increases. When raising

Figure 1.15—Violent ebullition causes changes in drum water level and unacceptable discharge of water in the steam. The upper sketches show two conditions registering normal water level. In the right drum the true water level is higher because it is being forced up by violent steam action of steam in the water causing an undetected water level offset. In the left sketch a low steam generation rate does not offset the water level from the observed level. In the lower sketch, longitudinal mounding near the steam nozzle is caused by violent steam formation aided by low steam pressure in the steam system piping. Also, dynamic forces may set up a sloshing action causing localized high water and wave like motion contributing to priming.

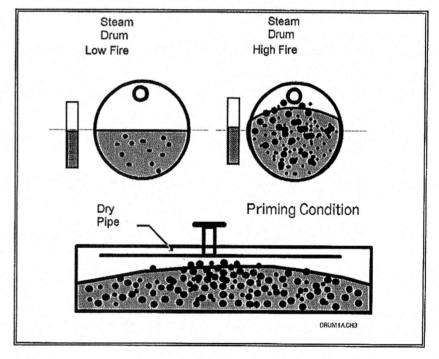

the steam pressure on a boiler not connected to a header system, the steam line should be warmed up along with the boiler by the operation of drain valves to remove condensation and create the desired flow of warming steam.

In bringing a boiler on the line with other boilers on a header system, certain precautions are necessary to avoid water hammer and excessive temperature gradients in the piping. Adequate drainage and warming of the piping will eliminate the risk of water hammer. The judicious use of bypass valves around main header valves will avoid steep temperature gradients. Header drains should be operated. The steam line from the boiler to the· header should be brought up to temperature by operating bypass and drain valves. When up to temperature and line pressure, the header valve may be opened wide and the bypass closed. The stem for the non-return valve should be back off to a position corresponding to about 25% open until the boiler begins to supply steam to the header, after which it can be

set at the wide open position. In the absence of a non-return valve, the boiler stop valve should be opened slowly when the pressure in the boiler and header are approximately equal.

WARM-UP

During the period when steam pressure is initially being raised, the boiler, especially the boiler drum which is fabricated from thick metal, can experience some unusual and possibly damaging stresses. To avoid thermal stress, temperature changes should be limited. Figure 1.16 is a typical chart governing pressure raising in drum boilers. These temperature limitations are very critical during the startup and shutdown of the boiler. They are necessary to minimize thermal stress in the steam drum. Temperature changes can be controlled by controlling the firing rate on startup and controlling the cooling down of the boiler when shutting down. Your boiler may be

Figure 1.16—Thermal stress can cause serious damage to a boiler. Most manufacturers provide a curve similar to this which limits the rate at which pressure can be raised. Water temperature differences can also cause thermal stresses so it has been added to this graphic as a reminder of its possible influence. Usually a standard of I 00 F or 38 C per hour is used as a general limit for bringing a cold boiler up to operating pressure.

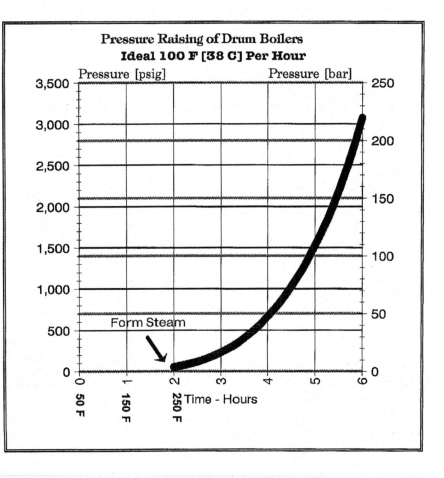

different so the manufacturers technical manual should be consulted.

FLAME SAFETY CONTROL

A furnace explosion is the ignition and almost instantaneous combustion of highly flammable gasses in the boiler. It can cause great explosive force on the boiler, minor explosions, called puffs or blowbacks can blow flames from doors and covers. Anyone in the flame path· can be seriously burned. Major explosions can shatter observation ports, blow off doors, and demolish boilers and buildings.

> **THE MAIN CAUSES OF EXPLOSIONS ARE:**
>
> 1. Flame failure with no resulting fuel isolation
> 2. Faulty regulating controls
> 3. Failure to purge a boiler with at least 4 air volume changes after shutdown or before each attempted light-off.
> 4. Fuel shut-off valve leakage
> 5. Loss of furnace draft

> **AN OVER PRESSURE CONDITION CAN OCCUR IN ONE OF TWO WAYS:**
>
> A gradual increase in pressure due to improper operation or malfunctioning equipment.
>
> A sudden and rapid increase in pressure caused by an immediate loss of steam flow.

The flame safety control is designed to prevent any condition which could result in a boiler explosion. Its basic function is to prevent the accumulation of fuel in the furnace. The fuel oil safety shut off valves will automatically shut when flame can not be detected in the furnace.

BOILER SAFETY VALVES

Safety valves are devices that protect the steam and water circuits against accidental over pressurization.

Safety valves are required on every pressure vessel. The boiler and pressure code also requires that the safety valves have a total steam relieving capacity at least equal to the rated full load steam flow of the boiler.

A safety valve setting table is supplied in the instruction manual for each unit showing the prescribed valves and settings.

There are a variety of safety valves installed on systems from over pressurization.

A. There are spring type safety valves supplied with the boiler main steam and reheat steam piping.

B. There are drum safety valves connected directly to the steam drum.

C. Superheat safety valves are connected directly to the main steam: line to the high pressure turbine.

D. Safety valves are connected directly to the reheat inlet piping.

E. Safety valves are connected directly to the reheat outlet piping.

There are electrically or pilot operated pressure relief valves located on the superheater outlet lead downstream from the spring type safety valves. These relief valves are not included in the total relieving capacity of the safety valves because they can be isolated from the boiler. They are set to open at a pressure below the popping pressure of the lowest set superheater safety valve. This is to prevent the safety valves from opening except on major over pressures.

RATING BOILER CAPACITY

Sometimes one can become confused by the different ways boiler ratings are described. For example firetube boilers are rated by horsepower, commercial boilers are rated in Btus per hour and industrial boilers are rated in thousands of pounds of steam per hour. This complicated system has a historical background. Originally firetube boilers were used in applications where horsepower was a term commonly where fire tube boilers were first used in mining· operations and ship propulsion. In buildings; heating and air conditioning requirements are calculated in Btus and this is commonly used for their boiler specifications. Many industries use pounds or tons of steam per hour as a way of expressing energy requirements. Figures 1.17 and 1.18 provide additional information.

TIME—TEMPERATURE—TURBULENCE

The type of fuel burned determines the size of the combustion chamber and ultimately the overall size of the boiler. If gas or lighter grades of distilled fuel oil are used, the burn out is faster than for heavier residual grade oils and coal. The key is the carbon content of the fuel. The higher the carbon content, especially fixed carbon an opposed to carbon gasses, the longer the fuel burnout takes requiring larger and larger furnaces. The key to good combustion relies on the three Ts: *Time, Temperature and Turbulence*. See Figure 1.19.

TURNDOWN RATIO

A measure of the quality of a burner and control system lies in its capability to modulate through a firing range, this is known as turndown.

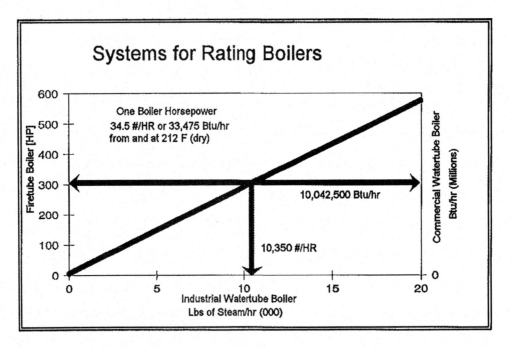

Figure 1.17—The systems for rating boilers has developed with usage over the years. Firetube boilers were first used to power ships and heavy duty equipment where a horsepower convention was in use. Industrial facilities adopted the use of thousands of pounds of steam per hour and building architects and engineers calculate heating needs in terms of Btus and called for boilers rated in terms of Btus. The EPA has issued rules using millions of Btus per hour.

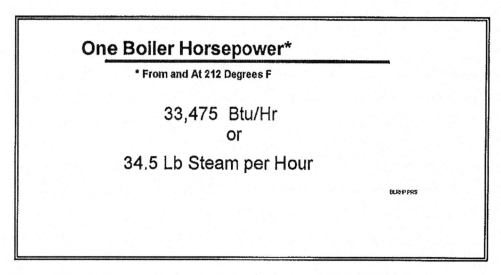

Figure 1.18—Conversion factors for boiler horsepower. In smaller boilers the capacity is measured in steam produced at 212 F. Larger boiler capacity is usually given in (000)lb of steam evaporated per hour at specified conditions of pressure and temperature. A recent trend for rating large boilers is kW or mw of the turbine generator.

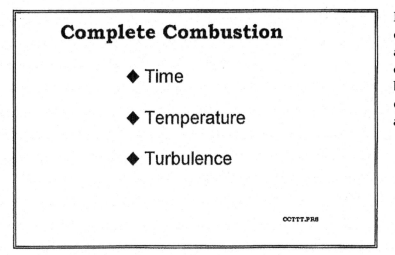

Figure 1.19—The three Ts of combustion: Time, Temperature and Turbulence. As the carbon content of fuel increases the 3 Ts become more important in the design and operation of burners and firing equipment.

Turndown is the measure of the maximum firing rate compared to the minimum firing rate. [See Figure 1.20.] With larger boilers taking a boiler off line can be a complex and time consuming affair. Being able to reduce its firing rate provides an economy in fuel, man-hours and maintenance. Each time a boiler is taken out of service and restarted, it must be purged with at least four changes of air cooling it down, wasting stored heat and causing some thermal stresses from the cold air. The fuel used to bring it back to the point where steam is being exported again is wasted because the only work it does is replace the heat lost to the environment during the shutdown.

When working with other boilers and in responding to variable plant steam demands a large turndown ratio has an advantage. Commercial boilers are not required to respond to large load ranges and because of economic reasons, first cost competitiveness, the turndown is low about, 2-3 to 1. Firetube boilers, because of their large water capacity, can store a lot of energy allowing them to respond to large short term demands very quickly. This and the fact they are not designed for heavy

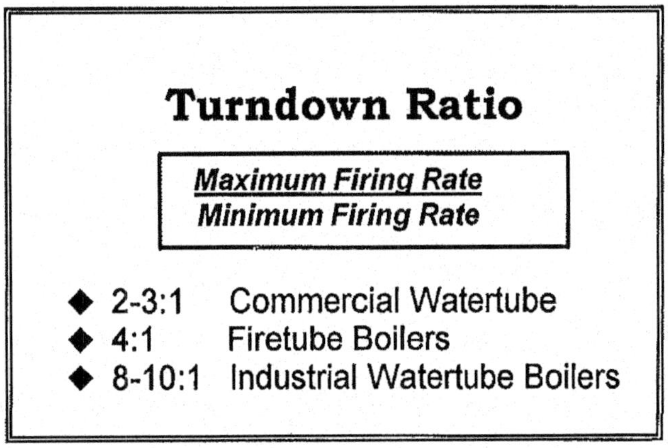

Turndown Ratio

$$\frac{\textbf{\textit{Maximum Firing Rate}}}{\textbf{\textit{Minimum Firing Rate}}}$$

◆ 2-3:1   Commercial Watertube
◆ 4:1   Firetube Boilers
◆ 8-10:1   Industrial Watertube Boilers

Figure 1.20—The turndown ratio of a boiler is important for efficiency and operations. If a boiler has to cycle on and off because of low load conditions energy will be lost during the shutdown and purge cycles before and after the firing sequence.

industrial applications allows them to have a turndown around 4 to 1. Large industrial boilers must respond to large demand requirements and changes so a large turndown is desirable.

BOILER CASUALTIES

The insurance companies insuring boilers and property are very interested in the causes of boiler casualties as they have millions of dollars at risk in many plants. So, when a major loss occurs, they investigate the cause very thoroughly. Here is the result of one such study:

BOILER CASUALTIES

Percentage	Cause
%	
54	Maintenance
10	Operational
7	External
5	Construction
4	Design
4	Application
2	Repair
14	Undetermined

HIGH EFFICIENCY BOILER OPERATIONS

In this book you will find a chapter with over 60 ways to improve boiler efficiency, the following list contains the key items to reducing the energy costs of boilers.

THE KEY TO EFFICIENT BOILER OPERATIONS

1. [2M] Measure and Manage. Change from an operational to management philosophy.

2. Operate with minimum excess air levels.

3. Reduce boiler exhaust temperature to minimum levels corresponding to fuel sulfur content.

4. Capture latent heat energy from boiler exhaust with condensing heat recovery technology.

Chapter 2

Optimizing Boiler Plants
Establishing the Ideal Scene

Optimizing the performance of boiler plants and distribution systems is a broad and rewarding subject. On the surface you might think it requires a great deal of knowledge and insight. It covers many engineering subjects and involves an understanding of the dynamics of operating systems composed of many subsystems and components. If the job is approached on a systematic basis, it is much simpler and you will be more successful. This book is organized for a systematic and simple approach to boiler plant optimization.

As-found efficiency is the efficiency for a boiler in its existing state of repair and maintenance. This efficiency will be used as the baseline for any later efficiency improvements. It is very important that the as-found efficiency be recorded because it will serve as a benchmark to estimate the value of a Boiler Optimization program

The beginning point for Boiler Plant Optimization Program is to establish the *asfound efficiency* for each boiler and system. This information will serve as a datum to show the expected benefits of the program.

It may be that millions of dollars can be saved by improving plant and distribution system efficiency. The as-found conditions will serve as a datum for establishing this fact and also illustrate the folly of neglecting boilers and their energy distribution systems to management and decision makers.

Also, if the plant is found to be efficient, it will bring credit to those responsible for the plant. In any case the as-found efficiency is important for

economic evaluations and justifying additional personnel and modifications to existing systems.

The next step will be to tune-up the boiler and accomplish any maintenance and repair identified during the initial testing to bring it back, as close as reasonably possible, to its original design performance level.

Tuned-up efficiency is the efficiency after operating adjustments, lowering excess air, and minor repairs have been completed. This will be the baseline efficiency for estimating all future savings.

There are many minor problems which can develop during the life of a boiler and distribution systems that cause non-optimum performance. Over time, they can waste a great deal of energy.

To make an honest and accurate evaluation of the potential savings available, all equipment must be in a normal state of repair with the boiler air/fuel ratios at designed levels. If this is not the case, estimates of savings and justifications for new equipment and modifications will contain false information leading to poor decisions.

It is very important that the tuned-up efficiency be accurate, because the economic benefits of improvement options will be estimated from this efficiency.

Next, there should be an estimate of the savings if the efficiency were to be improved to a reasonable level. This information is used to determine if further work on a boiler is justified and to set priorities for competing projects.

Maximum economically achievable efficiency is the efficiency that can be achieved, with efficiency improvement changes only, if it is economically justifiable.

Maximum attainable efficiency is the result of adding the best available efficiency improvement technology, regardless of cost considerations.

Table 2.1 lists this efficiency level for a wide range of boilers and fuels. Chapter 13 lists more than 100 possible ways to improve boiler plant efficiency.

Next, it might be necessary to know what the maximum efficiency of a boiler could be if considered. For example, knowing maximum efficiency would be useful for emergency planning if fuel were very scarce.

Also, it might be cheaper to raise efficiency (productivity) of a boiler rather than install another boiler, at high cost, to keep up with growing plant steam demands.

Table 2.2 shows maximum attainable efficiency for various size boilers and fuels.

Table 2.1— Maximum Economically Achievable Efficiency Levels

Fuel	Rated Capacity Million BTU's/HR		
	10 -16	16 - 100	100 - 250
Gas	80.1%	81.7%	84.0%
Oil	84.1%	86.7%	88.3%
Coal Stoker	81.6%	83.9%	85.5%
Pulverized	83.3%	86.8%	88.8%

Table 2.2—Maximum Attainable Efficiency Levels

Fuel	Rated Capacity Million BTU's/HR		
	10 -16	16 - 100	100 - 250
Gas	85.6%	86.2%	86.5%
Oil	88.8%	89.4%	89.7%
Coal Stoker	86.4%	87.0%	87.3%
Pulverized	89.5%	90.1%	90.4%

THE FIRST STEP

The first step is to find the *as-found efficiency* of each boiler.

Efficiency improvement potential is based on the *as-found efficiency*. For example, it can be used to show the value of testing the performance of boilers on a regular basis by identifying losses caused by the drop in efficiency from a well-maintained or tuned up condition.

This testing can indicate if your maintenance program needs improvement and will show the dollar losses if efficiencies are allowed to drop off because of neglected maintenance.

An efficiency monitoring system is also a good way to check the work of contractors and engineering consultants to insure they actually improved efficiency or to find out if new problems have been created.

When one speaks of boiler efficiency, a degree of generality is present unless the term efficiency is further defined. Boilers normally operate over a range of efficiency (Figures 2.1 through 2.3).

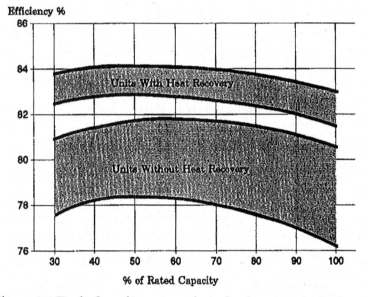

Figure 2.1 Typical performance of gas fired water tube boilers.

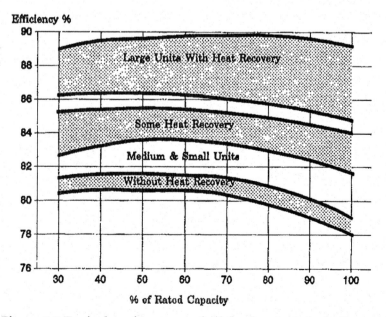

Figure 2.2 Typical performance of oil fired water tube boilers.

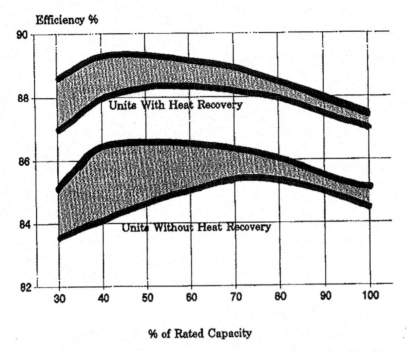

Figure 2.3 Typical performance of pulverized coal fired boilers.

The efficiency of a boiler usually falls off as the bottom end of the of the turn-down ratio is approached. This is because the volume of air through the burner is greatly reduced affecting the performance of air fuel mixing. The excess air must be increased to compensate for this problem to prevent smoking and incomplete combustion which could lead lower efficiencies.

As the boiler approaches its maximum firing rate, it's capability to recover all of the heat from the combustion process diminishes and stack temperature rises. This is the primary reason for the fall off of efficiency as the maximum firing rate is approached.

Due to the highly diverse range of boiler designs over many years, any curve of performance which might be presented can only be considered for general illustration. Figures 2.1 through 2.3 are the result of an extensive survey sponsored by the Department of Energy and are to a large degree typical for the aggregate boiler population in the U.S.

A curve of efficiency over the entire load range should be developed for each specific boiler. This curve can serve as an accurate statement of a boilers efficiency but further examination will be necessary to establish its real operating efficiency. Figures 2.4 through 2.6 show typical characteristics that may be found when plotting these curves.

Where does the boiler operate typically? Does it stay at full load or some other fixed load most of the time? Does the load vary quite a bit, just what is the average load and average efficiency for a particular boiler? Does it cycle on and off a lot? The more accurate the data, the closer the estimation of efficiency to actual conditions.

These factors argue against arriving at boiler efficiency easily or quickly using the heat loss method. This being the case, judgments and estimates are usually applied unless comprehensive metering exists for accurate input-output efficiency monitoring.

Long-term monitoring with good instruments and modern data recording and storage is the best way to go.

LOSS PREVENTION

A further refinement of boiler optimization comes under the heading of "LossPrevention" which is a system of identifying key performance

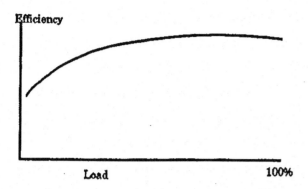

Figure 2.4—Efficiency curve showing high excess air losses below half load.

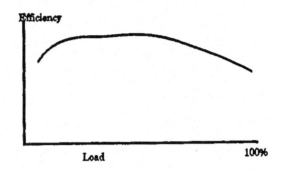

Figure 2.5—Efficiency curve showing high stack temperature losses above half load.

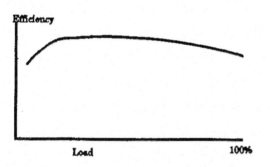

Figure 2.6—A typical efficiency curve showing moderate excess air losses at low firing rates and heat exchange losses at high firing rates.

parameters and identifying them with annual fuel or dollar losses when they vary from the ideal.

Once the ideal scene has been defined, loss prevention can be indexed to these and other key indicators:

a. Stack temperature
b. Excess air
c. Percent make up water
d. Blowdown losses
e. Condensate return temperature losses

A loss prevention program is an essential part of boiler plant optimization as it deals with the control of specific elements of boiler and system operation and maintenance.

Thus, two ideal scenes are appropriate to any plant. The first is ideal design. Efficiency measurements can be used to justify projects such as economizers,

oxygen trim systems, heat recovery options and such to achieve the ideal design for the plant.

The second ideal scene deals with how the plant is operated and maintained relative to unnecessary energy losses within the control of plant personnel.

OPTIMIZING BOILER MANAGEMENT

If tests show an annual fuel savings of $100,000 were possible just by tuning up a boiler and correcting simple maintenance problems uncovered during a tune up, this fact would present a convincing argument for additional personnel, training or technical support to insure that these savings are maintained. Hidden costs from maintenance problems can be a very expensive oversight.

The term management is not often discussed in connection with boilers and distribution systems, probably because they are operated, but not managed in the sense of maintaining optimum performance.

A good many managers feel that the efficiency of a plant should never change. What is missed is that they are a very dynamic energy consuming monster that must be managed. If not prevented from wasting fuel, they will surely be feasting on your utility dollars.

Plant engineering and maintenance staff are often too busy or undermanned or don't understand the magnitude of losses involved in boiler plants and distribution systems to be concerned about this very real problem.

Too often plants don't have instrumentation to measure what is happening with their data, the assumption is made that boiler and steam distribution system losses are negligible.

ACCOUNTABILITY

If you want to be successful in an energy conservation program, accountability must play a key role, there must be someone who is accountable for energy. This is just not someone who checks the utility bills. It may take a professional who can spot problems and knows how to correct them.

Surveys of many plants have shown very conclusively that there is usually a lot more attention on "operating" than on "managing" in most plants.

This is clearly evident from examining plant instrumentation, most of it is for plant operations. There usually aren't any instruments or means to measure stack losses, blowdown losses, condensate system losses or to measure over-all system efficiency. Existing instrumentation is used for estimating these values, however typical instrumentation errors can produce large errors compounding the problem.

MEASURE TO MANAGE

The 2M system "Measure to Manage" has been a very successful approach to reducing plant energy costs. It is too easy to waste energy without it.

There is usually no accountability system except in the some office where the utility bills get checked and paid. Here, the math of the bills is reviewed and perhaps comparisons of last years bills to this years are made. Plant personnel are often left out of the loop and this doesn't create any incentive to manage efficiently or to save energy.

Savings more than 20% have been achieved just by installing a system to measure and man-

age energy in a plant to find out what was happening.

Once you have complete and accurate data, it's easy to see what is happening in any plant or distribution system. Establishing a system to get data, interpret it and then controlling unnecessary losses is the real challenge.

AN EFFECTIVE
ENERGY CONSERVATION PROGRAM

The essential elements of an effective energy conservation program are:

1. Management commitment
2. Boiler and system testing
3. Economic evaluation of energy conservation projects
4. Assignment of project priorities
5. Final plan
6. Plan implementation
7. Loss control management program
8. Monitoring results

1. Management Commitment

One of the most important aspects of an energy conservation program, often missing, is management support. If nobody seems to be interested in the efficiency or productivity, things can get very bad.

It was just a few years ago when a vice president of a major U.S. corporation addressed a group of energy engineers at a national conference after receiving an award for being "The Energy Manager of the Year." He stated that the most important contributing fact to his very successful energy conservation program was the full backing of the president of his company. He said, "If the boss shows an interest, something is going to get done."

2. Boiler and System Testing

In the first place, the plant or system may not have been designed for optimum efficiency. Design omissions, deviations from plans by contractors, modifications over the years, changes in operational requirements and maintenance modifications may have all influenced efficiency. It is wise not to make any assumptions about performance. Testing and actual measurements will tell the story about how the system is performing *now*.

3. Economic Evaluation of Energy Conservation Projects

Testing and evaluation will form a good basis for judging the opportunities for saving energy with the boiler or system as it operates in "to days" environment. Payback period, return-on-investment, life cycle costing and other economic evaluation methods are important measurements of the real worth of a project to modify a boiler or improve a systems performance. When faced with many competing projects, economic evaluation will show which is best and which projects won't pay.

4. Assignment of Project Priorities

There is usually competition for limited funding to improve efficiency and save energy. Management must have data on which choices will be most beneficial. The economic evaluation process makes it possible to rank various projects on the strength of their merits.

5. Final Plan

A wise man once said, "If you don't have a plan, then you plan to fail." A final plan infers that all facts have been considered and a smart and orderly course will be followed.

6. Plan Implementation

The plan has now become a working tool. Like many other tools, plan implementation will get the job done. Responsibility and control are key words here. The design must be right, the contractor must install the changes properly, maintenance personnel can change things so they don't work well and operators can make improper adjustments. So, someone needs to "close the loop" by asserting responsibility and control over the project.

7. Loss Control Management

Assign an annual loss value in dollars to indicators of system performance. Each operating parameter that will have a significant impact on operating cost and energy losses should be monitored.

Examples of possible items for loss control management are:

(a) for each 10°F rise in stack temperature there will be a $10,000 annual loss if not corrected,

(b) for each 10% change in excess air fuel costs will go up $15,000 annually,

(c) For each 10°F the condensate returns below the ideal temperature, $40,000 will be lost in a year and so on. Each plant will have its own values.

8. Monitoring Results

The use of computers and the ease of acquiring data from plant operations provides new opportunities to actually monitor performance on a full time basis rather than guessing or making assumptions. Its easy to claim that equipment will improve performance or repairs were made properly, a monitoring system will tell the story and can be a valuable tool in managing an efficient plant.

In the following chapters we will go through a step by step program for a typical facility where you will be shown how to identify and deal with the many problem areas and establish the "ideal scene" for your plant.

Chapter 3

Energy Management Basics for Boilers and Systems

ENERGY BALANCE AND LOSSES

Understanding how energy works in systems is essential to an effective energy management program, so lets examine some basics.

The purpose behind managing an efficient plant is to get as much energy as possible from the fuel. How much would this be?

A good starting point for establishing the "Ideal Scene" for an efficient plant is knowing how much energy is being delivered to the system? The term used to express the energy delivered to a plant in fuel is Higher Heating Value or (HHV—Table 3.1).

THE HIGHER HEATING VALUE (HHV)

The total heat obtained from the combustion of a specified amount of fuel.

The HHV represents the total amount of heat energy released from the fuel when it is completely burned, at 60 Deg F [15.6C] when combustion starts and the combustion products of which are cooled to 60 Deg F [15.6C] before the quantity of heat released is measured.

For practical reasons, flue gas is not cooled down to 60 F because acid will form which can cause severe corrosion problems. Stack temperature is kept above the acid dew point or acid formation temperature which wastes some of this heat energy.

Fossil fuel heating values are expressed in terms of Btu/lb in tables and for standard calculations.

THE BTU (BRITISH THERMAL UNIT)

"The quantity of heat required to raise the temperature of one pound of water one degree Fahrenheit"

For scientific purposes this is measured at, or near, its point of maximum density of water 39.1 Deg F [3.9C].

INPUT-OUTPUT EFFICIENCY

The starting point for finding efficiency begins with accurate knowledge of the amount of energy "input" into the process. Two things must be known, how much fuel is being fired and how many Btu's are in each pound of fuel. The range of heating values for various fuels is given in Table 3.1.

Boiler efficiency is the percentage of the fuel's higher heating value which is converted to steam. This is sometimes called the Fuel-To-Steam Efficiency. The formula for Input-Output efficiency is:

$$\text{Efficiency} = \frac{\text{Btu Output} \times 100}{\text{Btu Input}}$$

Figure 3.1 Efficiency defined

Efficiencies are basically limited by the acid dew point arid excess air. The acid dew point is the temperature at which acid begins to form on

Table 3.1—Higher Heating Value (HHV) of various fuels.

	As Fired Heating Value	
Fuel	Btu/Lb	Other Units
Natural Gas	20,000 - 23,500	950-1,100 Btu/SCF
#1 Fuel Oil	19,670 - 19,860	134-135 KBtu/gal
#2 Fuel Oil	19,179 - 19,750	138-142 KBtu/gal
#4 Fuel Oil	18,280 - 19,400	142-151 KBtu/gal
#5 Fuel Oil	18,100 - 19,020	142-149 KBtu/gal
#6 Fuel Oil	17,410 - 18,990	143-156 KBtu/gal
Bituminous Coal	11,500 - 14,500	NA
Subituminous Coal	11,500 - 14,500	NA
Lignite	5,000 - 8,300	NA
Wood-moisture free	8,000 - 10,100	NA

colder metal in the boiler exhaust. This minimum temperature does not allow for recovery of all the heat of the HHV of the as-fired fuel. Also, because extra air is required to insure complete combustion of the fuel at the burner the extra volume of exhaust gasses carry away additional waste heat.

WHY STEAM IS USED

Steam is very useful because it automatically flows to the point of use and does not need to be pumped. It moves through the piping system to a point of lower pressure where steam is condensing into water as it gives up its latent heat.

STEAM PROPERTIES

The steam has three properties that can be put to work; pressure, temperature and volume. Pressure and volume combine to drive machinery and temperature is used in processes using heat. In the conversion of water from its liquid phase to steam, its vapor phase, heat is added to initially increase the water temperature to the boiling point.

Water heated under constant pressure will rise in temperature until a certain temperature is reached when it begins to evaporate into steam or boil. The temperature at which water boils is known as saturation temperature and depends on the pressure under which the water is heated.

If pressure increases the saturation temperature rises, if pressure decreases the saturation temperature falls. Each pressure has a corresponding saturation temperature. The temperature of the steam and water remain constant at each pressure (See Tables 3.2 and 3.4.)

Sensible heat is a rise in the measurable temperature of the water before it boils. This is what thermometers indicate. This is approximately one Btu per pound of water for degree of temperature rise.

When the water gets to its boiling point a change occurs, the water begins to evaporate with no change in temperature. This phase change from liquid to vapor absorbs energy known as the *Latent Heat of Evaporation*.

Steam that is not fully vaporized is called *wet steam* and the weight of water droplets in the wet steam compared to the weight of the steam is known as *% moisture*. Also, when heat is taken away from saturated steam, condensation takes place creating moisture.

For example: Steam at 100 psi has a total heat of 1190 Btu/lb and water has 309 Btu/lb. The steam has 3.85 times as much Btu content. If you are conducting a heat balance test and don't measure the percent moisture in the steam, a significant error will be introduced.

Steam that is heated above the saturation temperature is called *superheated steam*. This steam is dry and does not contain water droplets, and it can have any temperature above the saturation temperature. This is usually done by applying heat to steam after it has been removed from contact with the water.

This phenomena also occurs when steam pressure is reduced by a pressure reducing valve in the steam distribution system.

ACCOUNTING FOR HEAT IN STEAM AND WATER

The heat content of water and steam is expressed in British thermal units (Btu) per pound and is known as Enthalpy expresses by the symbol "h" in tables and formulas.

The *"Steam Tables"* contain the basic information on the energy properties of steam. The information in this book was obtained from works of Keenan and Keys "Thermodynamic Properties of Steam" published by John Wiley and Sons, Inc.

Enthalpy of Saturated Liquid (h_{fg}) is the heat required to rise the temperature of one pound of water from 32°F to the saturation temperature. This property is sometimes known as the *Heat of the Liquid*.

Enthalpy of Evaporation (h_{fg}) is the amount of heat of heat required to change one pound of water at the saturation temperature to dry saturated steam at the same temperature. This is also known as the *Latent Heat of Evaporation*.

Enthalpy of Saturated Vapor (h,) is the heat required to change one pound of water at 32 F into dry saturated steam. It is the sum of the enthalpies of saturated water and evaporation. It is also known as the *Total Heat of Steam*.

Table 3.1 lists the HHV for fuels and Table 3.2 shows the Btu values for steam and Table 3.3 the values for water. This information combined serves as a common basis for Input-Output energy calculations.

The steam tables also show the saturation temperature changes with pressure. At atmospheric pressure water boils at 212 Deg F [100C] and at 250 psig the boiling temperature goes up to

400 Deg F [204C]. The working temperature of the steam is raised by increasing pressure {Table 3.4)

ENERGY RECOVERY IN THE CONDENSATE SYSTEM

On the condensate recovery side of the steam system, there is an important rule of thumb to know about. Heating water for use in the boiler uses a significant percentage of the total energy in the steam, so it is important to get condensate back to the plant as hot as possible.

The water being fed to the boiler must be heated to the boiling point to drive off oxygen and carbon dioxide which can cause severe pitting and corrosion to the boiler and piping systems.

The hotter the returning condensate is, the higher the efficiency of the system.

The Rule of Thumb for feedwater heating:

Every 11 Deg F [6 Deg C) that has to be added to the boiler feed water reduces efficiency by 1 %.

Starting with the cold water side of the system 50 Deg F [10C], water is raised to approximately 220 Deg F [104C] in the feed water heater, depending on the type of system, for injection into the boiler which takes one Btu per pound per degree F. In this case 170 Btus per pound of water. This amounts to 15% of the energy in the steam.

Unfortunately, there can be another reason for boiling hot water returning to the boiler room. It is a sure sign that steam traps are leaking steam.

CALCULATING FUEL SAVINGS AND LOSS BASED ON EFFICIENCY CHANGE

There is an important difference between efficiency improvement and fuel savings. An efficiency increase from 80% to 81% is a 1% efficiency improvement. It is a proportional increase of 1% out of 80% (1180%) or 1.25. This represents a 1.25% fuel savings from the 1% efficiency improvement.

Table 3.2 Typical Steam Table information

Abs Press Lb/Sq in	Saturation or Boiling Temperature (Degrees F)	Specific Volume (Cu. Ft/Lb.)	Heat Content Above 32 degrees F		
			Sensible Heat or Heat of Liquid (Btu/lb)	Latent Heat or Heat of Evaporation (Btu/lb)	Total Heat (Btu/lb)
P	t	V_g	h_f	h_{fg}	h_g
14.696	212	26.80	180	970	1151
20	228	20.90	196	960	1156
25	240	16.31	209	952	1161
30	250	13.75	219	945	1164
35	259	11.90	228	939	1167
40	267	10.50	236	934	1170
45	274	9.40	244	929	1172
50	281	8.52	250	924	1174
55	287	7.79	256	920	1176
60	293	7.18	262	916	1178
65	298	6.66	268	912	1180
70	303	6.21	273	908	1181
75	308	5.82	278	905	1182
80	312	5.47	282	901	1184
85	316	5.17	287	898	1185
90	320	4.90	291	895	1186
95	324	4.65	295	892	1187
100	328	4.43	299	889	1188
110	335	4.05	306	884	1190
120	341	3.73	313	879	1191
130	347	3.46	319	874	1193
140	353	3.22	325	869	1194
150	358	3.02	331	864	1195
160	364	2.84	336	860	1196
170	368	2.68	341	856	1197
180	373	2.53	346	852	1198
190	378	2.41	351	848	1199
200	382	2.29	356	844	1199
250	401	1.84	376	826	1202
300	417	1.54	394	810	1204
350	432	1.33	410	795	1205
400	445	1.16	424	781	1206
450	456	1.03	437	768	1206
500	467	0.93	450	756	1205
600	486	0.77	472	732	1204
700	503	0.66	492	711	1202
800	518	0.57	510	690	1199
900	532	0.50	527	670	1196
1000	545	0.45	542	650	1192
1500	596	0.28	612	557	1169
2000	636	0.19	672	464	1136
2500	668	0.13	731	361	1091
3000	696	0.08	803	213	1016
3203.6	705	0.06	903	0	903

Table 3.3—Thermal properties of water

Water Temperature (Degrees F)	Saturation Pressure (Inches of Mercury Vacuum or (psig)	Specific Volume (Cu. ft./lb.)	Density (lb./cu. ft.)	Weight (lb./gal.)	Specific Heat (Btu/lb. — Degrees F — Hr.)	Specific Gravity
32	29.8	.01602	62.42	8.345	1.0093	1.001
40	29.7	.01602	62.42	8.345	1.0048	1.001
50	29.6	.01603	62.38	8.340	1.0015	1.000
60	29.5	.01604	62.34	8.334	.9995	1.000
70	29.3	.01606	62.27	8.325	.9982	.998
80	28.9	.01608	62.19	8.314	.9975	.997
90	28.6	.01610	62.11	8.303	.9971	.996
100	28.1	.01613	62.00	8.289	.9970	.994
110	27.4	.01617	61.84	8.267	.9971	.991
120	26.6	.01620	61.73	8.253	.9974	.990
130	25.5	.01625	61.54	8.227	.9978	.987
140	24.1	.01629	61.39	8.207	.9984	.984
150	22.4	.01634	61.20	8.182	.9990	.981
160	20.3	.01639	61.01	8.156	.9998	.978
170	17.8	.01645	60.79	8.127	1.0007	.975
180	14.7	.01651	60.57	8.098	1.0017	.971
190	10.9	.01657	60.35	8.068	1.0028	.968
200	6.5	.01663	60.13	8.039	1.0039	.964
210	1.2	.01670	59.88	8.005	1.0052	.960
212	0.0	.01672	59.81	7.996	1.0055	.959
220	2.5	.01677	59.63	7.972	1.0068	.956
240	10.3	.01692	59.10	7.901	1.0104	.947
260	20.7	.01709	58.51	7.822	1.0148	.938
280	34.5	.01726	57.94	7.746	1.020	.929
300	52.3	.01745	57.31	7.662	1.026	.919
350	119.9	.01799	55.59	7.432	1.044	.891
400	232.6	.01864	53.65	7.172	1.067	.860
450	407.9	.0194	51.55	6.892	1.095	.826
500	666.1	.0204	49.02	6.553	1.130	.786
550	1030.5	.0218	45.87	6.132	1.200	.735
600	1528.2	.0236	42.37	5.664	1.362	.679

Table 3.4—Saturation or boiling temperature change with boiler pressure.

Pressure Psig	Temperature °F	°C
5	227	108
10	240	116
15	250	121
20	259	126
25	267	131
50	298	148
75	320	160
100	338	170
150	353	178
200	388	198
250	406	208
300	422	217
400	448	231
500	470	243
600	488	253

The actual fuel savings percentage is always larger than the efficiency increase. Similarly the fuel loss percentage is always greater than the corresponding efficiency decrease.

The formula for fuel savings or loss resulting from the change in efficiency is:

$$\text{Savings} = \frac{\text{New Efficiency} - \text{Old Efficiency}}{\text{New Efficiency}}$$

CALCULATING PERFORMANCE DEFICIENCY COSTS

The cost benefit of maintaining boiler efficiency at a high level is easily calculated with this formula:

$$S = \frac{W_f \times E_n - E_I \times C_f \times Hr}{E_n}$$

S is the potential fuel savings per year.

W_f is the fuel use rate in million Btu/Hr

E_I is the ideal efficiency

E_n is the new or existing efficiency

C_f is the cost of fuel per million Btu

Hr is operating hours per year

Because boiler efficiency usually changes with load, the potential for fuel savings will change with the typical load on the boiler based on the ideal or reference efficiency at the load being considered. This formula can be used for estimating energy saving at the typical load under study.

Example: a boiler is firing at 100 million Btu/Hr and its efficiency had dropped from the ideal of 83% to 78%. The cost of fuel is currently $6.00 per million Btus and it fires at this load for 6,000 hours a year.

This loss of efficiency will cost an estimated $230,400 in wasted fuel for the year if this loss of efficiency remains uncorrected.

CARBON DIOXIDE REDUCTION

The basic data on carbon dioxide emissions from different fuels published by the Office of Global Change of the Environmental Protection Agency:

Fuel	Lbs per Million Btu
Natural Gas	117
Fuel Oil	173
Coal	215

Quick Reference Charts

It is useful to have a quick way to estimate potential fuel savings based on improvement in either net stack temperature or excess-air. Figures 3.2 and 3.3 can be used as a quick reference for estimating savings potential

The efficiency improvement for each· one percent change in excess air varies with the stack temperature (Fig 3.2). To estimate efficiency change, multiply the factor (left) corresponding to the stack temperature times the change in excess air.

Example, what will the efficiency change be with a stack temperature of 500°F if the excess air is reduced by 50%? The factor from Fig 3.2 at 500°F is .075; this multiplied by 50% excess air reduction is a 3.75% improvement in efficiency.

The efficiency improvement for each 10 degrees F change in stack temperature varies with the excess air level (Fig 3.3). To estimate efficiency change, multiply the factor (left) corresponding to the excess air level times each 10°F change in stack temperature.

Example: What will the efficiency change be with an excess air level of 60% if the stack temperature is reduced by 100°F? The factor from Fig 3.3 at 60% excess air is 0.325; this multiplied by 100°F/10 a 3.25% improvement in efficiency.

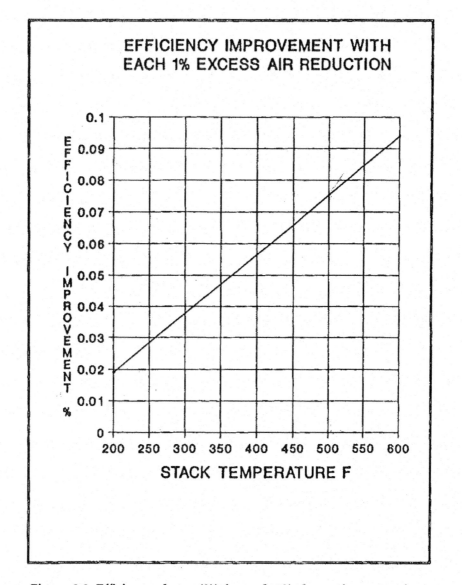

Figure 3.2, Efficiency change(%) for each 1% change in excess air at various stack temperatures.

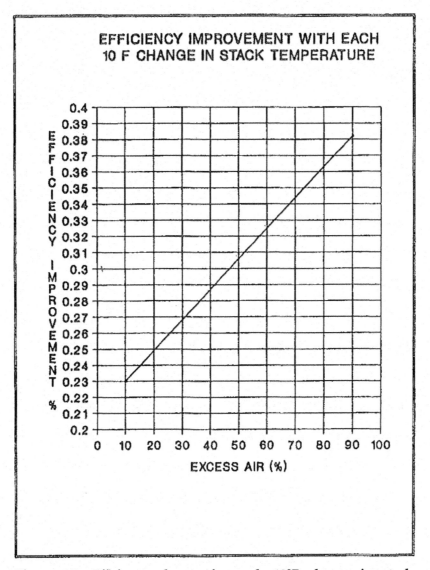

Figure 3.3—Efficiency change for each 10°F change in stack temperature for various levels of excess air.

IMPROVING RADIANT HEAT TRANSFER

The radiation section or that section of the boiler exposed to direct radiation from combustion, is the most effective heat transfer zone of a boiler. Figure 3.4 shows that 8% of the heat exchange surface of a boiler is absorbing about 48% of the energy from combustion. The convection section, which comprises 34% of the heat exchange surface only absorbs about 20% of the heat of combustion.

FLAME TEMPERATURE

Table 3.5 shows that as the excess air is reduced, flame temperatures rise. Combustion flame temperatures are very complex to analyze and this chart should be treated only as an approximation. It can be seen in Table 3.5, at 0% excess air, the flame temperature is about 3,400 F [1,871C] for natural gas and at 100% excess air this temperature drops to about 2,000 F [1,093C]. For number 6 oil these numbers are 4,000 F [2,204C] for 0% excess air and 2,300 F [1,260C] for 100% excess air.

The basic equation for radiant heat transfer, in the flame zone section, is:

$$Q = \rho ST^4$$

Q is Btu/Hr

ρ is the Stefan-Boltzman constant,
1.17 x 10^{-9} Btu/sq ft, hr, T^4

S is surface area, sq ft

T is absolute temperature F + 460

It can be seen from the mathematics involved, the higher the temperature of the flame, the more intense the heat transfer. It becomes obvious that if the temperature of the radiant heat transfer section of the boiler is lowered by lower flame temperatures, there will be lower radiation heat transfer. This will ultimately increases flue gas exit temperature and decrease efficiency.

Stack temperatures have been observed to change over 50 F [28C] due to this phenomena. Changes in excess air directly affects flame temperature, and radiation heat transfer in the combustion zone.

Table 3.5—Excess air flame temperature relationship

Excess Air%	Temperature Deg F C			
	Natural Gas	Propane	#2	#6
0%	3,400	3,700	3800	4,000 F
	1,871	2,038	2,093	2,204 C
25%	2,900	3,100	3,200	3,400 F
	1,593	1,704	1,760	1,871 C
50%	2,500	2,600	2,800	2,900 F
	1,371,	,424	1,538	1,593 C
75%	2,300	2,300	2,400	2,600 F
	1,260	1,260	1,316	1,427 C
100%	2,000	2,000	2,200	2,300 F
	1,093	1,093	1,204	1,260 C

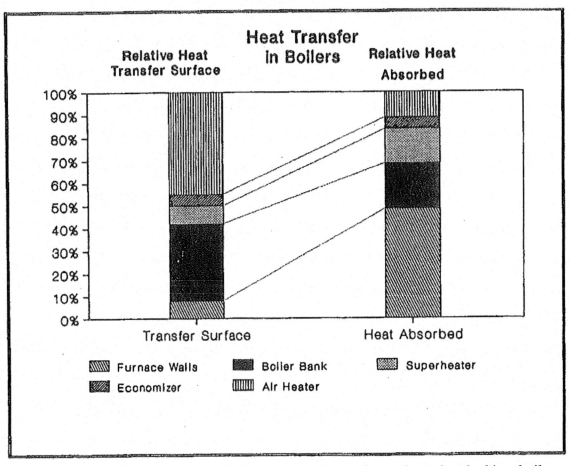

Figure 3.4—Graph of the ratio of relative heat transfer surface to heat absorbed in a boiler.

Chapter 4

Basic Boiler Plant Efficiency

This chapter is designed to show the logical steps that should be followed to find basic boiler plant efficiency. It covers the basic steps for plant survey and testing to identify sources of losses in the plant and distribution system.

In figure 4.1 the typical cost of 1,000 pounds of steam [approximately 1 million Btus] is shown. Boiler plant operations are fairly fixed. Optimizing boiler efficiency will often offer a good opportunity for improved efficiency and fuel dollar savings. The typical distribution system wastes a great deal of energy, about double a boiler's losses. When the distribution system wastes energy, the boiler load goes up accordingly compounding

the losses by another 20%. It can get quite expensive when this circle of losses is escalated by poor conditions in the distribution system.

Figure 4.2 points out an interesting fact about the real value of fuel savings. When considering where the money comes from to pay fuel bills, boiler and distribution system optimization should have a very high priority because the money that pays the fuel bill comes from company profits. It is not unusual to be able to identify a savings of $50,000 a year in a typical plant. The real question is, how does this relate to profits? If there is a 5% profit one million dollars will have to be invested in raw materials, production, mar-

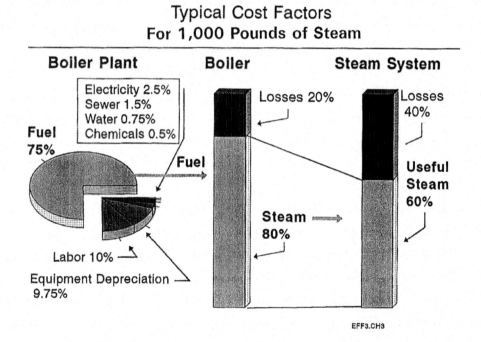

Figure 4.1—Typical cost factors for 1,000 pounds of steam. The boiler plant operational costs are fairly constant and predictable. The steam system losses are unpredictable and can be very large depending on conditions.

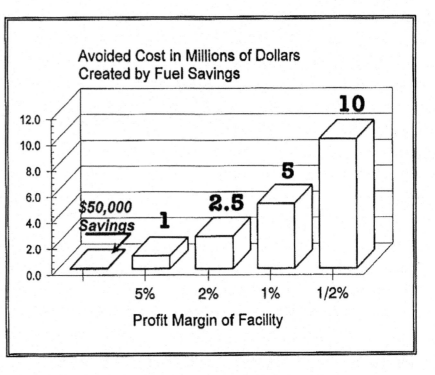

Figure 4.2—What is the true value of fuel savings? As the money for fuel must be earned through production and marketing of some product, profitability enters the equation. An annual fuel savings of $50,000 could be worth 1 million dollars in avoided cost at 5% profit to 10 million dollars at 1/2% profit.

keting and sales to earn the $50,000 in fuel.

If profit falls off to 1%, $5 million in product sales will be required. Anything with this kind of an influence on the bottom line of a company's profits deserves further inspection.

How would one go about investigating the potential for improving profits and reducing energy losses in boilers?

Essential Information
Fuel Consumption and Costs
Boiler Losses

FUEL CONSUMPTION

The first requirement is to gather accurate fuel use information for each boiler. Figure 4.3 shows the basic information required.

Each boiler should have a separate fuel meter so fuel use can be accurately tracked for each boiler. The 181.7 million Btu/hr represents the maximum input for this boiler. Without a fuel meter, actual firing rate can not be determined whether 120% [dangerous] or 60% of maximum rated capacity. Measuring the actual fuel rate

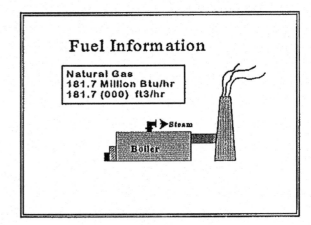

Figure 4.3—Fuel use information

with a stopwatch and good fuel meter will tell the story. Also, making a plot of the fuel delivery rate verses control system position can uncover problems with control system linearity. Three basic instruments can be very valuable in estimating and tracking performance: a totalizing fuel flow meter, an steam or feedwater flow meter and a simple timer

FUEL COST

Figure 4.4 shows the basic data on fuel costs necessary for financial evaluations. This information is important for the analysis of dollar losses and for evaluating the dollar value of fuel saving opportunities. The $727 per hour represents the maximum firing rate.

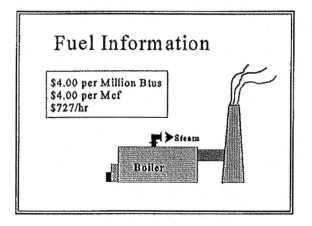

Figure 4.4—Fuel cost information

ANNUAL FUEL COSTS

Figure 4.5 gives annual fuel use information for our typical boiler, this data is essential for evaluating potential savings and ranking energy conservation options based on annual fuel consumption. It will also provide data for evaluating the effectiveness of energy conservation options after they have been identified.

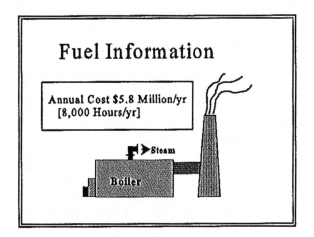

Figure 4.5—Annual fuel costs

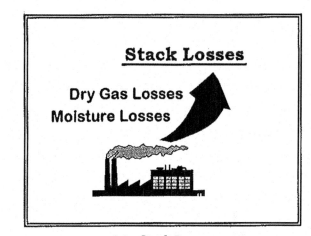

4.6—Stack Losses

ANALYZE PERFORMANCE

Analysis of plant performance requires identification of the principal losses. Stack Losses are usually the largest single loss in boiler operations.[Figure 4.6] By testing the exit gas temperature and excess air levels, the sensible and latent heat being lost up the stack can be established. A rough estimate of boiler efficiency is often obtained by subtracting stack losses from 100 percent. It must be noted that many other losses are involved in operating boilers and this method of using just one loss for the efficiency determination has limited value. There are a number of very good portable analyzers on the market that give stack loss efficiency directly, they also show other important information like excess air, stack tem-

perature, carbon monoxide, combustibles and oxygen readings that can be very useful in identifying the sources of boiler problems. Other chapters in this book cover how to measure these losses in more detail.

STACK LOSSES

As shown in Figure 4.7, stack losses are 18.6% which provides a rough efficiency of 81.4%. Stack loss is not the only operational boiler loss. Because it is the largest, it provides rough information on efficiency. It involves two very important parameters, excess air and stack temperature which are items that should be given first priority in any boiler optimization program.

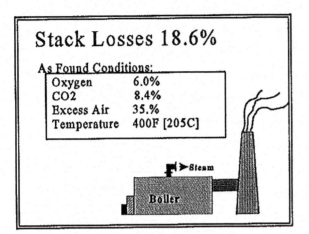

Figure 4.7—Stack losses measured using the Heat Loss method.

BLOWDOWN LOSSES

Blowdown losses are often overlooked because they are hard to measure and water chemistry work is not always fully understood by plant personnel. As shown in Figure 4.8., this boiler has a blowdown rate of 4% of its total steam output. Hot water is being dumped into a drain or sewer system and excess heat is being vented to the atmosphere as flash steam. This amounts to 6,000 lb/hr of very hot water which is responsible for an energy loss of 1.57%.

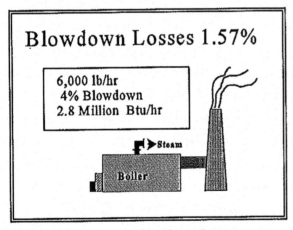

Figure 4.8—Blowdown losses

SURFACE LOSSES

Another significant loss is the boiler surface loss (Figure 4.9), caused by radiation and convec-

tion from the boiler surface and valves and piping in the boiler plant. Usually this loss is between 1 and 3 percent but this number can not be generalized.

The percentage of this loss grows as boiler load decreases. When the boiler load starts to drop off, this percentage continues to go up because the boiler is still losing the same amount of heat but the output is less so the loss is a bigger percentage at low firing rates could be 8 to 10 percent. This becomes a real significant problem of their design output capacity. Figure 4.9 shows the percentage loss of this boiler at 100% of its full rated output.

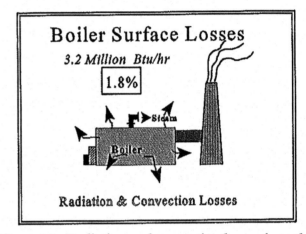

Figure 4.9—Radiation and convection losses from the boilers surface

Figure 4.10 illustrates a method of surveying a boiler for surface losses using infrared measurements. These measurements of both surface temperature and surface heat losses are useful for estimating energy losses and for planning maintenance. Changing surface temperatures will indicate casing and insulation failures. A modern spread sheet can be used to determine total and average heat losses and record changes over time.

Figure 4.11 illustrates how this radiation and convection losses increase as the boiler load goes down. The lower plot represents radiation and convection losses from the boilers exterior surface, the upper curve shows the effect of on/off operation at the same loads. When a boiler stops firing a purge of at least 4 air changes is necessary to sweep out any remaining fuel vapors. This same purge is required on light off.

These cold air purges act to cool down the boiler wasting energy.

The cause of unusually hot boiler rooms usually this heat from hot surfaces. Good design and maintenance can reduce this loss significantly.

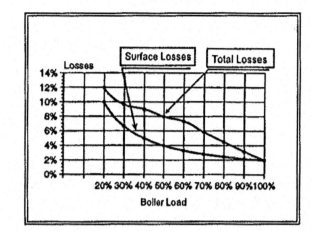

Figure 4.11—As boiler load decreases the percentage of losses increase. These losses remain relatively constant but because the percentage loss is a function of the actual load, the percentage loss increases as the load goes down. A 2% surface loss is illustrated above to show the relationship more dramatically.

AS-FOUND EFFICIENCY

The largest boiler losses are stack losses, boiler surface losses and blowdown losses. In our example boiler, shown in Figure 4.12, they add up to 22% giving an efficiency of 78%. These losses are not all of the losses but they represent the major losses that can be measured easily and corrected economically.

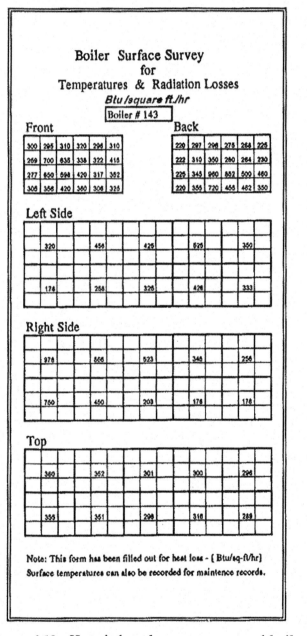

Figure 4.10—How infrared measurements of boiler surface temperatures and radiation heat loss measurements can be used evaluate conditions. This is a chart of one such survey.

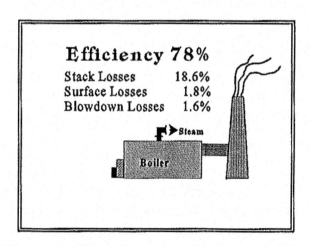

Figure 4.12—When the percent losses are subtracted from 100% an approximation of efficiency is possible.

TUNE-UP

Figure 4.13 shows the first step in dealing with boiler losses. The Tune-Up is almost always found to be necessary. Problems in the burner system, control system and other components are often "corrected" by plant personnel by increasing excess air.

When a boiler is having problems it usually begins to smoke. Problems have often been "fixed" by increasing excess air. If no one is monitoring and correcting the excess air level, this situation can cost a lot of wasted fuel. Be advised that problems that have been "fixed" by increasing excess air will reappear when a tune-up is attempted.

By reducing excess air from 35.5% to 9.3% in the example boiler, fuel losses have been reduced by 1.8%. This is a typical fuel savings for many boilers.

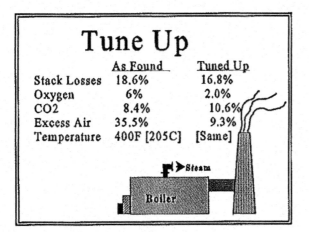

Figure 4.13—The first step in improving efficiency is to reduce excess air losses by tuning up a boiler.

REDUCING STACK TEMPERATURE

Figure 4.15 shows that a fuel savings of 2.7% can be produced by reducing net stack temperature from 400F [205C] to 250F [121C]. This high as-found temperature may be caused by soot or scale deposits in the boiler or by hot gas short circuits through defective baffles. Figure 4.16 is a list of things to check when boiler exit temperatures are too high.

Sources of Air Fuel Ratio Errors

1. Fuel: gravity, temperature, Btu content
2. Fuel System: pressure variations & wear
3. Combustion air: pressure, temperature, humidity fan performance, dampers & ducting
4. Exhaust system: back pressure variations, boiler fouling, stack effect & wind effect
5. Control System: fuel & air positioning accuracy, alignment, synchronization, repeatability & linearity

Figure 4.14—A troubleshooting table for reducing excess air.

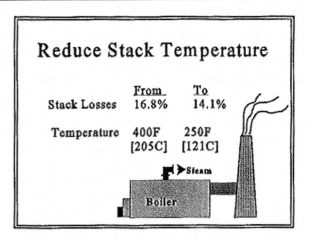

Figure 4.15—Reducing outlet temperature produces higher efficiency.

Stack Temperature Check List

1. Check for over firing
2. Firesides deposits of soot and slag
3. Waterside deposits of scale and baked on sludge
4. Gas pass short circuits
5. Install economizer
6. Install air preheater
7. Reduce Excess Air
8. Install turbulators
9. Install condensing heat recovery

Figure 4.16
Check list for reducing stack temperature.

BLOWDOWN HEAT RECOVERY

Figure 4.17 shows that by using energy from the blowdown system to heat incoming makeup feed water, losses can be cut from 1.6% to 0.13%. Heat recovery equipment in this case can reduce blowdown losses from 475 Btu/lb to 60 Btu/lb of blowdown water.

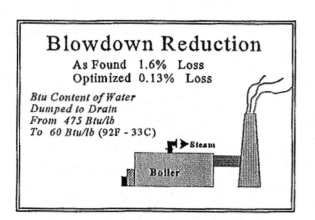

Figure 4.17
Blowdown heat recovery improvements.

OPTIMIZED EFFICIENCY

Figure 4.18 shows that by taking the standard actions of tuning up the boiler, recovering blowdown losses and reducing stack temperature, efficiency has been increased from 78% to 85%

Figure 4.18—Optimized efficiency

COST OF STEAM

Figure 4.19 shows one of the benefits of improving efficiency. The cost of steam has been reduced from $5.09 to $4.70 per million Btus produced.

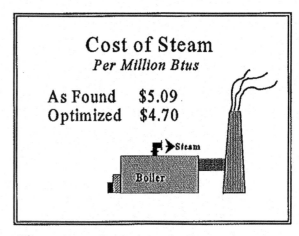

Figure 4.19—Optimizing boiler performance reduces steam cost.

BOILER PRODUCTIVITY

Figure 4.20 shows how the productivity of the boiler has been increased from the "as found" output of 142.8 million Btu/hr to an optimized 154.5 million Btu/hr, an increase of 7.6%. This represents an increase in steam production from 135.5 to 146.6 thousands of pounds of steam per hour.

Increased boiler productivity can save money in two additional ways; first if additional boilers are being kept on the line at reduced firing rates to pick up demand surges, this additional capacity may provide a safety margin which will allow the extra boilers to be shut down. Also, if there is a need for more capacity because of plant expansion or production requirements, increasing the efficiency and productivity of existing boilers may eliminate the need for high capital investment and pollution compliance costs for new boilers.

DOLLAR SAVINGS

Figure 4.21 shows the dollar savings can be quite impressive. Total savings are $403,989 a

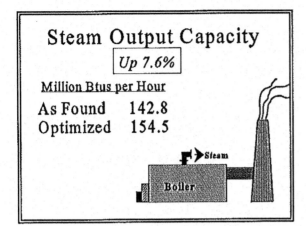

Figure 4.20 Increased boiler capacity by 7.6% is a direct result of optimizing performance.

year. The tune up will save $104,673, the lowered stack temperature $157,000, the shell loss reductions $58,786 and reduced blowdown $83,520. Each plant is different but this example is not unusual.

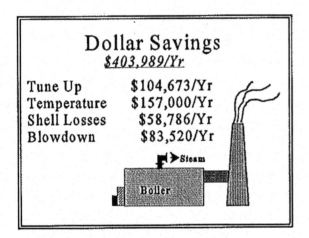

Figure 4.21—Fuel savings from boiler optimization

ADDITIONAL LOSSES

As this chapter is an introduction to major boiler energy losses many additional losses were not included in this chapter and are covered following chapters.

Energy Losses:
- Fuel Oil Atomization
- Soot Blowing
- Oil Heating
- Carbon Monoxide
- Combustibles

Operating Losses:
- Partial load operations (large losses)
- Blowers Fans and Pumps
- Make Up Water
    - Chemical Treatment
    - Cost of Water
    - Water Heating
- Blowdown (additional considerations)
- Oil
    - Oil Pumping
    - Fuel Oil Additives
    - Heating
    - Inventory Investment
- Fuel Handling [Coal & Biomass]
- Environmental Compliance
- Maintenance and Repair

Chapter 5

Distribution System Efficiency

STEAM SYSTEM LOSSES

The steam system is ideal for moving energy through piping. No pumps are needed because when the latent energy in the steam is expended it condenses to approximately 1/700 of its original volume causing instant replacement of the use steam. This replacement steam flows to fill the volume of the steam which has condensed, often reaching velocities of 8,000 to 12,000 feet per minute on the way.

The "pure" water from the condensed steam still contains valuable energy and for efficient operation it should be recovered and transported back the boiler plant where it is reheated into turned back into steam and recycled in the system once again.

Rule of Thumb

Every 11 deg F [6 deg C] that has to be added to boiler feed water costs 1% in efficiency.

Boilers and their distribution systems have a closely interrelated efficiency. The losses involved in distribution systems have a significant impact on boiler operations and efficiency. For example, if the hot condensate from the recovery system does not return to the boiler plant it must be replaced with cold makeup water which must be heated and also receive chemical treatment. Every Btu wasted in the distribution system must be replaced by the boiler which wastes about 20% additional energy. This cycle feeds on itself, the more energy lost in the distribution system, the more energy wasted by the boiler replacing it.

When energy losses in the steam distribution system are corrected, boiler plant losses are reduced too.

The steam distribution system moves the steam from the boiler to perform work at some distant location. They do not usually exist separately and they work together to establish a combined efficiency.

Example

The distribution system for our example has 10,000 feet [3,050 meters] of high pressure piping at 600 psig [120 bar] connected to 4,000 feet [1,220 meters] of 175 psi [12 bar] sub-system piping. Figure 5.1 shows a typical steam distribution and condensate recovery system which will serve for our example.

In this chapter we will examine how a typical steam system loses energy and becomes inefficient. We will examine the following topics:

---

1. Trap flash losses

2. Insulation losses

3. Condensate system losses

4. External steam leaks

5. Internal steam leaks

6. Steam trap leakage

7. Inactive steam piping.

Figure 5.2—major steam system losses which will be covered.

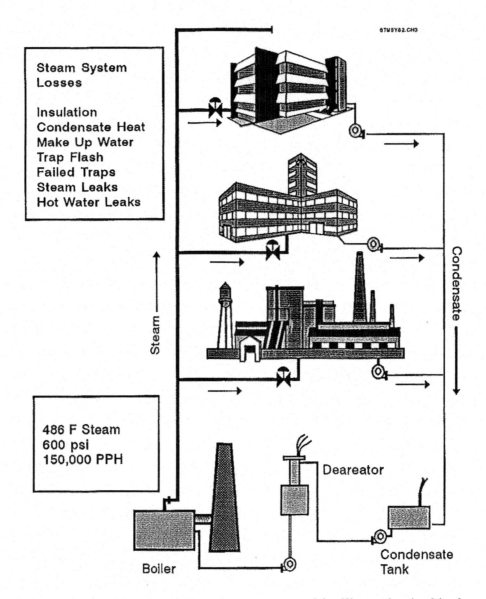

Figure 5.1—A typical steam distribution system used for illustration in this chapter

The steam load on the system in this case is 100,000 pounds [50 tons] of steam per hour. Total energy being delivered to the distribution system is 105 million Btus per hour. The cost of steam is $4.70 per million Btus.

The 600 psi steam has an enthalpy of 1204 Btu per pound. The make-up water for this system is 62 F [17C], containing 30 Btus per pound. The energy value for the steam is actually 1174 Btu/lb when we subtract the heat value of our "heat-sink"; the 62 F makeup water.

One must be aware of the fact that conditions in a boiler plant and distribution system are changing constantly and that instruments either don't exist or have become inaccurate over years in a harsh environment. It was once said by an old and seasoned plant engineer when commenting on his instrumentation: *"accurate to two hammer handles—more or less.* Numbers have been rounded off in this chapter to communicate important information in a less complex way.

FLASH STEAM

The largest energy loss in steam systems is flash steam in open condensate recovery systems. When the hot condensate passes through a trap from a high energy level to a lower energy level and percentage of the water flashes into steam. Figure 5.3 illustrates this energy balance across a steam trap.

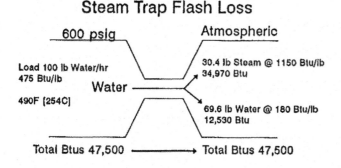

Steam Trap Flash Loss

Figure 5.3—Flash steam forms on the low pressure side of the steam trap, in this case with almost 3 times as much energy as the condensate.

The formula for flash steam is simple; the difference in enthalpy [Btu/lb] of the water on either side of the steam trap is divided by the Btus required to form a pound of steam at the lower pressure.

$$\frac{475 - 180}{970} = 30.4\% \text{ Flash Steam}$$

All steam used by the system must pass through steam traps, so the whole system is subject to this loss if atmospheric vents are used. Closed pressurized condensate recovery systems are used in well designed plants to avoid this problem.

Flash steam percentages vary with pressure drop across steam traps, for example exhausting to an open (atmospheric) system:

600 psi trap has a 30% flash
175 psi trap has a 17% flash
50 psi trap has a 4% flash

This is why high pressures are used in distribution systems and pressure reducing valves (PRV) are used close to the point of use of the steam.

Let's look at an open system, in this case the 175 psi [12 Bar] steam system in our example which is delivering steam to the points of use. Assuming there is a load on this system of 75,000 pounds of steam per hour passing through steam traps, the flash rate is about 17%. This will produce 12,000 pounds of flash steam per hour costing over $480,000 a year. This is an 11% system loss.

This is a basic problem of these systems, flash steam losses as great as total boiler plant losses. High temperature hot water systems do not have a number of problems associated with steam systems.

HIGH TEMPERATURE HOT WATER SYSTEMS

High temperature water distribution systems are divided into three categories:

1. High Temperature Hot water systems have a temperature range from 350 F [177C] to 450 F [232C].

2. Medium Temperature Hot water systems have a temperature between 250 F [121C] and 350F [177C].

3. Low Temperature Hot water systems are usually below 250F [121C].

It is interesting to note that one cubic foot of hot water at 350F contains 64 times as much heat as a cubic foot of saturated steam at 350F [177C]. When high temperature hot water cools from 350F to 25°F, it releases 40 times more heat than a cubic foot of steam.

Other features of high temperature water systems are:

1. High Temperature Water systems are closed systems and should only use one half of one percent make up water compared to boilers which use quite a lot; in some cases 100%.

2. They do not have the usual losses associat-

ed with steam systems, such as steam trap flash losses, steam trap malfunctions [often as high as 40%], packing gland leakage, water hammer damage and high temperature stresses caused by rapid temperature changes.

3. Because of the higher heat content, piping sizes are smaller saving on construction costs.

4. On the down side, pumps are required to circulate the hot water throughout the system.

INSULATION LOSSES

All sections of hot pipe, valve bodies, unions, flanges and most steam traps should be insulated. Insulation will pay for itself quickly, less than one half year, as well as prevent the overheating of working spaces and the elimination of hot and dangerous surfaces.

Steam users should be fully aware of the need to insulate hot surfaces but it takes constant vigilance to keep steam systems in good condition.

In our example we have 10,000 ft [3,050 meters] of well insulated piping which has an average heat loss of 200 Btus per lineal ft per hour, this works out to 2 million Btus per hour which is 1.9% of the system load costing $82,344 a year.

Our system also includes 10,000 feet [3,050 meters] of condensate return piping plus a number of tanks and pumps. The criteria for surface temperature of insulated piping us generally agreed to be about 140F [60C], so in the case of condensate return piping insulation is often not used because it is uneconomical and the surface temperature will not normally exceed 140 F [60C]. Under these conditions, surface losses will be about 60 Btu per lineal ft per hour. For our example, this loss will be 0.6 million Btus an hour costing 24,700 a year.

In the 4,000 ft [1,220 meter] system at 175 psi, because of lower temperatures and Btu values, the losses are less; 120 Btu per lineal ft. The insulation losses in this section are about half a million Btus per hour with an annual cost of about

$20,000. Flash steam losses are much less at 72,000 Btu per hour, worth about $3,000 a year.

FLASH STEAM LOSSES FROM INSULATION LOSSES

There is another significant loss resulting from the insulation losses. When steam in the piping loses energy through the pipe wall, it condenses then goes through the steam trap as water. In this case of the 600 psi steam, the water temperature will drop from 489F to 212F across the trap.

The Btu content of the water at 600 psi is 475 Btu per pound before the trap and is only 180 Btu/lb on the low pressure side of the trap. At atmospheric pressure, 970 Btus/lb are required to form a pound of steam. This results in about 30. 4% flash steam.

With an average surface loss of 200 Btu per lineal ft for 10,000 ft [3050 meters] of piping, insulation losses from the 600 psi system will form 2,743 pounds of water per hour. One pound of water forms for each 729.1 Btu lost from the pipe. So, with the 30.4% flash rate, 834 pounds of steam per hour will be produced. This steam will have a heat value of 1120 Btu/lb. If this flash steam is lost through atmospheric vents it will cost $39,522 a year.

BARE PIPING LOSSES

If the 600 psi [40 Bar] piping system were not insulated but bare, it would lose about 3,400 Btu per lineal foot rather than 200 Btu per ft, a factor of 17 greater. In the 175 psi [12 Bar] system the value for bare pipe is 1200 Btu per lineal ft per hour compared to 120 Btu for economical insulation. The increase of losses from bare pipe is a factor of 10 greater than economical insulation.

The North American Insulation Manufacturers Association (NAIMA) has developed software applications to calculate heat losses from bare pipe and for various combinations of insulation. This software solves these problems rather handily and often predicts that insulating bare pipe will pay off in less than a month.

MEASURING INSULATION LOSSES

A useful way to measure insulation losses is to weigh the condensate produced by a section of piping under study. This is sometimes called the barrel test as it requires a barrel, a scale and a drain connection to the discharge end of the steam trap. If the piping being tested is typical of the entire system, then an overall estimate of system performance is possible. The effectiveness of insulation under different conditions can also be studied by the change in condensate formed when the piping is exposed to variable conditions of weather such as changing temperatures, humidity, rain and wind conditions.

CONDENSATE SYSTEM LOSSES

Recycling hot condensate for reuse as boiler feed water is an important way to maintain the efficiency of the system. The energy used to heat cold make-up water is a major part of the heat delivered for use by the steam system.

For example, in the system under study, the condensate should return at 182 F [83 C], but if it returns to the boiler plant at 127 F [53 C]; 55 F [30 C] cooler, it will cause a loss of 5 million Btu/hr, a 4.8% energy loss costing $242,000 a year.

If no condensate is returned to the plant and 62 F [17 C] make-up water is used instead, this will cause a 11.4% loss costing $581,000 a year.

In this example we can see that heating cold make-up water requires a significant energy investment; 11.7% of the heat in the steam.

The formula below illustrates this point:

a. Heat loss 120 Btu/lb, using 62 F water rather than 182 F condensate return.

b. Heat in Steam 1204 Btu/lb

c. Boiler Efficiency 85%

$$\frac{120}{1204 \times 0.85} = 11.7\%$$

EXTERNAL STEAM LEAKS

External steam leaks are usually visible and the length of the plume can be measured by eye. The following chapter on steam systems has more information on this. In our example system there are 4 visible steam leaks, two 6 foot plumes and two 8 foot plumes. This works out to an $84,000 a year loss plus the cost of water, chemicals and fuel wasted heating make up water to operating temperature.

INTERNAL STEAM LEAKS

Internal steam leaks are hard to detect but common sense indicates there could be many such leaks. Steam traps are installed with bypass valves which can be opened in the event there is a problem with the trap. The bypass valves are often opened on start up or when water hammer becomes a problem in the system. Steam driven equipment such as turbines and pumps often have drains and warm up systems and these can be left open too. In our example 10 traps had partially open bypass valves averaging 75 pounds per hour. This was costing $26,500 a year.

STEAM TRAP MALFUNCTIONS

It is not unusual to find that half of the steam traps in a plant aren't functioning properly. They work in a harsh environment and have an uncertain lifetime measured from 3 to 5 years. In our example we have uncovered 35 defective traps wasting on the average of 50 pounds of steam per hour costing about $62,000 a year. [See chapter on steam traps]

*When conditions are not observable or known,
the assumption is often made that
they are negligible.*

*In steam systems most of the losses
are not observable.
Unfortunately, the losses are quite
large and jar from negligible.*

In this chapter we have examined information about various ways steam distribution systems control and ultimately waste energy including various energy loss mechanisms and their relative costs.

For the planner and designer, eliminating life cycle system losses should be a primary concern right from the drawing board. For the maintenance force, pipe insulation, steam trap testing and repair, steam leak prevention, and condensate recovery system maintenance can have a significant impact on operating costs. In the boiler plant monitoring the amount of make up water and the condensate return temperature are important indicators of the health of the distribution system. If the pure hot water is not brought back to the plant by the condensate recovery system, then the expense of new water, chemical treatment and heating it to operating temperature will just add to avoidable plant expenses.

Loss Summary

Distribution System Losses	Percent	Dollars/yr
Insulation Losses		
600 psi (10,000 ft.)	1.90%	$82,344
175 psi (4,000 ft)	0.46%	$19,763
Condensate Recovery sys (10,000 ft)	0.57%	$24,703
Flash Steam Load		
600 psi (surface losses)	0.91%	$39,522
175 psi (surface losses &	11.14%	$481,514
75,000 pph load)		
External Steam Leaks	1.94%	$84,000
Internal Steam Leaks	0.61%	$26,556
Steam Trap Failures	1.43%	$61,964
	18.51%	$820,360

What If

		Percent	Dollars/yr
(1)	All steam piping were bare:		
	600 psi system surface losses	32.38%	$1,399,848
	600 psi flash losses	33.32%	$1,440,585
	175 psi system surface losses	4.57%	$197,626
	175 psi flash losses	15.75%	$680,869
		86.02%	$3,718,928
(2)	Condensate Recovery System		
	A 55 F [13C] drop		
	in system Temperature	4.76%	$242,131
(3)	No condensate returns		
	100% make up @ 62F [17C]	11.43%	$581,115

Chapter 6

Efficiency Calculation Methods

A key part of a boiler efficiency improvement program is knowing the operating efficiency of the boiler and the corresponding increase in efficiency from as-found conditions to the final optimized condition. This may require several efficiency tests over an extended period.

The following paragraphs discuss the various test methods and computational procedures available for measuring efficiency.

ASME COMPUTATIONAL PROCEDURES

The basis for testing boilers is the American Society of Mechanical Engineers Power Test Code 4.1.

Figure 6.1 shows the ASME Test Form for Abbreviated Efficiency Test or so called "ASME Short form" which is used for both the Input-Output and heat loss methods. Figure 6.2 is the calculation sheet for the abbreviated ASME efficiency test.

Both Heat Loss and Input-Output boiler efficiency calculations are included in the ASME short Form. This power test code has become the standard test procedure in many countries. It neglects minor efficiency losses and heat credits, considering only the Higher Heating Value of the input fuel.

COMPARISON OF THE INPUT-OUTPUT AND HEAT LOSS METHODS

Both methods are mathematically equivalent and would give identical results if the required heat balance (or heat loss) factors were considered in the corresponding boiler measurements could be performed without error.

When very accurate instrumentation and testing techniques are used, there is reasonably good agreement between the two calculation procedures. However, for practical boiler tests with limited instrumentation, comparisons between the two methods are generally poor. The poor results are primarily from the inaccuracies associated with the measurement of the flow and energy content of the input and output streams.

The efficiencies determined by these methods are "gross" efficiencies as opposed to "net" values which would include as additional heat input the energy required to operate all the boiler auxiliary equipment (i.e. combustion air fans, fuel pumps, fuel heaters, stoker drives, etc.)

These "gross" efficiencies can be considered essentially as the effectiveness of the boiler in extracting the available heat energy of the fuel. It is important to take complete data when using this test form to fully document the test results no matter what procedure is used.

THE INPUT-OUTPUT METHOD

$$\text{Efficiency (percent)} = \frac{\text{Output}}{\text{Input}} \times 100$$

This method measures the heat absorbed by the water and steam and compares it to the total energy input of the higher heating value (HHV) of the fuel.

This method requires the accurate measurement of fuel input. Also, accurate data must be available on steam pressure, temperature and flow, feed water temperature, stack temperature, and air temperature to complete energy balance calculations.

Figure 6.3 illustrates the envelope of equipment included in the designation of "Steam Generating Unit."

Figure 6.4 shows the relationship between input, output, credits and losses.

ASME TEST FORM
FOR ABBREVIATED EFFICIENCY TEST

SUMMARY SHEET

PTC 4.1-a (1964)

		TEST NO.		BOILER NO.	DATE
OWNER OF PLANT		LOCATION			
TEST CONDUCTED BY		OBJECTIVE OF TEST			DURATION
BOILER MAKE & TYPE		RATED CAPACITY			
STOKER TYPE & SIZE					
PULVERIZER, TYPE & SIZE		BURNER, TYPE & SIZE			
FUEL USED	MINE	COUNTY	STATE		SIZE AS FIRED

PRESSURES & TEMPERATURES — FUEL DATA

	PRESSURES & TEMPERATURES				COAL AS FIRED PROX. ANALYSIS	% wt		OIL	
1	STEAM PRESSURE IN BOILER DRUM	psia							
2	STEAM PRESSURE AT S. H. OUTLET	psia		37	MOISTURE		51	FLASH POINT F°	
3	STEAM PRESSURE AT R. H. INLET	psia		38	VOL MATTER		52	Sp. Gravity Deg. API°	
4	STEAM PRESSURE AT R. H. OUTLET	psia		39	FIXED CARBON		53	VISCOSITY AT SSU° BURNER SSF	
5	STEAM TEMPERATURE AT S. H. OUTLET	F		40	ASH		44	TOTAL HYDROGEN % wt	
6	STEAM TEMPERATURE AT R H INLET	F			TOTAL		41	Btu per lb	
7	STEAM TEMPERATURE AT R.H. OUTLET	F		41	Btu per lb AS FIRED				
8	WATER TEMP. ENTERING (ECON.) (BOILER)	F		42	ASH SOFT TEMP.° ASTM METHOD			GAS	% VOL
9	STEAM QUALITY % MOISTURE OR P.P.M.				COAL OR OIL AS FIRED ULTIMATE ANALYSIS		54	CO	
10	AIR TEMP. AROUND BOILER (AMBIENT)	F		43	CARBON		55	CH₄ METHANE	
11	TEMP AIR FOR COMBUSTION (This is Reference Temperature) †	F		44	HYDROGEN		56	C₂H₂ ACETYLENE	
12	TEMPERATURE OF FUEL	F		45	OXYGEN		57	C₂H₄ ETHYLENE	
13	GAS TEMP. LEAVING (Boiler) (Econ.) (Air Htr.)	F		46	NITROGEN		58	C₂H₆ ETHANE	
14	GAS TEMP. ENTERING AH (If conditions to be corrected to guarantee)	F		47	SULPHUR		59	H₂S	
	UNIT QUANTITIES			40	ASH		60	CO₂	
15	ENTHALPY OF SAT. LIQUID (TOTAL HEAT)	Btu/lb		37	MOISTURE		61	H₂ HYDROGEN	
16	ENTHALPY OF (SATURATED) (SUPERHEATED) STM.	Btu/lb			TOTAL			TOTAL	
17	ENTHALPY OF SAT. FEED TO (BOILER) (ECON.)	Btu/lb			COAL PULVERIZATION			TOTAL HYDROGEN % wt	
18	ENTHALPY OF REHEATED STEAM R.H. INLET	Btu/lb		48	GRINDABILITY INDEX°		62	DENSITY 68 F ATM. PRESS.	
19	ENTHALPY OF REHEATED STEAM R.H. OUTLET	Btu/lb		49	FINENESS % THRU 50 M°		63	Btu PER CU FT	
20	HEAT ABS/LB OF STEAM (ITEM 16 – ITEM 17)	Btu/lb		50	FINENESS % THRU 200 M°		41	Btu PER LB	
21	HEAT ABS/LB R.H. STEAM (ITEM 19 – ITEM 18)	Btu/lb		64	INPUT-OUTPUT EFFICIENCY OF UNIT %		ITEM 31 × 100 / ITEM 29		
22	DRY REFUSE (ASH PIT + FLY ASH) PER LB AS FIRED FUEL	lb/lb			HEAT LOSS EFFICIENCY			Btu/lb A. F. FUEL	% of A. F. FUEL
23	Btu PER LB IN REFUSE (WEIGHTED AVERAGE)	Btu/lb		65	HEAT LOSS DUE TO DRY GAS				
24	CARBON BURNED PER LB AS FIRED FUEL	lb/lb		66	HEAT LOSS DUE TO MOISTURE IN FUEL				
25	DRY GAS PER LB AS FIRED FUEL BURNED	lb/lb		67	HEAT LOSS DUE TO H₂O FROM COMB. OF H₂				
	HOURLY QUANTITIES			68	HEAT LOSS DUE TO COMBUST. IN REFUSE				
26	ACTUAL WATER EVAPORATED	lb/hr		69	HEAT LOSS DUE TO RADIATION				
27	REHEAT STEAM FLOW	lb/hr		70	UNMEASURED LOSSES				
28	RATE OF FUEL FIRING (AS FIRED wt)	lb/hr		71	TOTAL				
29	TOTAL HEAT INPUT (Item 28 × Item 41)/1000	kB/hr		72	EFFICIENCY = (100 – Item 71)				
30	HEAT OUTPUT IN BLOW-DOWN WATER	kB/hr							
31	TOTAL HEAT OUTPUT (Item 26×Item 20)+(Item 27×Item 21)+Item 30 /1000	kB/hr							

FLUE GAS ANAL. (BOILER) (ECON) (AIR HTR) OUTLET

32	CO₂	% VOL	
33	O₂	% VOL	
34	CO	% VOL	
35	N₂ (BY DIFFERENCE)	% VOL	
36	EXCESS AIR	%	

° Not Required for Efficiency Testing

† For Point of Measurement See Par. 7.2.8.1-PTC 4.1-1964

Figure 6.1—ASME test form summary sheet for abbreviated efficiency testing.

PTC 4.1-b (1964)

ASME TEST FORM

CALCULATION SHEET FOR ABBREVIATED EFFICIENCY TEST *Revised September, 1965*

	OWNER OF PLANT		TEST NO.	BOILER NO.	DATE

| 30 | HEAT OUTPUT IN BOILER BLOW-DOWN WATER = LB OF WATER BLOW-DOWN PER HR × $\dfrac{\text{ITEM 15} - \text{ITEM 17}}{1000}$ = ... kB/hr |

| 24 | *If impractical to weigh refuse, this item can be estimated as follows*

 DRY REFUSE PER LB OF AS FIRED FUEL = $\dfrac{\% \text{ ASH IN AS FIRED COAL}}{100 - \% \text{ COMB. IN REFUSE SAMPLE}}$

 CARBON BURNED PER LB AS FIRED FUEL = $\dfrac{\text{ITEM 43}}{100} - \left[\dfrac{\text{ITEM 22} \times \text{ITEM 23}}{14,500}\right]$ = ... | NOTE: IF FLUE DUST & ASH PIT REFUSE DIFFER MATERIALLY IN COMBUSTIBLE CONTENT, THEY SHOULD BE ESTIMATED SEPARATELY. SEE SECTION 7, COMPUTATIONS. |

| 25 | DRY GAS PER LB AS FIRED FUEL BURNED = $\dfrac{11CO_2 + 8O_2 + 7(N_2 + CO)}{3(CO_2 + CO)}$ × (LB CARBON BURNED PER LB AS FIRED FUEL + $\frac{3}{8}$ S)

 = $\dfrac{11 \times \text{ITEM 32} + 8 \times \text{ITEM 33} + 7 \left(\text{ITEM 35} + \text{ITEM 34}\right)}{3 \times \left(\text{ITEM 32} + \text{ITEM 34}\right)}$ × $\left[\text{ITEM 24} + \dfrac{\text{ITEM 47}}{267}\right]$ = ... |

| 36 | EXCESS AIR † = $100 \times \dfrac{O_2 - \frac{CO}{2}}{.2682 N_2 - (O_2 - \frac{CO}{2})}$ = $100 \times \dfrac{\text{ITEM 33} - \frac{\text{ITEM 34}}{2}}{.2682 (\text{ITEM 35}) - (\text{ITEM 33} - \frac{\text{ITEM 34}}{2})}$ = ... |

	HEAT LOSS EFFICIENCY	Btu/lb AS FIRED FUEL	$\dfrac{\text{LOSS}}{\text{HHV}} \times 100 =$	LOSS %
65	HEAT LOSS DUE TO DRY GAS = $\dfrac{\text{LB DRY GAS}}{\text{PER LB AS FIRED FUEL}}$ × C_p × ($t_{lvg} - t_{air}$) = ITEM 25 × 0.24 (ITEM 13) – (ITEM 11) = ...	......	$\dfrac{65}{41} \times 100 =$	......
66	HEAT LOSS DUE TO MOISTURE IN FUEL = $\dfrac{\text{LB } H_2O \text{ PER LB}}{\text{AS FIRED FUEL}}$ × [(ENTHALPY OF VAPOR AT 1 PSIA & T GAS LVG) – (ENTHALPY OF LIQUID AT T AIR)] = $\dfrac{\text{ITEM 37}}{100}$ × [(ENTHALPY OF VAPOR AT 1 PSIA & T ITEM 13) – (ENTHALPY OF LIQUID AT T ITEM 11)] = ...	......	$\dfrac{66}{41} \times 100 =$	......
67	HEAT LOSS DUE TO H_2O FROM COMB. OF H_2 = $9H_2$ × [(ENTHALPY OF VAPOR AT 1 PSIA & T GAS LVG) – (ENTHALPY OF LIQUID AT T AIR)] = $9 \times \dfrac{\text{ITEM 44}}{100}$ × [(ENTHALPY OF VAPOR AT 1 PSIA & T ITEM 13) – (ENTHALPY OF LIQUID AT T ITEM 11)] = ...	......	$\dfrac{67}{41} \times 100 =$	......
68	HEAT LOSS DUE TO COMBUSTIBLE IN REFUSE = ITEM 22 × ITEM 23 = ...	......	$\dfrac{68}{41} \times 100 =$	......
69	HEAT LOSS DUE TO RADIATION* = $\dfrac{\text{TOTAL BTU RADIATION LOSS PER HR}}{\text{LB AS FIRED FUEL}}$ – ITEM 28	......	$\dfrac{69}{41} \times 100 =$	......
70	UNMEASURED LOSSES **	......	$\dfrac{70}{41} \times 100 =$	......
71	TOTAL	......	......	......
72	EFFICIENCY = (100 – ITEM 71)	......	......	......

† For rigorous determination of excess air see Appendix 9.2 – PTC 4.1-1964
* If losses are not measured, use ABMA Standard Radiation Loss Chart, Fig. 8, PTC 4.1-1964
** Unmeasured losses listed in PTC 4.1 but not tabulated above may by provided for by assigning a mutually agreed upon value for Item 70.

Figure 6.2—ASME calculation sheet for abbreviated efficiency testing.

Because of the many physical measurements required at the boiler and the potential for significant measurement errors, the Input/Output method is not practical for field measurements at the majority of industrial and commercial boiler installations where precision instrumentation is not available.

Large errors are possible because this method relies on the difference in large numbers. If the steam flow is off by 2-3% and other instrumentation have a similar level of error, then the cumulative error can become unacceptable, producing false information.

The Input-Output test method is also very labor intensive. Precision instrumentation must be specified and installed. Test runs are usually more than four hours and must be rejected for any inconsistent data. Trial runs are often required to check out instrumentation and identify problems with the boiler as well as to train test personnel and observers.

Often plants cannot support testing for long periods because of load considerations. For example, a plant may not be able to provide either full load or partial load conditions for extended periods for various reasons which can cause premature curtailment of tests.

Problems like inconsistent water level control or variations in outlet steam pressure can prevent the stable thermal balance required for accurate test information.

THE HEAT LOSS METHOD

Efficiency(%) = 100 % – Heat Loss %

The Heat Loss method subtracts individual energy losses from 100% to obtain percent efficiency. It is recognized as the standard approach for routine efficiency testing, especially at industrial boiler sites where instrumentation quite often is minimal.

The losses measured are:
1. Heat loss due to dry gas.
2. Heat loss due to moisture in fuel.
3. Heat loss due to Hp from the combustion of hydrogen.

4. Heat loss due to combustibles in refuse. (for coal)
5. Heat loss due to radiation.
6. Unmeasured losses.

This procedure neglects minor efficiency losses and heat credits and only considers the chemical heat (Higher Heating Value) of the fuel as energy input.

In addition to being more accurate for field testing, the heat loss method identifies exactly where the heat losses are thus aiding energy savings efforts.

This method might be termed the flue gas analysis approach since the major heat losses considered by this method are based on measured flue gas conditions at the boiler exit together with an analysis of the fuel composition.

This method requires the determination of the exit flue gas excess O_2 (or CO_2), CO, combustibles, temperature and the combustion air temperature.

The heat loss method is a much more accurate and more accepted method of determining boiler efficiencies in the field provided that the measurements of the flue gas conditions are accurate and not subject to air dilution or flue gas flow stratification or pocketing.

COMBUSTION HEAT LOSS TABLES

Tables for stack gas heat losses for different types of fuels have been prepared separately and are available for determining flue gas losses due to dry gas and moisture losses.

HEAT LOSS DUE TO RADIATION

Radiation loss, not associated with the flue gas conditions can be estimated from the standard curve given in the ASME Power Test Code.

Radiation Joss can also be measured using a simple direct method, using an infrared instrument which has specially designed to detect radiation losses from the boiler surface and gives a read out in Btu/sq-ft/hr.

PTC 4.1 – 1964

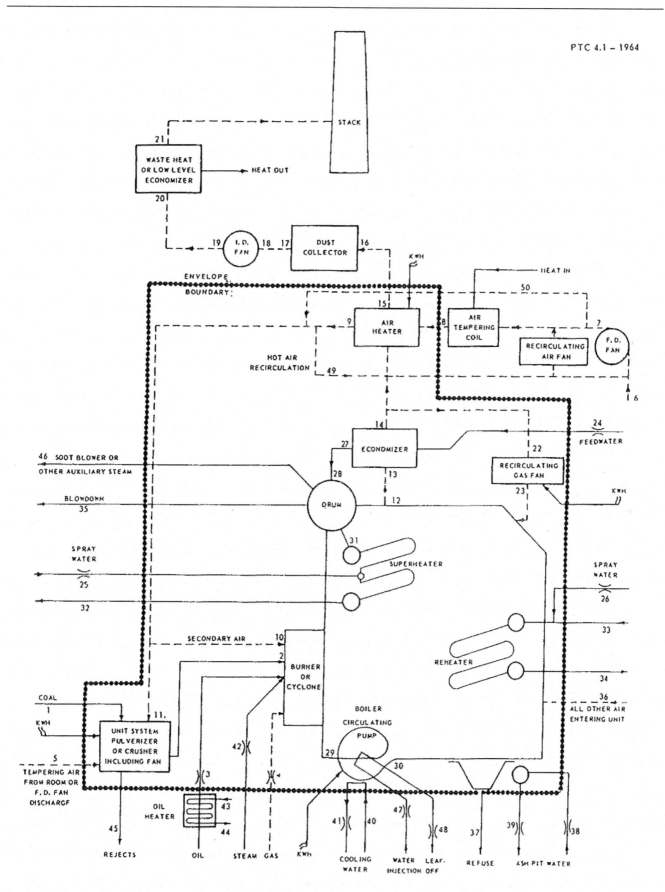

Figure 6.3—ASME designated envelop for steam generating unit.

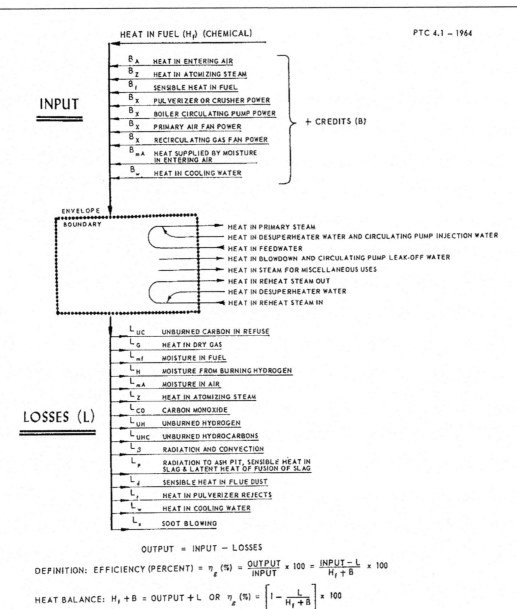

Figure 6.4—ASME Heat Balance for steam generators.

The technique is to make a drawing of the boiler surface dimensions and make grids of equal areas with the average measured heat loss for each grid then adding up the losses for all the surface grids. [See Figure 4.10]

COMPARING METHODS FOR MEASURING BOILER EFFICIENCY

Input-Output Method
   a.   Most direct method
   b.   Difficult and expensive to measure accu-

rately
   c.   Does not locate energy losses

Heat Loss Method
   a.   Indirect method (100% – energy losses)
   b.   Simple and accurate
   c.   locates and sets magnitude of energy losses and therefore is a key to efficiency improvement efforts. d. Allows assessment of potential efficiency improvements and energy savings.

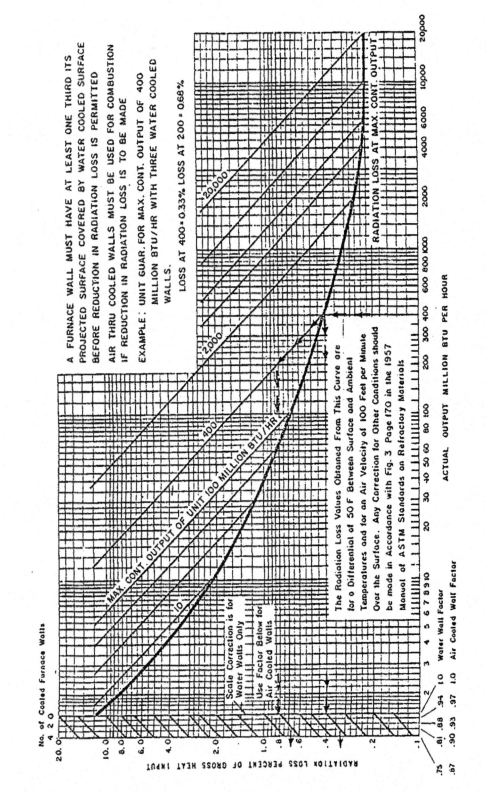

Standard Radiation Heat Loss Chart. (Courtesy of the American Boiler Manufacturers Association.)

Stack Gas Heat Loss Solution

To calculate the stack heat loss using the ASME Short Form, it is necessary to make separate calculations of the heat losses due to dry gas, moisture in the fuel and moisture in the flue gas due to hydrogen in the fuel. The three heat losses are then totaled to form the stack gas heat loss.

Calculation of Stack Gas Heat Loss

Heat loss due to Dry Gas $= P_f \times 0.24 \times (T_g - T_a) =$ BTU/LB (As Fired Fuel)

$$P_f = \frac{LB\ Dry\ Gas}{LB\ as\ fired\ fuel} = \frac{11\ CO_2 + 8\ O_2 + 7(CO + N_2)}{3(CO2 + CO)} \times \left(\frac{\%C}{100} + \frac{\%S}{267} \right)$$

T_g = Flue gas temperature leaving unit, degrees F

T_a = Combustion air temperature entering unit, degrees F

%C, %S = Percent by weight in fuel analysis of carbon and sulfur

O_2, CO_2, CO, N_2 = Percent by volume in flue gas.

Heat Loss Due to Moisture formed from H_2 in fuel $= \frac{9\left(H_2 \right)}{100} \times (h_g - h_a) =$ BTU/LB (As Fired Fuel)

Heat Loss Due to Moisture in Fuel $= \%\ Moisture \times (h_g - h_a) =$ BTU/LB (As Fired Fuel)

H_2 = % hydrogen in as fired fuel by weight

% Moisture = % moisture in as fired fuel by weight

h_g = Enthalpy of water vapor at 1 psia and T_g

h_a = Enthalpy of liquid water at T_a

Excess Air Solution

$$Excess\ Air = 100 \times \frac{O_2 - CO/2}{.2682N_2 - (O_2 - CO/2)}$$

Chapter 7

Combustion Analysis

COMBUSTION ANALYSIS

Understanding the combustion process is very important for safe and efficient operation of a plant. Perfect combustion is the proper mixture of fuel and air under exacting conditions where both the oxygen and the fuel are completely consumed in the combustion process. Having just the right amount of oxygen (no more, no less) is called the stoichiometric point, simply the ideal air to fuel ratio for combustion.

Anything above the ideal amount of air supplied to the combustion process is called Excess Air, and is wasteful.

A CO_2 analysis alone does not provide a safe indication of the combustion air/fuel setting. Additional requirements of either smoke or CO is recommended as the same CO_2 measurements can occur on either side of stoichiometric.

Excess air is the preferred term to describe the combustion setting on the safe side of stoichiometric. Using oxygen measurements is the best way find excess air.

By understanding a few simple instruments most of the potentially hazardous conditions can be reduced. The key is to properly measure smoke, oxygen, carbon monoxide, draft and gas pressure.

CAUTION

A most dangerous approach, when dealing with combustion systems is thinking the systems will always be correct and not considering that it can be affected by small and seemingly unrelated external forces.

The following is a list of key parameters considered as important safety measurement areas:

- Input gas pressure
- Draft (over fire and boiler exhaust)
- Carbon Monoxide
- Stack temperature
- Smoke
- Combustibles

Too much smoke is one of the most common indicators of excessive fuel wastage and can cause major problems.

STACK FIRES

Buildup of combustibles in the exhaust system and chimneys which can cause fires and explosions. To a lesser extent a build up of soot in the exhaust system can block the normal passage of flue gases further restricting the amount of oxygen supplied for combustion progressively compounding the problem.

SOURCES OF PROBLEMS

Four basic combustion zone conditions that prevent clean, efficient combustion are:

- Insufficient combustion air applied to the flame to permit clean combustion at an acceptable combustion efficiency

- Non-uniform delivery of fuel/or combustion air to the combustion zone.

- Insufficient temperature of the combustion zone to permit proper burning of the fuel.

• Insufficient flame turbulence or inadequate mixing of fuel and air during the vaporization and burning.

These points are precisely why the smoke test is an indispensable part of oil burner servicing, accurate smoke testing takes less than a minute.

THE SMOKE SPOT TEST

The smoke test method has long been recognized as the acceptable standard for oil burners. It is used by Underwriters Laboratories in approval testing of oil burners and it is the test method specified by the National Oil Fuel Institute and by the U.S. Department of Commerce, in their standard for testing materials; (02156-65) and also in the European standard DIN 51402.

All three components of the smoke test set (pump, scale and filter paper) must meet the U.S. commercial standards (Figure 7.1).

The ten spots on the scale range in equal photometric steps from white to black to completely cover any smoke condition which may be experienced.

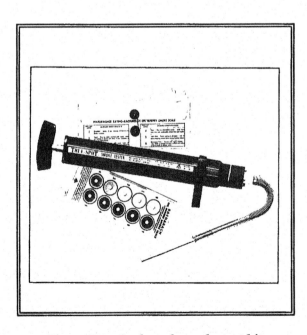

Figure 7.1—Bacharach smoke test kit.

Use of the Smoke Tester, Three Simple Steps (Figure 7.2)

1. Clamp the filter strip into the test pump and insert into the flue.

2. Pump strokes.

3. Remove filter paper strip and place between smoke scale and white reflective and compare smoke test spot on filter paper with smoke spots on scale. The smoke reading is the closest match.

Interpreting the Results

Not all kinds of oil burners will be equally affected by the same smoke content in the flue gas, this fact is shown in Figure 7.3 which interprets smoke scale readings in terms of sooting produced. Depending on the construction of the heat exchanger or the boiler, some units will accumulate soot rapidly with a number 3 smoke spot number, while accumulation of soot on other units at the same smoke spot reading may be relatively slower.

POSSIBLE CAUSES OF SMOKE CONDITIONS

Condition

If the Smoke Spot Number is too high, soot deposits in furnace and on heat exchange surfaces will lead to poor efficiency.

Cause:
1. If excess air is low 5%-20%
   a. Over firing
   b. Too little excess air
2. If excess air is high above 50%
   a. Faulty nozzle inefficient atomization of fuel.
   b. Combustion chamber trouble
   c. Chilling of the combustion process before complete combustion occurs.

HIGH STACK TEMPERATURE

A high stack temperature may indicate any of the following conditions and 'should be immediately checked and remedied:

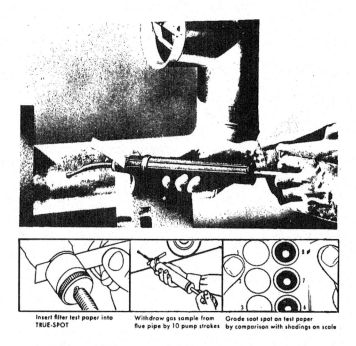

Insert filter test paper into   Withdraw gas sample from   Grade soot spot on test paper
TRUE-SPOT   flue pipe by 10 pump strokes by comparison with shadings on scale

Figure 7.2—Smoke spot testing procedure.

Effect of Smoke on Burner Performance		
Smoke Scale Number	Rating	Sooting Produced
1	Excellent	Extremely light if at all
2	Good	Slight sooting that may not increase stack temperature appreciably
3	Fair	May be some sooting but will rarely require cleaning more than once a year
4	Poor	Borderline condition. Some units will require cleaning more than once a year
5	Very Poor	Sooting occurs rapidly and heavily

Figure 7.3—Smoke spot rating scale.

1. Soot deposits on heat exchange surfaces.

2. Short circuits of hot combustion gases due to problems with baffles.

3. Over firing, check fuel rate.

4. Water side scale deposits from improper water treatment.

5. High excess air which reduces combustion chamber heat transfer.

LOW STACK TEMPERATURE

If the stack temperature drops lower than normal, lower than 250 F for natural gas or 275-300 F for oil.

1. Possible under firing.

2. Incomplete combustion of gas with dangerous carbon monoxide production.

3.   Possible ruptured boiler tube or other component which is cooling the gases with steam or water.

4.   In negative draft units, cold air may be entering boiler through open door or defective wall or skin.

THE DANGERS OF CARBON MONOXIDE

An example of the dangers of carbon monoxide is evident in a report from Canada. Between 1973 and 1983 there were 293 reports of carbon monoxide poisoning, including 145 deaths. Also, combustion systems caused 238 deaths during the same period.

What is Carbon Monoxide?

It is the product of incomplete combustion and is a flammable colorless and odorless gas. Carbon monoxide is about the same density as the air that we breath. Therefore, easily mixes to form a deadly atmosphere.

The major hazards of carbon monoxide are it's toxicity and flammability. Carbon monoxide becomes a combustible gas when its concentration reaches 12.5% by volume (125,000 ppm).

Carbon monoxide is classified, however, as a chemical asphyxiant which produces a toxic action by preventing the blood from absorbing oxygen. Since the affinity of carbon monoxide is 200-300 times that of oxygen in blood, even small amounts of carbon monoxide in the air will cause toxic reactions to occur.

If breathed for a sufficiently long time, a carbon monoxide concentration of only 50 ppm will produce symptoms of poisoning. As little as 200 ppm will produce slight symptoms like a headache or discomfort in just a few hours. A concentration of 400 ppm will produce a headache and discomfort in two to three hours. The effect at higher concentrations may be so sudden that a person has little or no warning before collapsing. It should be noted that all of these values are approximate and vary as to the individual.

To prevent over-firing and under-firing the fuel supply pressure must be tested from time to time to insure the firing rate has not shifted because of a pressure change.

In small boilers or furnaces, a daft measurement is necessary to guard against a gas reversal where exhaust gases are escaping to the environment, which is potentially toxic. A small draft gage will indicate problems, like building exhaust fans pulling gases back down the chimney, obstructions in the exhaust system, down drafts from high wind conditions and defective stack covers and other problems.

CHIMNEY EFFECT

When taking draft readings, insure that the system is warmed up to normal operating temperatures. As air is warmed, it expands and the same weight of air will take up more space becoming lighter. This warm "light" air will rise up the chimney decreasing the furnace pressure. If this low pressure is not established, combustion products may escape. If the draft is too high, and the hot gases are creating too much negative draft heat will be lost up the stack.

CONDENSING FLUE GASES TO IMPROVE EFFICIENCY

Fuel is a hydrocarbon which means that it is made up of hydrogen and carbon. Carbon burns dry but each pound of hydrogen that enters into the combustion process forms about 9 pounds of water. Now, at the 2,000 to 3,000 degree combustion temperature this water is in the form of steam and it carries a considerable amount of latent heat. If this latent heat can be extracted from the exhaust gases, there is an opportunity to raise efficiency by 10% or more.

Roughly there will be about 970 Btus available from each pound of water that is condensed in the flue gas. In flue gas condensing systems the exit temperature is typically about 100 F.

SULFUR IN FUEL FORMS ACID

Sulfur in some fuels can end up as sulfuric acid when the flue gas temperatures drop too low. Boiler damage and corrosion from sulfuric acid has been a problem and a challenge for many years, causing large (energy wasting) safety margins in stack temperature to be used to avoid damage.

In the past, temperatures were maintained above the approximate levels listed below to prevent formation of SO_2 and SO_3 which combines with moisture to form acids.

- Natural Gas   250 F
- No. 2 Heating Oil   275 F
- No. 6 Fuel Oil   300 F
- Coal   325 F
- Wood   400 F

As the chief concern with acid formation is cold surfaces, minimum metal temperatures are used as a more precise means to control corrosion Figure 7.4 shows minimum metal temperature guidelines for air heaters.

Air Heater Comparison

Minimum Metal Temperature

Regenerative

Oil/Gas	< 1%	Sulfur	210 F	99 C
Natural Gas	Sulfur	Free	150 F	66 C
Coal			155 F	69 C

Recuperative

Residual Oil	< 1%	Sulfur	240 F	116 C
Distillate Oil	< 1%	Sulfur	220 F	105 C
Natural Gas	< 0.1%	Sulfur	200 F	93 C
Coal	<1%	Sulfur	160 F	71 C

Figure 7.4
Minimum metal temperatures for air heaters.

COMBUSTIBLES

Because combustibles in the flue gas are unburned fuels, this represents fuel flowing out of the stack.

Scientists have observed on occasion that combustibles are composed of equal parts carbon monoxide and hydrogen. Hydrogen has a heating value of 61,100 Btu/lb. Carbon monoxide has a heating value of 4,347 Btu/lb.

COMBUSTION EFFICIENCY

In practice combustion efficiency is thought of as the total energy contained per pound of fuel minus the energy carried away by the hot flue gases exiting through the stack, expressed as a percentage.

Combustion efficiency is only part of the total efficiency. Radiation loss from hot exposed boiler surfaces, blowdown losses and electrical losses in pumps and fans are examples of other kinds of losses that must be considered in determining total efficiency. However in most fuel burning equipment, the most effective way to reduce wasted fuel is to improve combustion efficiency. To do so, it is necessary to understand the fundamentals of combustion.

STOICHIOMETRIC COMBUSTION

The three essential components of combustion are fuel, air and heat. In fossil fuels, there are really only three elements of interest: carbon, hydrogen and sulfur. During combustion, each reacts with oxygen to release heat:

$$C + O_2 \rightarrow CO_2 + 14,093 \text{ Btu/lb.}$$
$$H + {}_{1/2}O_2 \rightarrow H_2O + 61,095 \text{ Btu/lb.}$$
$$S + O_2 \rightarrow SO_2 + 3,983 \text{ Btu/lb.}$$

Pure carbon, hydrogen and, sulfur are rarely used as fuels. Instead, common fuels are made up of chemical compounds containing these elements. Methane, for example, is a hydrocarbon gas that burns as follows:

$$CH_4 + 2O_2 \rightarrow$$
$$CO_2 + 2H_2O + 1{,}013\ Btu/Ft^3$$

Pure oxygen is also rarely used for combustion. Air contains about 21 percent oxygen and 79 percent nitrogen by volume and is much more readily available than pure oxygen:

$$CH_4 + 2O_2 + 7.53N_2 \rightarrow$$
$$CO_2 + 2H_2O + 7.53N_2 + 1{,}013\ Btu/Ft^3$$

In this example, one cubic foot of methane (at standard temperature and pressure) will burn completely with 9.53 cubic feet of air containing 21 percent oxygen and 79 percent nitrogen. This complete burning of fuel, with nothing but carbon dioxide, water, and nitrogen as the end product is known as stoichiometric combustion (Figure 7.5). The ratio of 9.53 cubic feet of air to one cubic foot of methane is known as the stoichiometric air/fuel ratio. The heat released when the fuel burns completely is known as the heat of combustion.

EXCESS AIR

The Importance of Excess Air

As most combustion equipment operators know, it is extremely undesirable to operate a burner with less-than-stoichiometric combustion air. Not only is this likely to result in smoking but it will significantly reduce the energy released by the fuel.

If a burner is operated with a deficiency of air, carbon monoxide and hydrogen will appear in the products of combustion. CO and H_2 are the result of incomplete combustion and are known as combustibles. Anything more than a few hundred parts per million of combustibles in flue gas indicates inefficient burner operation.

In actual applications, it is impossible to achieve stoichiometric combustion because burners can not mix fuel and air perfectly. To insure that all of the fuel is burned and little or no com-

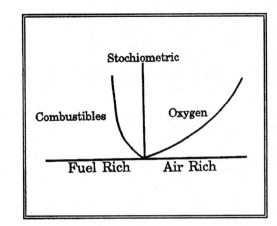

Figure 7.5—Stoichiometric point and air rich and fuel-rich combustion shown as a function of the air fuel ratio.

bustibles appear in the flue gas, it is common practice to supply some amount of excess air. In the era of cheap energy it was not uncommon to run a burner with a large amount of excess air in order to avoid smoking. Today this is becoming known for the highly wasteful practice it really is.

HOW DO YOU ACHIEVE OPTIMUM COMBUSTION EFFICIENCY?

Too little excess air is inefficient because it permits unburned fuel, in the form of combustibles, to escape up the stack. But too much excess air is also inefficient because it enters the burner at ambient temperature and leaves the stack hot, thus stealing useful heat from the process. This leads to the fundamental rule:

"Maximum combustion efficiency achieved when the correct amount of excess air is supplied so the sum of both unburned fuel loss and flue gas heat loss is minimized."

Measuring Combustibles

Combustible analyzers are available to accurately measure CO and H_2 concentrations in flue gas to (+/−) 10 ppm or less. Carbon monoxide analyzers are often used in control systems because of their greater accuracy and calibration stability.

FLUE GAS HEAT LOSS

Flue gas heat loss is the largest single energy loss in every combustion process. It is generally impossible to eliminate flue gas heat losses because the individual constituents of flue gas all enter the system cold and leave at elevated temperatures. Flue gas heat loss can be minimized by reducing the amount of excess air supplied to the burner.

Flue gas heat loss increases with both increasing excess air and temperatures. As both the carbon dioxide and oxygen level in flue gases are directly related to the amount of excess air supplied, either a CO_2 or an O_2 flue gas analyzer can be used to measure this loss. However, in recent years, CO_2 analysis has fallen out of favor.

There are a number of problems when CO_2 is used for analysis. CO_2 can be measured on both sides of the stoichiometric mix bringing about confusion about air deficiency or excess air (Figure 7.6). Also CO_2 readings may not be correct when different fuels having different hydrocarbon ratios are used.

Both of these errors are unacceptable in modern combustion control systems. The development of improved oxygen analyzers has all but eliminated the use of carbon dioxide flue gas analyzers.

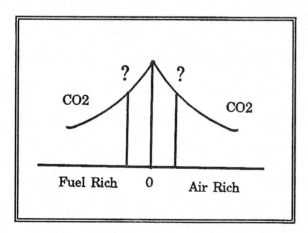

Figure 7.6—CO_2 can be measured on both sides of stoichiometric. This can lead to questions about the actual air-fuel ratio being either air-rich or fuel-rich.

FLUE GAS MEASUREMENT
ORSAT TESTING

One of the earliest methods of measurement is still in use today. The Orsat test is a manually-performed test in which a flue gas sample is passed successively through a series of chemical reagents. The chemicals each absorb a single gas constituent, usually carbon dioxide, oxygen, and carbon monoxide. After the sample passes through each reagent, its volume is accurately measured. The reduction in volume indicates the amount of gas that was originally in the sample.

There are several disadvantages to using the Orsat flue gas testing apparatus:

a.   It is slow, tedious work

b.   Its accuracy is affected by the purity of the reagents

c.   Operator skill is very demanding and under field conditions expert control is necessary to prevent data scatter and test rejection.

d.   Most important, the Orsat test measures only small samples and an unacceptable amount of time can go by before the unit is ready to analyze another sample, quite possibly missing information on the actual dynamics of the combustion systems operation.

e.   All data must be hand recorded. Computerized data systems are becoming increasingly important in combustion system analysis because of their continuous flow of data and automatic record keeping capabilities.

Measuring Carbon Dioxide (CO_2) and Oxygen (O_2) using chemical absorption instruments.

Because of the complexity and the operator skills demanded by the Orsat flue gas analyzer, simpler less complex devices have been developed using the chemical absorption process.

Notable among the test instruments now in daily use in many plants using this technique are instruments like the Bacharach Fyrite CO_2 and O_2 indicators (Figure 7.7).

The primary difference between the O_2 and CO_2 indicators are the chemicals used to absorb

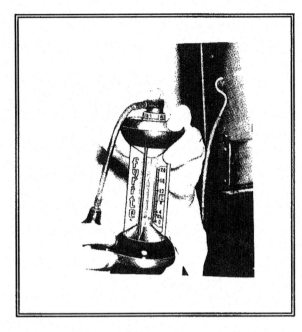

Figure 7.7—Fyrite CO$_2$ or O$_2$ indicator

on which is in use. The volume change, percent CO$_2$ or O$_2$, can be read in percent on a convenient scale.

FINDING STACK LOSSES

By measuring the net stack temperature (Stack temperature—combustion air temperature), combustion tables or slide rule (Figure 7.9) can be used to determine flue gas losses.

Measuring Flue Gas Oxygen on a Continuous Basis

There are three common methods of measuring oxygen in flue gas on a continuous basis. The paramagnetic sensor, the wet electrochemical cell and the zirconium oxide ceramic cell.

these gases. They are a very practical, simple, rugged and economical approach to combustion testing. Also, they are very useful for backing-up the more complex instruments which may develop errors and faults from time to time.

To operate the Fyrite O$_2$ and CO$_2$ indicators, a flue gas sample is extracted from an appropriate point, using the hand operated sampling assembly. The rubber cap is placed over a spring loaded plunger valve which opens when pressure is applied (Figure 7.8). The aspirator rubber bulb is squeezed about 18 times in succession to clear the sampling apparatus, insuring an undiluted flue gas sample. When the plunger is released, the flue gas sample is trapped in the instrument.

The FYRITE is now inverted twice, thoroughly mixing the flue gas sample with the chemical reagents which absorb of either the O$_2$ or CO$_2$, depending

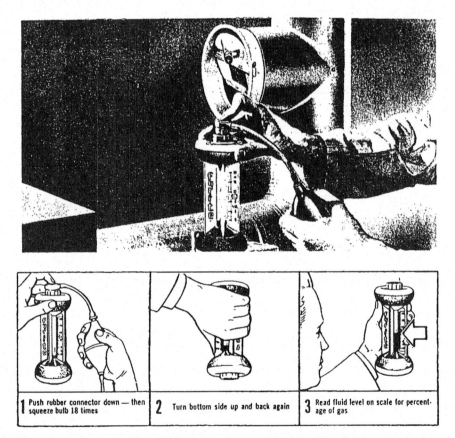

Figure 7.8—Use of the Fyrite illustrated for testing for carbon dioxide (CO$_2$) and Oxygen (O$_2$)

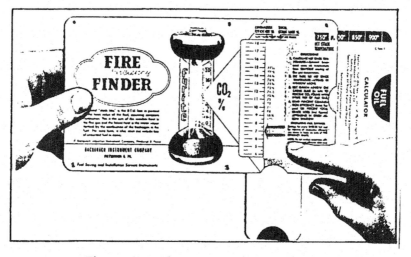

Figure 7.9—The combustion slide rule

PARAMAGNETIC

The paramagnetic sensor takes advantage of the fact the oxygen molecules are strongly influenced by a magnetic field. Because of this and because other flue gas constituents, notably NO, NO_2 and certain hydrocarbons exhibit appreciable paramagnetic properties this instrument is usually limited to the laboratory.

WET ELECTROCHEMICAL INSTRUMENTS

Wet electrochemical cells, of which there are many designs, all use two electrodes in contact with an aqueous electrolyte. Oxygen molecules diffuse through a membrane to the cathode where a chemical reaction occurs.

The electrochemical cell is essentially a battery with an electrical current that is directly proportional to the flow of oxygen through the membrane.

These cells are designed to be replaced easily, however as the flue gas sample is extracted from the stack and brought to the sensor, sample conditioning is required as well as periodic maintenance on the sampling system which becomes fouled with combustion products and a high moisture level. The high maintenance required for these sensors is a definite drawback.

Zirconium Oxide Cell

In recent years, The zirconium oxide cell has become the most common oxygen sensor for continuous monitoring of flue gases. The sensor was developed in the mid-1960s in conjunction with the U.S. space program and because of its inherent ability to make oxygen measurements in hot, dirty gases without sample conditioning, it was quickly accepted by industrial users. The sampling element itself is a closed end tube or disk made from ceramic zirconium.

The zirconium oxide cell has several significant advantages over the other oxygen sensing methods. First, since the cell operates at high temperatures, there is no need to cool or dry flue gases before it is analyzed. Most zirconium oxide analyzers make direct oxygen measurements on the stack with nothing more than a filter to keep ash away from the cell. The cell is not affected by vibration and unlike other techniques, the output actually increases with decreasing oxygen concentration. In addition the cell has a virtually unlimited shelf life.

IN SITU VS. CLOSE-COUPLED EXTRACTIVE ANALYZERS

There are two basic types of flue gas oxygen analyzers that employ the zirconium oxide cell: *in situ* and closed coupled extractive.

In situ Analyzer

In an *in situ* analyzer, the zirconium oxide cell is located at the end of a stainless steel probe that is inserted directly into the flue gas stream. A small heating element encompasses the cell, and a thermocouple provides feedback to an external temperature control circuit.

Close-coupled Extractive Analyzer

A close-coupled extractive analyzer is designed somewhat differently. The zirconium oxide element and temperature controlled furnace are housed in an insulated enclosure mounted outside, but immediately adjacent to, the flue gas stack or duct.

While *in situ* analyzers are limited to flue gas temperatures of about 1100 F [593 C] or less, close-coupled sensors can be used with high temperature probe materials up to 3,200 F [1760 C].

In general, close-coupled units respond much faster to changes in the flue gas stream because they do not rely on diffusion to carry the sample to the sensing cell. A close coupled sensor can be fitted with a catalytic combustibles sensor in the same flow loop as the oxygen cell, thus making a combination oxygen/combustibles analyzer.

NET OXYGEN VS. GROSS OXYGEN MEASUREMENTS

As burners cannot mix fuel and air perfectly, both oxygen and unburned combustibles are in the flue gas. zirconium oxide analyzers indicate net oxygen; i.e., the oxygen left over after burning whatever combustibles are present on the hot zirconium oxide cell. Orsat, paramagnetic and wet cell oxygen analyzers measure gross oxygen.

Usually the difference between net and gross measurements are small since combustibles are generated in the parts per million range. Occasionally conditions may occur where net and gross readings are significantly different. Differences may also occur because zirconium oxide measures oxygen on a wet basis; i.e. the flue gas contains water vapor. The other measuring techniques all require cool, dry samples and are said

to measure on a dry basis. The difference between wet and dry measurements can result in readings that may differ by as much as 0.5 percent oxygen.

If the condition occurs where the combustibles concentration increases to a point where there is no net oxygen in the flue gas, it becomes a sensor of net combustibles. The voltage generated by the cell increases sharply as the flue gas changes from a net oxygen to a net combustibles condition. This property of a zirconium cell is extremely useful on some combustion processes because it permits measurement on both sides of stoichiometric combustion, either excess air or excess fuel.

RELATIVE HUMIDITY

Relative humidity can change the amount of oxygen in air at 70 F [21 C] from 20.9% at 0% RH to 20.4% at 100% RH. This 0.5% change will effect excess air settings and efficiency (0.2% or more).

OXYGEN DEFICIENCY AND SAFETY

In measuring oxygen O_2, here are some important points to keep in mind. Normal air contains 20.9% oxygen and 79.1% nitrogen; the usual alarm point is 19.5% unconsciousness occurs at 15%; brain damage at 10%, and death at 5%. (These are approximate figures and may vary with the individual.

CARBON MONOXIDE AND COMBUSTIBLES MEASUREMENT METHODS

There are three prevalent methods for on line monitoring of flue gas combustibles: wet electrochemical cell, catalytic element, and non-dispersive infrared absorption.

Electrochemical cell

The wet electrochemical cell technique is used only for carbon monoxide. It works on the principle that current flowing between the anode and cathode is directly proportional to the flow of carbon monoxide through the membrane. There are problems that occur with this type, flow rate as affected by the ambient pressure, temperature and humidity. Furthermore, the membrane can become coated with flue gas condensation, thus reducing its effectiveness. Because of this, these sensors are prone to zero and span drift.

CATALYTIC COMBUSTIBLES SENSOR

Catalytic element sensors have been widely used for detecting combustible gases in ambient air in mine shafts, parking garages, and other closed areas. High quality sensors with carefully selected elements, can be used to measure carbon monoxide in flue gas. Catalytic sensors are available with full scale ranges as sensitive as 0-2000 ppm combustibles and with accuracies of (+/−) 100 ppm or better.

The principle behind all catalytic sensors is the same, if combustibles and oxygen are both present in a gas stream, they will not normally burn together unless the temperature is elevated something above 1000 F [638 C] However, if the same gas mixture comes in contact with a solid catalyst, such as platinum, combustion will occur at temperatures as low as 400 F [204 C].

There are two elements present, one with a catalyst in an inert binder the other is inert. The entire housing is heated to over 400 F, when the flue gas sample containing both oxygen and combustibles pass through the housing, combustion occurs on the active element but not on the reference element. This causes the temperature of the active element to rise and its resistance to change. Some close-coupled extractive oxygen analyzers have been modified to incorporate a catalytic combustibles sensor in addition to the oxygen sensor.

The catalytic sensor has the advantage of being both low cost and sensitive to hydrogen and carbon monoxide. These sensors make it ideal for flue gas monitoring and recording. However, the zero and span stability of the sensor is not as good as that of the infrared sensor. Combustion

systems requiring CO/combustibles measurement as an active input generally utilize infrared carbon monoxide analyzers.

INFRARED CO MEASUREMENT

Carbon monoxide is one of many gases that are known to absorb infrared energy at specific discrete wavelengths. The amount of energy absorbed is a measure of the concentration of carbon monoxide.

There are two types of carbon monoxide analyzers: off-stack (sampling) and across the stack (*in situ*).

OFF-STACK CO ANALYZERS

Off-stack analyzers are housed in enclosures suitable for the environmental conditions and are usually located at easily accessible places near the combustion process. In most cases a sampling system is required to clean, dry, and cool the sample before it enters the analyzer. Provision for the introduction of calibration gases are usually an integral part of the design of the sample conditioning system.

ACROSS-THE-STACK CO ANALYZERS

Across-the-stack CO analyzers are based on the same technical principles as off-stack analyzers but their design is somewhat different. The infrared source is housed in an enclosure that mounts directly on the stack or duct. The infrared beam generated by the source passes completely through the stack into a similar enclosure mounted on the other side.

There are two major advantages of the across-the-stack systems. First, the speed of response is nearly instantaneous. Off-stack systems, conversely, can take several minutes to respond to a change in flue gas conditions. Second, across-the-stack systems provide a measurement of the average CO concentration in the stack.

Unlike off-stack analyzers, which sample from a single point, they are unaffected by stratification or stagnation of flue gases in various areas of the stack.

The measurement of carbon monoxide or combustibles is an important part of achieving maximum combustion efficiency. The result will be less fuel wasted and more money saved.

OPACITY

Smoking with oil and coal fuels indicates the presence of flue-gas combustibles or unacceptable flame conditions, and always should be avoided. Some boilers, especially larger ones, are equipped with smoke detectors, which can indicate poor stack conditions. Ultimately, stack conditions should be checked by visual observation.

Accurate spot check type smoke measurements can also be made with the inexpensive, portable hand pump with filter paper testers described above.

These devices use the smoke spot number or ASTM (American Society for Testing & Materials) smoke scale(standard D-2156), and can be very helpful in establishing optimum boiler conditions.

STACK TEMPERATURE

Deposits and fouling of external tube surfaces with soot, ash and other products inhibit the absorption of heat in the unit and lead to lower efficiencies. Deposits are indicated by flue-gas temperatures that are high compared to clean conditions. The efficiency loss resulting from dirty tubes can be estimated with the RULE OF THUMB:

EVERY INCREASE OF 40 DEG F IN STACK TEMPERATURE REDUCES EFFICIENCY BY ABOUT 1%.

Waterside deposits caused by improper water treatment also can lead to high stack temperatures, but tube failures due to overheating generally occur before any substantial efficiency losses are evident from these internal tube deposits.

Rising Stack Temperatures Indicate a Problem

Stack-temperature measurements are an easy and effective means for monitoring boiler-tube fouling. This is done by comparing the present temperature to the start up temperature or a temperature recorded when the boiler was in a clean condition.

Since stack temperatures usually increase with firing rate and excess air, make your comparisons at similar boiler operating conditions.

In the absence of previous data, flue gas temperatures normally are about 150 to 200 F above steam temperature for a boiler producing saturated steam at high firing rates.

Boilers equipped with economizers and air preheaters should be judged by observing the flue gas temperature immediately after leaving the last of these heat recovery units.

ACCURACY OF SAMPLING TECHNIQUES

For oxygen, CO and smoke analyzers the portion of the gas analyzed must be representative of the total gas stream. Location of the sampling site can be as important as the selection of the proper measurement device.

To illustrate: on negative draft boilers, the gas sampling point should be upstream of the air preheater, if one is installed, or upstream of any known air leaks. Reason is that air leakage into the gas ducts can dilute flue gas and resultant measurements won't give a true indication of furnace conditions. Air leakage in preheaters poses the same problem.

Sample conditions immediately downstream of bends, dampers, or induced fans should be avoided. Gases in such areas can stratify or form pockets leading to errors, especially when samples are withdrawn from a single point in a duct.

When a single-point probe is to be used, compare several readings in the duct first, to find the most representative probe location. When existing ports are not satisfactory, drill or cut out new ports and run traverse measurements. Remember, unless you get truly representative data, your testing program will be of little value.

Flue-gas temperatures are subject to stratification in ducts and a representative location of thermometers or other temperature sensors should be verified.

Position them close to the boiler outlet, because thermal losses can occur in the flue gas duct, especially in uninsulated sections.

USING CARBON MONOXIDE
TO MEASURE PERFORMANCE

On gas fired boilers carbon monoxide is the primary indicator of incomplete combustion and usually determines the lowest practical level of oxygen. The concentration of CO in the flue-gas should not exceed 400 ppm or .04% (the limit established by many state and city ordinances, industry codes and insurance companies).

Once the final adjustments are made. It is wise to observe the operation of a boiler for an extended period to insure that your adjustments are final and there is no condition present that can increase the CO level above the acceptable limit.

When performing tests, occasional CO levels of up to 1000-2000 ppm may be encountered. Adequate boiler monitoring and flame observation are very important to assure stable conditions. Use caution at these levels because even a slight lowering of excess air can cause the CO level to skyrocket, which can lead to smoking, flame instability, furnace pulsation and possibly an explosion.

The situation is further complicated by the possibility of some CO monitoring instruments becoming insensitive and going off scale requiring a waiting period for them to come back to operating range. During this period you may be blind to what is actually happening to the CO level. One precaution that may be taken is to use a combustibles analyzer along with the CO instrument, the combustibles analyzer is less sensitive and will indicate the actual situation over a wider range.

Carbon monoxide measurements on oil and coal fired equipment is less often used because smoking or excessive carbon carryover usually precedes the formation of large quantities of CO. This is not always the case, however.

High CO levels have been measured on units where burner equipment had deteriorated or malfunctioned, impellers had burned off, oil tips plugged, over fire air was too low, etc. Also, CO can be caused by chilling the combustion process before the fuel is completely burned, two ways this can happen is chilling of the flame with excessive (cold) concentrations of combustion air in part of the flame and through flame impingement on the (cooler) boiler tubes.

Knowing the CO level is very valuable, The CO analyzer should be capable of measuring from less than 100 to more than 2,000 ppm. While Orsat analyzers have traditionally been used to determine CO, difficulties in accurate reading of concentrations less than 1000 ppm have presented problems in the modern environment. Portable or permanently installed electronic type CO analyzers have the ability to measure CO continuously, having the advantage of indicating excursions in CO that may not be detected with occasional spot readings.

Chapter 8

The Control of Boilers

Controlling boilers can be a complex subject. The intent here is to summarize, simplify and clarify the subject of boiler controls. There are many very good companies specializing in boiler controls who have been continually improving control system technology for many years.

The field of boiler controls is an advancing rapidly especially since the introduction of distributed digital systems, the latest in a long string of advances. However, the basics of boiler control are straight forward and subject to less change.

STEAM PRESSURE CONTROL

The steam pressure is the balance point between demand for energy in the distribution system and the supply of fuel and air to the boiler for combustion.

As energy is used by the steam system, pressure drops, creating a demand for more energy.

Several interacting control loops are used to insure steam pressure is maintained and that the flows of water, air and fuel are managed to insure safe, efficient and reliable operation.

These control loops are:

- Boiler Safety Controls
- Combustion Control
- Feedwater Control
- Blowdown Control
- Furnace Pressure Control
- Steam Temperature Control
- Cold End Temperature Control for air heaters and economizers

---

SAFETY CONTROLS

The purpose of the boiler safety control system is to prevent explosions and other damage to the boiler. If an accumulation of unburned fuel vapor suddenly ignites, explosive levels of energy can be released and cause severe damage to the boiler and surrounding countryside.

If the water level in the boiler drops below a certain point, it could have a melt down. If there are disruptive fuel pressure changes, the ensuing flame instability could cause an explosion or heavy smoking. These and several other conditions must, for the sake of safety, be controlled at all times.

COMBUSTION CONTROL

The energy is supplied to the boiler by the combustion process and the combustion control system regulates the firing rate by controlling amount of air and fuel delivered to the burners.

Combustion control systems are regulated to maintain the desired steam pressure and they must be able to respond the many dynamic aspects of the burner, fuel and air control sub-systems in a coordinated way to maintain the steam pressure set point in spite of demands of the steam distribution system.

FEEDWATER CONTROL

The purpose of the feedwater control system is to maintain the correct water level in the boiler during all load conditions.

Feedwater control must be able to regulate

water level under very dynamic conditions when the heat rate changes in the boiler causes the water level to shrink and swell, due to steam bubble volume changes in response to firing rate changes.

The boiler feedwater control system must also respond to momentary changes in steam demand replacing steam that has left the boiler with feedwater.

If the feedwater control system fails, there may be serious problems. A high water level can cause severe damage to the distribution system and machinery like turbines. A low water level can allow the high flame temperatures to weaken and melt the boiler steel causing a catastrophic high energy release of steam from inside the boiler.

BLOWDOWN CONTROL

As water in the boiler is evaporated to produce steam, impurities are concentrated. Unless the concentration of these impurities is kept under control, severe scaling of the heat transfer surfaces and tube failure can occur.

Impurities can also be carried over into the distribution system and depending on conditions at the steam water interface, can cause surges of water into the steam system.

Blowdown is also controlled to eliminate the waste of energy by preventing more hot water than is necessary from being dumped from the operating boiler.

FURNACE PRESSURE CONTROL

Some larger boilers have been designed for balanced or negative draft and requiring a furnace pressure slightly below atmospheric for proper combustion and safe operation.

STEAM TEMPERATURE CONTROL

In boilers where superheaters are used to raise steam temperature above the saturation point, the temperature of this superheated steam must be regulated.

COLD END TEMPERATURE CONTROL

When economizers and air heaters are used to remove heat from flue gases, sulfur oxides can form sulfuric and sulfurous acids which can cause damage to boiler and exhaust system components.

SOOT BLOWER CONTROL

The accumulation of soot, fly ash and other deposits on heat exchange surfaces lowers the heat transfer and efficiency of a boiler raising the stack temperature (Figure 8.1). It is an economic consideration to keep boilers as clean as possible.

On many boilers soot blowing operations are automated and since the accumulation of soot is roughly proportional to the number of hours the boiler has been operated, automatic sootblowers are usually activated on a simple time clock mechanism.

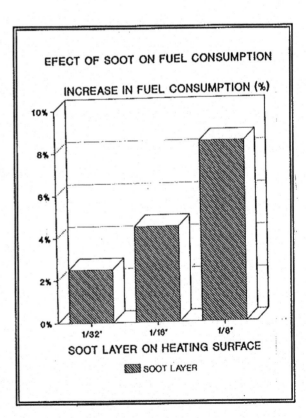

Figure 8.1—Effect of soot on fuel consumption

SAFETY VALVES

The purpose of safety controls is to prevent explosions. Safety devices are installed for the pressure vessel which usually consists of safety valves sized to carry away steam faster than it can be generated by the boiler in the event an over pressure situation exists. This is a mechanical system consisting of safety valves and exhaust piping to vent the escaping steam harmlessly through the roof.

Boiler controls also sense overpressure and override the main control system and shut off the flow of fuel to the boiler.

FLAME SAFETY CONTROLS

Flame safety controls continuously monitor the pilot and main flame zone to detect the uninterrupted presence of the flame (Figure 8.2). If unburned fuel is allowed to accumulate in the boiler there could be an explosion, so this system must have the capability to shut down the fuel system almost immediately. These systems were developed from studying lessons learned from boiler explosions and disasters over the years.

Flame safety controls are designed to shut off the fuel to prevent explosions that can occur when:

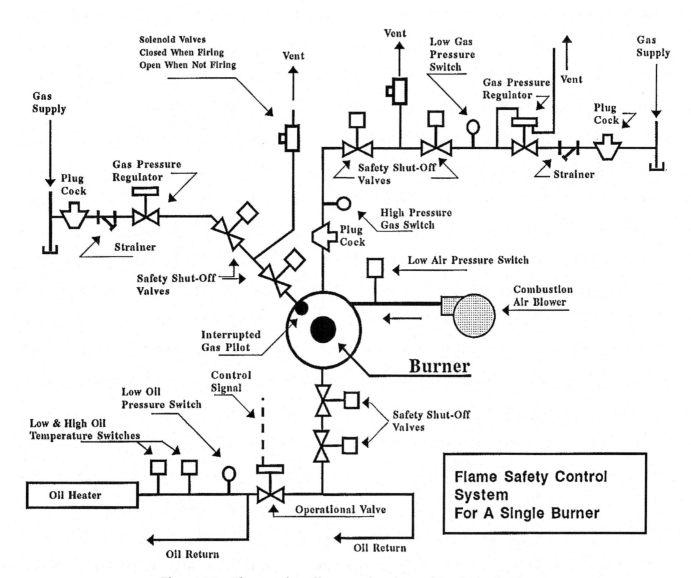

Figure 8.2—Flame safety diagram showing safety shut off valves

1. An ignition source could be introduced into a furnace that contains air and accumulated fuel vapors that could become an explosive mixture.

2. Fuel is discharged into the furnace during start up without proper ignition taking place.

3. The burner flame is extinguished during normal operation without the fuel supply being shut off.

4. A major malfunction in the burner or feedwater systems.

5. Variations exist in fuel pressure which could cause flame instability.

Purging Combustible Gasses on Light-off

The start-up condition presents perhaps the greatest danger from explosion. It is essential that the furnace be purged of all combustible gases before any source of ignition is introduced and that once fuel flow is initiated, ignition take place quickly.

Purging the furnace is usually accomplished with the combustion air fan. A purging air timer does not allow ignition until the fan has been operated for a specific period of time. For watertube boilers the purge airflow must be sufficient for at least eight air changes to occur.

For firetube boilers, the purge is conducted with wide open dampers and four air changes is normally required. In either case, air flow is verified by providing limit switches on the dampers and a pressure switch at the fan discharge, or by providing airflow measurement devices such as differential pressure switches.

Once fuel has been initiated to the pilot flame or the main burner, flame verification is usually achieved by means of a flame scanning device. These flame scanners may be either the infrared flicker type or the ultraviolet type.

Recently manufacturers of flame safety control components have introduced integrated flame safety systems based on microcomputer technology with self diagnostic features. They are programmed to perform boiler start up and shutdown sequences, alarm communication and perform energy conservation functions.

SAFETY INTERLOCKS

• Automatic start up sequencing has been designed to insure fuel air and combustion occur under controlled and safe conditions. Each boiler must be equipped with an approved safety system which has the capability to shut it down in the event an unsafe condition develops during start up. This system coordinates with safety interlocks for shutdown if an unsafe condition is detected. Shutdown is accomplished by stopping the start up sequence and closing the Safety Shut Off Valves (SSV).

• Loss of flame interlock; if there is no flame in the combustion zone, the boiler must be shut down immediately to prevent the accumulation of combustible vapors, which when conditions are right, could cause a powerful explosion. Pilot flame ignition, main flame ignition and main flame monitoring are all important to the prevention of combustibles build up. The Safety Shut Off Valves (SSV) close to stop fuel supply to burners when the flame scanners do not sense a flame at the burner.

• Purge interlock; the combustion products and possible accumulations of explosive fuel vapors must be purged from the boiler before each light off and after each shut down.

• Low air interlock; if air supply is less than required for safe combustion, the boiler must be shut down to prevent an accumulation of a fuel rich and possibly explosive mixture in the boiler.

• Low fuel supply interlock; loss of fuel delivery can cause flame instability, erratic combustion, poor atomization of some fuel oil equipment and the possibility of an explosion. If low fuel pressure is detected the boiler is shut down.

• Low water interlock; The combustion flame may be more than 3,000 F which is above the melting point for boiler steel. The pressure vessel is normally kept cool by the water and steam which carries away the energy from combustion. If the water level becomes low, metal temperatures can rise beyond a safe point, melting boiler components allowing a catastrophic release of energy from the steam and hot water contained in the pressure vessel.

• Other interlocks are used for specific boiler applications which override the control operation when necessary.

BASIC FIRING RATE CONTROL

There are several methods of controlling steam flow, the most common method uses steam pressure to generate a master control signal. The master control signal is usually utilized by one of two types of control methods: parallel control positioning or series control positioning.

PARALLEL POSITIONING CONTROL

Parallel positioning control is probably the most common control scheme for small industrial boilers. With this type of control the signal from the steam pressure sensor goes simultaneously to the fuel flow and air flow regulating devices. The position of these regulating devices is determined by the magnitude of the signal from the steam pressure sensor (Figure 8.3).

Parallel positioning systems offer the advantages of fast response and simplicity of operation and have been found to be very reliable. Individual components can be adjusted independently, so control system tuning is facilitated.

Parallel positioning systems do have a shortcoming. The master signal operates on feedback from of the actual steam pressure. The individual air and fuel regulators do not have a feedback loop to assure that the air to fuel ratios are in the desired range.

Parallel positioning control systems employ mechanical, pneumatic, electronic and digital control elements.

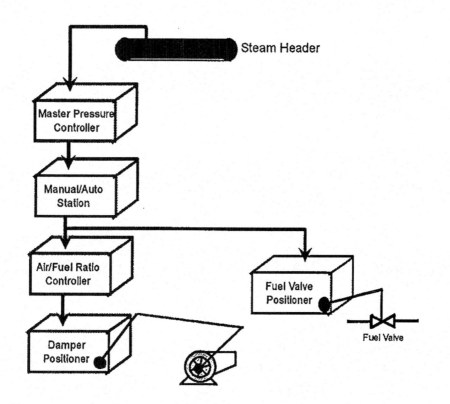

Figure 8.3—Parallel positioning control system

JACKSHAFT CONTROL SYSTEMS

The jackshaft control system (Figure 8.4) is a simple form of parallel positioning control. Figure (8.5) is a functional schematic of the logic of this type of system. This logic path becomes more sophisticated with the increase of boiler size where response time of different elements such as fuel and air require special consideration.

SERIES POSITIONING CONTROL

In a series positioning control system (Figure 8.6), the control signal from the steam pressure sensor is not simultaneously sent to the fuel and air regulating devices. Rather, the signal is sent to only one of the regulating devices. The displacement of this first device is then measured and used to control the position of the second device.

By using the actual displacement of one regulating device to control the other, series posi-

tioning control provides a margin of safety that is not available in parallel positioning control.

For example, if the damper positioner is the first controlled device, then the fuel flow cannot increase unless the air damper actually opens. This prevents a hazardous condition from occurring should the damper drive fail to operate correctly.

Control of the air and fuel regulators in a series positioning system is open-loop, just as in the parallel type system. A series positioning system can develop a temporary upset in the fuel to air ratio during rapid load changes. This upset can be corrected by adding "cross-limiting" or "lead-lag" to the control.

CROSS LIMITING CONTROL FEATURES

Cross limiting control prevents a fuel rich mixture on load changes. This is needed because fuel response is much faster than air system re-

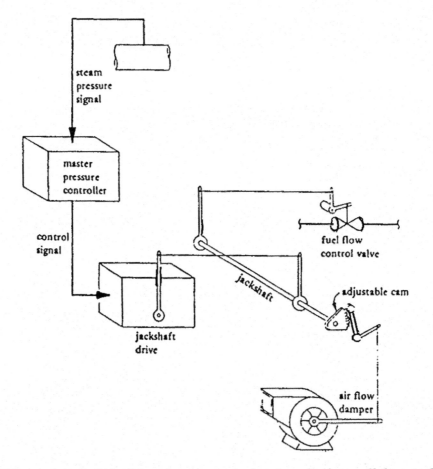

Figure 8.4—Jackshaft type burner control system, basic parallel control

Control of Boilers

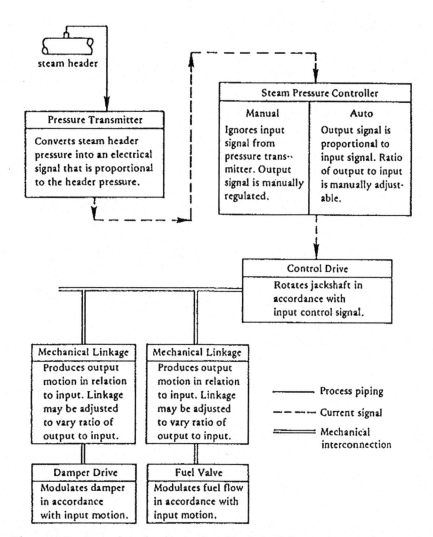

Figure 8.5—Functional schematic of jackshaft burner control system

sponse to load changes. The lag in response of air delivery is caused by the compressible nature of air and the slow response of damper positioners and the speed change characteristics of some types of blowers. Low and high selectors are used to insure there is more than enough combustion air available to insure there is no smoking on load changes.

For example, as in Figure 8.7, the high selector [>] receives the firing demand signal [60%] and actual fuel flow signal [55%]; it sends the higher value to the air damper positioner. The low selector [<] compares the firing demand signal [60%] and the air flow signal [55%] and sends this lower signal to control the fuel flow valve.

METERING CONTROL SYSTEMS

Metering control systems overcome both the shortcomings of parallel and series positioning control, but the added cost and complexity of metered control has generally restricted their use to large boilers (Figure 8.8).

With metering control, the control of the fuel and air regulators is a closed loop. The steam pressure is measured and feedback is provided to the master controller which adjusts the fuel and air flows. The fuel and airflows are also measured and feedback is provided to their control devices to insure they are in accord with the master controller.

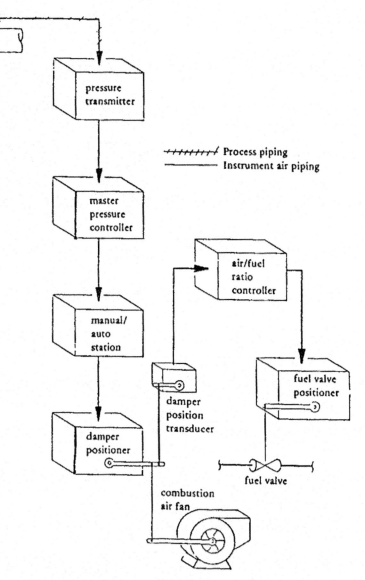

Figure 8.6—Series positioning control system

FEEDWATER CONTROL

The purpose of feedwater control is to maintain the correct water level in the boiler during all load conditions. The simplest control systems use one variable, water level as the input to the control system as illustrated in Figure 8.9.

The controllability can be improved by adding information about steam flow and feedwater flow rates. This is called a three element feedwater control system (Figure 8.10), the flow rates of steam and water are compared to overcome any momentary false water levels caused by the shrink and swell affects due to rapid load changes.

With the "shrink-swell" phenomena it is difficult to measure drum level accurately. As steam demand increases, the steam pressure decreases lowering drum pressure. This decrease in pressure plus the increased firing rate to meet this higher demand increases the volume of steam bubbles in the drum water, increasing its water level temporarily. This swelling of water level sends a false signal to the drum level controller indicating a need for less water rather than more, as the circumstances actually require. Conversely, as steam demand decreases, drum pressure increases and there is less combustion heat for steam formation. This contracts the volume of

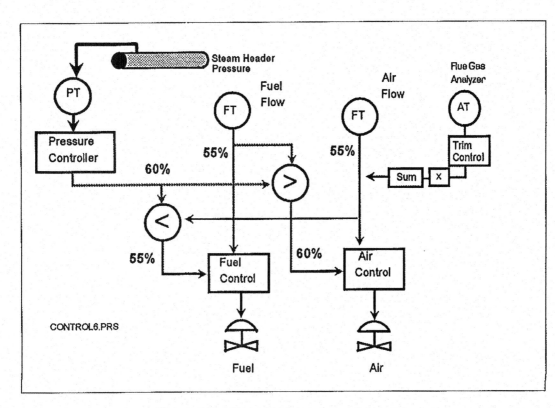

Figure 8.7—Cross limiting control features

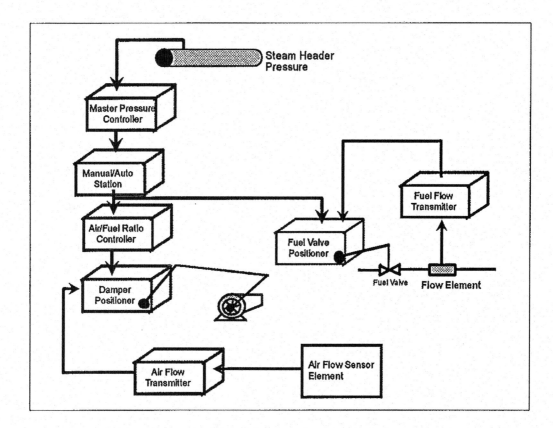

**Figure 8.8—Metering control system using flow
sensing elements to insure accuracy of fuel and air positioners.**

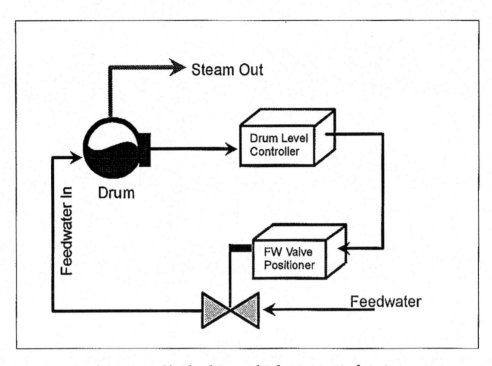

Figure 8.9—Single element feedwater control system

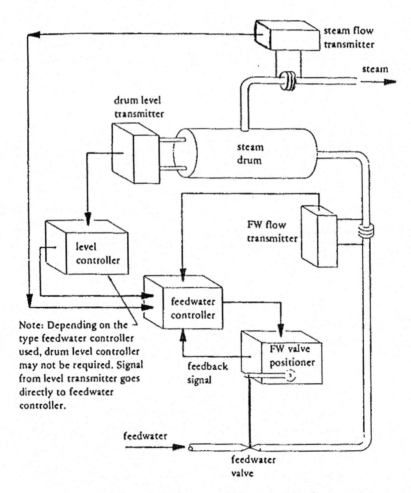

Note: Depending on the type feedwater controller used, drum level controller may not be required. Signal from level transmitter goes directly to feedwater controller.

Figure 8.10—Three-element feedwater control system.

steam bubbles in the drum water which the drum level controller interprets as a need for more water, which is colder than the boiler saturation temperature causing further shrinking.

The three element control loop is very precise and stable because of the closed loop control of the feedwater valve. It measures steam flow and water flow as well as drum water level. This system can adjust for actual steam and water flow conditions balancing this information against measured water level.

Feedwater controllers are available in pneumatic, analog electric and digital electronic models.

BLOWDOWN CONTROL

Blowdown is done either intermittently or continuously. On many smaller boilers blowdown is performed manually after an analysis of the boiler water indicates that the water is exceeding impurity limits.

If too much hot boiler water is blown down from the boiler an excessive amount of energy is wasted. If the concentration of impurities in the boiler is allowed to get to high, carry-over and scale formation can occur.

Control systems are available which will automatically perform the operations necessary to blow down the boiler.

Automatic blowdown controllers use electrochemical cells to measure the concentration of minerals by measuring electrical conductivity and acid (PH) levels.

The signal corresponding to the impurity level is compared to the setpoint and when exceeded, blowdown is initiated.

Boilers operated at a steady load may be equipped with a continuous blowdown system. A continuous blowdown system removes a continuous, usually small amount, of water from the steam drum to keep the concentration of undesirable chemicals at a safe level.

Manual blowdown to remove sludge which has settled in the mud (lower) drum is still necessary.

Boiler blowdown water is often passed through a heat exchanger to recover the energy which would otherwise be wasted. Heat from the boiler blowdown is usually used to heat the incoming boiler makeup feedwater.

FURNACE PRESSURE CONTROL

Furnace pressure conditions must be closely controlled to establish safe and efficient firing conditions. Air fuel mixing for the burner systems depends to a great degree on the differential pressure across the burner.

Different challenges apply to different sizes of boilers. Figure 8.11 shows a natural draft, balanced draft, induced draft and forced draft systems. Each configuration will require special control considerations.

Damper openings, forced draft fans, induced draft fans and stack effect all play an important role in various furnace pressure control systems.

Steam Temperature Control

Superheaters are used to raise the steam temperature above the saturation point for superheated steam. In doing so, the temperature of the superheated steam must be regulated. The main control loop is to regulate the superheated steam temperature by a desuperheater steam valve. This control loop is enhanced by a feed forward, usually air flow, signal into the loop plus an inner loop cascade signal for spray water flow to eliminate the effects of spray valve performance variations.

Cold End Temperature Control

Cold end corrosion occurs when metal temperatures fall below the dew point for sulfurous and sulfuric acids. The most critical point for the economizer is the feedwater entrance point to the heat exchanger where metal surfaces can be cooled below the acid dew point.

In air preheaters a steam or glycol heating system is used to regulate the cold end temperature. A single element control of the heating control valve may be adequate. However, the three

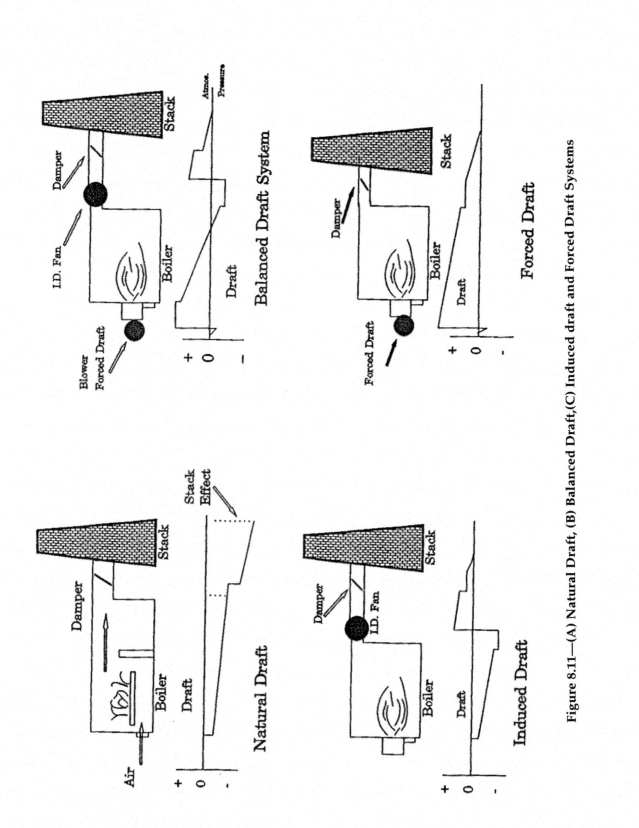

Figure 8.11—(A) Natural Draft, (B) Balanced Draft,(C) Induced draft and Forced Draft Systems

element air heater cold end temperature control may have to be used.

Controls to Improve Efficiency

To a great extent, the efficiency of a boiler is dependent on the design of the burner, heat exchanger and other design parameters which cannot be easily changed. However, changes in fuel composition, air temperature, pressure and humidity, boiler load and equipment condition all introduce changes in boiler performance which can be accommodated for with improved control.

In general "improved control" means better control of the air fuel ratio. The simple parallel positioning and series positioning systems have open-loop control of the fuel and airflows. There is no feedback into the control system which tells the system what the actual fuel and airflows are. Feedback
is provided only by the steam header pressure.

If the header pressure is at the control set point but the fuel to air ratio is nowhere near the desired value, the control system will take no action to correct it. For this reason it is highly desirable to modify a boiler control system to include feedback of information on the air to fuel ratio. This is done by adding either an oxygen trim or carbon monoxide trim system to the boiler controls.

The effect on air/fuel ratio on flue gas analysis is shown in Figure 8.12, knowing the relationship of measured flue gas oxygen, carbon dioxide, carbon monoxide, combustibles, special instruments can be used to measure and control the boiler for efficient air fuel ratios.

OXYGEN TRIM

The idea behind oxygen trim controls is to maximize boiler efficiency by operating at the point where the combined efficiency losses due to unburned fuel and excess air losses is minimized.

An oxygen trim system measures the excess

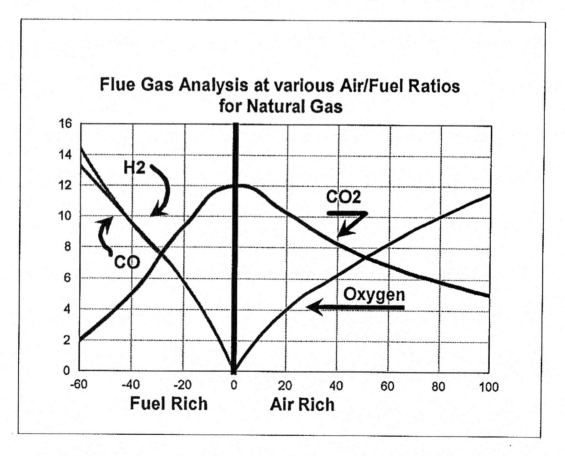

Figure 8-12—Composition of boiler flue gases with changes in air/fuel ration for natural gas.

oxygen in the combustion products and adjusts the airflow accordingly for peak combustion efficiency (Figure 8.13).

CARBON MONOXIDE TRIM

Carbon monoxide (CO) trim systems are also used to excess air. Carbon monoxide trim systems, in fact offer several advantages over oxygen trim systems.

In the carbon monoxide system trim system, the amount of unburned fuel (in the form of CO) in the flue gasses is measured directly and the air to fuel ratio control is set to actual combustion conditions rather than pre-set oxygen levels. This way the carbon monoxide trim system is continuously searching out the point of maximum efficiency (Figure 8.14).

Other benefits of carbon monoxide trim sys-

tems are that it is independent of fuel type and it is almost unaffected by air infiltration common in the negative draft type of boilers.

Carbon monoxide systems must be viewed with caution because the carbon monoxide level is not always a measure of excess air. A dirty burner, poor atomization, flame chilling, flame impingement on boiler tubes and poor fuel mixing can also cause a rise in carbon monoxide level.

DIGITAL CONTROL

The latest developments in automatic boiler control systems have been in the technology of distributed digital control (DDC) of boilers. DDC systems are based on microcomputer technology and control the boiler by means of "software" rather than fixed interconnections or "hardwired

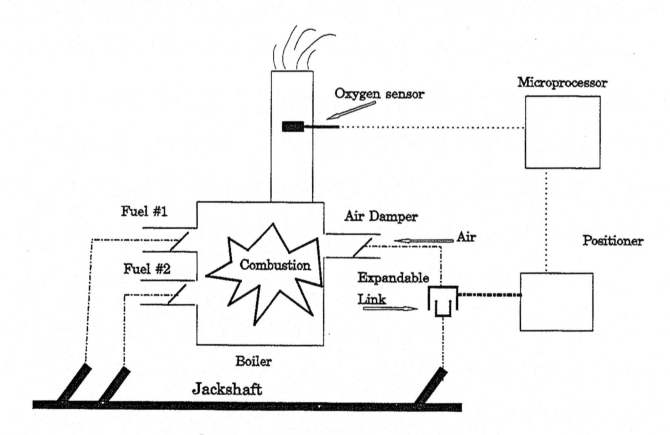

Oxygen Trim System

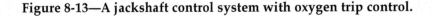

Figure 8-13—A jackshaft control system with oxygen trip control.

logic." Digital control systems often combine the functions previously performed by separate hardware systems such as combustion control, safety controls and interlocks, and monitoring and data acquisition.

In a typical DDC system, input devices such as sensors, switches and position encoders are accessed at very short intervals insuring any problem is dealt with almost immediately.

The advantages of distributed digital based control systems include:

1. The control instructions and set points can be easily changed.

2. The systems are easily expanded to include more control features or more boilers.

3. Low cost control redundancy.

4. Boiler operation diagnostics and performance data acquisition.

5. Control system self-diagnostics.

6. Easy access to important realtime or historical information by telecommunication links.

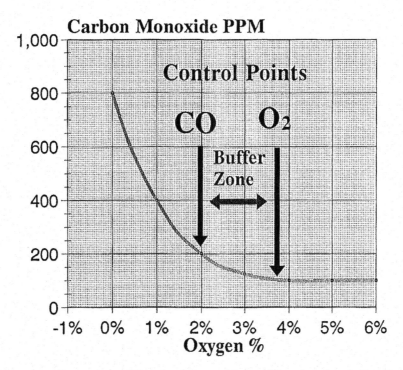

Figure 8.14—A carbon monoxide trim system is capable of maintaining the most efficient air fuel ratio because it does not need a buffer zone. It locates the most efficient operating level automatically.

Chapter 9

Boiler Tuneup

[The Importance of Operating Boilers with Minimum Excess Air]

Reducing excess air is one of the most effective boiler improvement techniques one can apply without high capital cost. When excess air is reduced, several things are accomplished:

- When hot combustion gases leave a boiler, they have the potential to carry away a lot of waste energy. The less volume of exhaust gas, the less loss there is.

- Flue gas velocity is reduced increasing the time available for heat transfer in the boiler.

- Flame temperature is raised, increasing radiant heat transfer in the combustion zone walls. Heat transfer in the combustion or radiant heat transfer zone is very efficient and becomes more efficient as flame temperature goes up. This increase of heat exchange efficiency reduces stack temperatures.

- Pollution is reduced because less fuel is required to meet the same demands.

The increase in efficiency available from tuning up a boiler has three direct and related benefits:

(1) it saves fuel dollars,

(2) it reduces the cost of energy at the point of use and

(3) it increases the steam output available from a boiler (increased productivity).

When evaluating boiler efficiency improvement projects, cost and benefit calculations must be based on the tuned up efficiency of the boiler to prevent false estimates of benefits. It doesn't make sense to attempt to correct a problem by adding something new to a boiler if it can be corrected by maintenance and repairs or a tuneup.

A TUNEUP STARTS WITH AN INSPECTION AND TESTING

Efficiency improvements obtained under a deteriorated state of the boiler can be substantially less than the improvements achieved under proper working conditions. Therefore, it is essential that the boiler be examined prior to testing and that necessary repairs and maintenance be completed.

One of the first questions in tuning up an operating boiler is whether or not is necessary to take the boiler out of operation and go through the expense of opening it up for a formal inspection.

A preliminary efficiency test and review of records might provide valuable information about a boilers operating condition and whether or not a more detailed inspection is necessary.

The condition of the burner system and combustion process can be judged by the excess oxygen level. Boiler start up records and records of previous tuneups provide a valuable reference point for your tuneup program. Even a call to the manufacturer can provide useful information on the boiler's expected performance characteristics and minimum expected excess air levels.

The two important goals of a boiler tuneup are minimum excess air and stack temperature. Figure 9.1 shows the zone of highest efficiency where excess air [oxygen] is lowest and just before carbon monoxide and combustibles, in the form of carbon monoxide, hydrogen and unburned hydrocarbons, begin to show up.

If this information is not available, then general information can be used based on information from typical minimum oxygen settings in similar boilers. The following general information on minimum excess oxygen is based on a large number of boiler tests and is applicable to high firing rates. As firing rate decreases burner performance falls off and more excess air may be needed for some burners.

Target Oxygen Levels
- Natural gas boilers,   0.5% to 3%
- Liquid fuels,   2% to 4%
- Pulverized coal,   3% to 6%
- Stoker fired coal,   4% to 8%

STACK TEMPERATURE

Stack temperature measurements are an easy and effective means for monitoring boiler tube cleanliness and the general effectiveness of the heat exchange process in a boiler. Existing temperatures can be compared to values obtained during start up or after maintenance and

cleaning, to identify any deviations from baseline levels. Since stack temperature usually increases with firing rate and excess air, make your comparisons at similar boiler operating conditions.

If previous information is not available or if temperatures seem excessive, the following graph (Fig 9.2) can be used for general estimates. Temperature readings to measure boiler performance must be taken before economizers or air heaters cool down the flue gases but overall performance is judged from the temperature after these units.

INSPECTING YOUR BOILER

Boiler components that you should inspect before conducting efficiency tests include burners, combustion controls and furnace. Typical things to look for are shown below (Figure 9.3 through Figure 9.5). Consult your boiler manufacturer for a more complete list appropriate to the specific equipment in your plant.

Oil burners
- Make sure the atomizer is suitable for your present firing conditions; for the type of oil being burned and burner geometry.

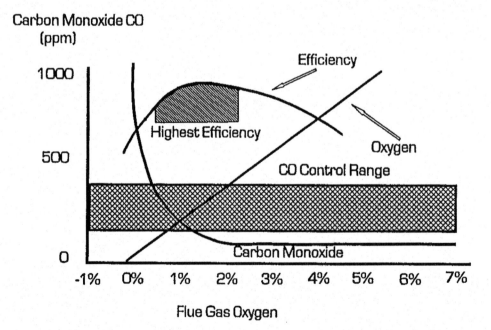

Figure 9.1—CO-Oxygen relationship showing CO control range and the effect of CO and oxygen levels on combustion efficiency.

Flue Gas Temperatures Above Steam Temperature

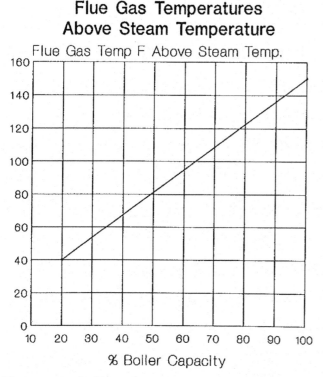

Flue Gas Temp F Above Steam Temp.

% Boiler Capacity

Figure 9.2—Boiler exhaust temperature above temperature of water-steam temperature in boiler.

- Verify proper flame pattern through the viewing ports located at the sides and back of the boiler, if installed.

- Inspect burners for warping or overheating, coke and gum deposits. Clean or replace parts as appropriate.

- Inspect oil tip passages and orifices for wear and scratches or other marks. Use proper size drill or machine gages for testing.

- Verify proper oil pressure and temperature at the burner. This may include resetting the fuel oil heater to the grade of oil presently being fired, calibrating gages and instruments and resetting pressures to manufactures values or values indicated in engineering records. If changes are made, careful observation of actual effects they produce is important in the case a problem develops.

- Verify proper atomizing-steam pressure. Also, be aware a defective steam trap may introduce unwanted water into the flame zone.

- Make sure that the burner diffuser (impeller) is not damaged, and is properly located with respect to the oil gun tip.

- Check to see that the oil gun is positioned properly within the burner throat, and the throat refractory is in good condition.

Gas burners
- Be sure that filters and moisture traps are in place, clean and, operating properly to prevent gas orifice plugging. Inspect gas-injection orifices and verify that all passages are unobstructed.

- Look for any burned off or missing burner parts. Confirm location and orientation of all parts. (Viewing ports are very helpful to identify faulty flame patterns or other problems)

Pulverized-coal burners
- Verify that fuel and air components, pulverizes, feeders, primary and tempering-air dampers etc are all working properly.

- Clear coal pipes of any coal and coke deposits.

- Check burner parts for any signs of excessive erosion or burn-off.

Spreader-stoker firing
- Check grates for wear. Check the stokers and the cylinder-reinjection system for proper operation.

- Confirm the proper positioning of all air-proportioning dampers.

- Verify proper coal sizing.

Combustion controls
- Be sure that all safety interlocks and boiler trip circuits operate.

- See that all system gages are calibrated and functioning.

- Eliminate play in all control linkages and air dampers. Also check to see if there is accurate repeatability when load points are approached from different directions.

Gas Fired Burners	Condition and cleanliness of gas injection orifices.
	Cleanliness and operation of filter & moisture traps.
	Condition of diffusers, spuds, gas cans, etc.
	Condition of burner refractory.
	Condition and operation of air dampers.
Oil Fired Burners	
	Condition and cleanliness of oil tip passages.
	Oil burning temperature.
	Atomizing steam pressure.
	Condition of impeller/diffusers.
	Position of oil guns.
	Cleanliness of oil strainer.
	Condition of burner throat refractory.
	Condition and operation of air dampers.
Pulverized Coal Firing	
	Condition and operation of pulverizes, feeders and conveyors.
	Condition of coal pipes
	Coal fineness
	Erosion and burnoff of firing equipment.
	Condition and operation of air dampers.
Stoker Firing	
	Wear on grates
	Position of all air proportioning dampers.
	Coal sizing
	Operation of cinder reinjection system

Figure 9.3—Preliminary Burner Equipment Checklist I

Combustion controls	Cleanliness and proper movement of fuel valves
	Smooth repeatable operation of all control elements.
	Adequate pressure to all regulators.
	Unnecessary cycling of firing rate
Flame Safety System	
	Proper operation of all safety interlocks and boiler trip circuits.
Furnace	
	Excessive deposits or fouling of gas-side boiler tubes.
	Proper operation of sootblowers.
	Casing and duct leaks.
	Clean and operable furnace inspection ports.

Figure 9.4—Preliminary Boiler Inspection Checklist II

COMMON CAUSES OF LOW CO₂ AND SMOKY FIRE ON OIL BURNERS

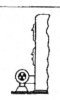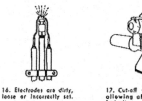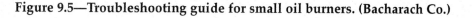

Figure 9.5—Troubleshooting guide for small oil burners. (Bacharach Co.)

- Check control elements for smooth accurate operation. Correct unnecessary hunting caused by improperly adjusted regulators and automatic master controllers.

- Inspect all fuel valves to verify proper movement, clean and repair as necessary.

FURNACE

- The firesides should be clean, check for sootblower cleaning efficiency. Consider periodic water-washing if firesides are not being kept clean by normal soot-blowing.

- Inspect and repair internal baffling. Defective baffling allows hot combustion gases to escape without giving up heat causing high stack temperatures. A traverse of the breaching with a temperature indicator may point out local hot spots behind baffle defects. Once hot spots are identified, the defects can be corrected.

- Repair any casing leaks and any cracked or missing refractory.

- Clean furnace-viewing ports and make sure that burner throat, furnace walls and leading convection passes are visible. Being able to see the condition of the flame, burner, refractory zone and furnace is essential to detecting and correcting problems.

FUNDAMENTAL COMBUSTION CHARACTERISTICS

The chemistry and physics of combustion very intricate and very difficult to define and describe. Combustion is notoriously elaborate and complex field of study. An attempt will now be made to put a simple face on the subject.

One of the main challenges faced by boiler and burner manufacturers is to design a compact and stable combustion system. Compactness, for lower first cost, and stability for trouble free and efficient operations. It has been well known for many years that a swirling motion of a burner flame will cause recirculation patterns which produce longer residence time, better air fuel mixing and stable flame conditions. Figure 9.6 is an example of the swirl pattern developed in register type burners.

Basic factors that must be considered when working with burners is burning velocity as influenced the air/fuel ratio and the use to toroids in the combustion zone. Figure 9.7 shows how burning velocity changes as the air/fuel ratio varies from stoichiometric conditions.

Figure 9.8 shows the use of toroids to stabilize the combustion zone. The recirculation patterns developed in the toroid zones have a great influence on combustion behavior. Swirl improves flame stability by forming toroidal recirculation zones that recirculate heat and unburned fuel constituents back to the base of the flame. At every point in the flame a balance exists between the velocity of flame propagation and the stream of incoming air and fuel. This also includes off stoichiometric mixtures of forming gases made up of partially burned mixtures of air and fuel and at varying levels of combustion activity. The trick is to get a complete but stable burn with no fuel energy wasted and the least amount of excess air possible. Figure 9.9 is a sketch of the final result of such a design.

Various flame types have been developed for different applications Figures 9.10 through Figures 9.16 The topic of NO_x control figures into these various flame types as shown in figure Figures 9.17, the peak in NO_x production belongs to the ball shaped flame which has extensive recirculation. It can be seen that as the surface area of the flame increases relative to volume, the temperature and NO_x production falls off.

FLAME APPEARANCE

The flame is the heart of the combustion process, if it isn't right you will have a serious challenge tuning-up a boiler.

The appearance of a boilers flame offers a good preliminary indication of combustion conditions. It is difficult to generalize the character-

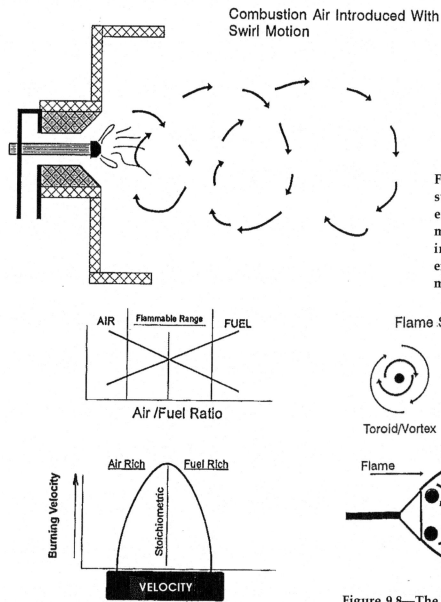

Combustion Air Introduced With Swirl Motion

Figure 9.6—An illustration of the swirling motion developed in burners to increase flame stability, better mixing and to shape the flame to fit into the combustion zone with out either rear wall or side wall impingement.

Figure 9.7—Illustrates the change in burning velocity or flame propagation rate relative to the air/fuel ratio. Maximum rate is at a stoichiometric mixing of air and fuel under proper conditions.

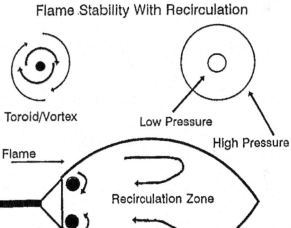

Flame Stability With Recirculation

Figure 9.8—The toroid or vortex is used extensively in burner design for flame zone stability. They are used to create recirculation patterns, to anchor flames and to recirculate partially consumed fuel back into the active combustion zone insuring good efficiency

istics of a "good" flame because of the variations due to burner design and operating conditions.

As the ideal situation is to operate with low-excess air, one must be familiar with the conditions this will create compared to higher excess-air conditions which may be favored by operators. Low excess-air operations demands that plant personnel pay close attention to the combustion process.

• Reduced oxygen levels leads to increased flame length because it takes more time to burn completely. It actually grows in size, filling the furnace more completely.

• It exhibits a lazy rolling appearance. Instead of intense, highly turbulent flames, low-oxygen flames may appear to move somewhat more slowly through the furnace.

Combustion Circulation Patterns

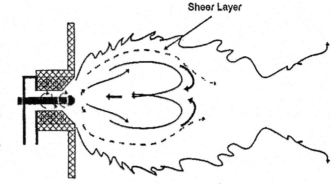

Figure 9.9—The combustion recirculation pattern established by main flame swirl to recirculate combustion products for more complete and stable combustion.

- It has an over-all color that may change as excess oxygen is decreased. Natural gas flames for instance, become more visible or luminous with yellow or slightly hazy, portions. Coal and oil flames become darker yellow and orange and may appear hazy in parts.

Although low excess-air operation is important, it is sometimes not possible to operate this way because of combustion related problems.

Observing oil flames provides important information concerning the combustion process.

The combustion problems which typically occur will be due to one or more of the following.

Observing oil flames provides important information concerning the combustion process. The combustion problems which typically occur will be due to one or more of the following.

Sources of Problems
1. Excess oxygen level.
2. Oil temperature or pressure.
3. Oil gun tip.
4. Air register setting.
5. Oil gun position.

Flame Types

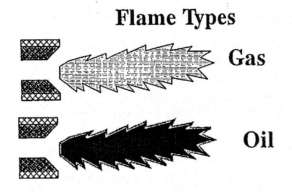

Gas

Oil

Conventional Feather Flame All Purpose

Figure 9.10—Conventional feather or "jet" flame for all purpose combustion chambers.

Flame Types

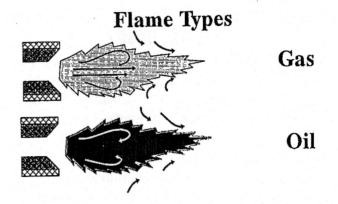

Gas

Oil

Headpin Flame Swirl Number 0.3

Figure 9.11—Headpin flame in which some slight swirl is introduced for stability especially at low firing rates.

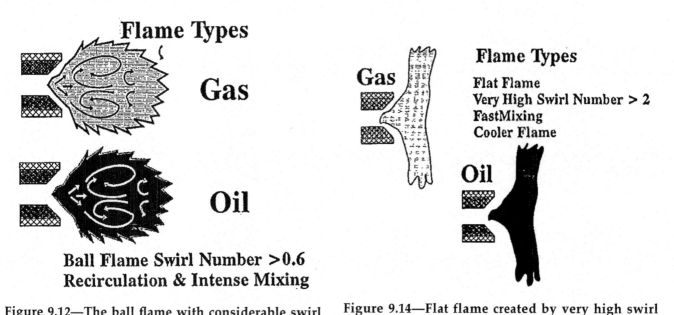

Flame Types

Gas

Oil

**Ball Flame Swirl Number >0.6
Recirculation & Intense Mixing**

Figure 9.12—The ball flame with considerable swirl for stability, used in cube type combustion chambers, produces high NO_x

Flame Types

Flat Flame
Very High Swirl Number > 2
FastMixing
Cooler Flame

Gas

Oil

Figure 9.14—Flat flame created by very high swirl number used to avoid flame impingement, to enhance wall radiation and to focus refractory radiation.

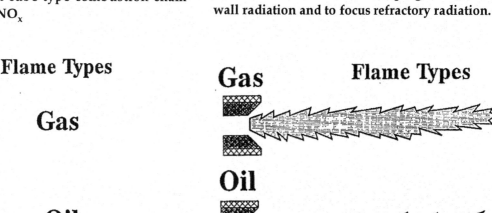

Flame Types

Gas

Oil

Conical Flame Swirl Number > 1 Cold Reverse Flow to Center & Intense Mixing

Figure 9.13—High swirl type flame with intense mixing.

Flame Types

Gas

Oil

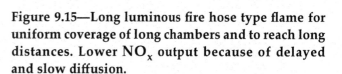

**Long Luminous Flame-No Recirculation
Delayed Diffusion
Uniform Coverage of Long Chambers**

Figure 9.15—Long luminous fire hose type flame for uniform coverage of long chambers and to reach long distances. Lower NO_x output because of delayed and slow diffusion.

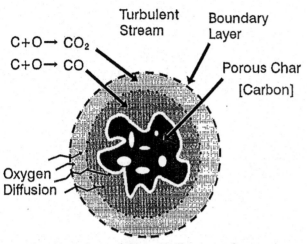

C+O → CO$_2$

C+O → CO

Turbulent Stream

Boundary Layer

Porous Char [Carbon]

Oxygen Diffusion

Combustion of Char in Furnace Gases

Figure 9.16—A governing reaction in flames and burners is the diffusion of oxygen into the fuel rich zone to sustain combustion. In low excess air operations the rate of evaporation may exceed the mixing of fuel and oxygen by turbulent diffusion yielding a process of rich limit ignition.

Nitric Oxide (NO) Change With % Swirl Of Combustion Air

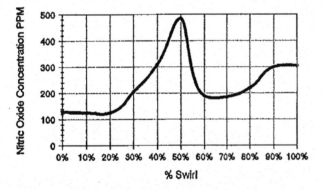

Figure 9.17—This graphic shows how nitric oxide concentration changes with swirl patterns, the peak point in this graph corresponds to the ball type flame in figure 9.12. On the left we have the long luminous or fire hose type flames and on the right the conical and flat type flames.

Figures 9.18 through 9.24 demonstrate various problems that may develop with typical burners.

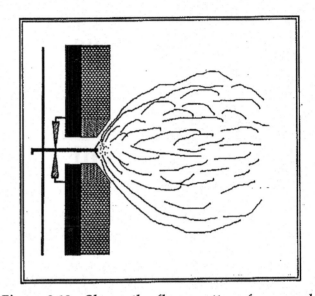

Figure 9.18—Shows the flame pattern for a nearly ideal flame geometry.

THE IDEAL FLAME

- Flame barely clears burner throat.

- Flame is bright and smooth with light smoke wisps on the end.

- Problems, none.

PROBLEMS THAT CAN OCCUR IN A BURNER FLAME OF IDEAL GEOMETRY

Flame contains ragged sparks
- Excess oxygen too high.

- Oil too hot.

- Ash in fuel.

Flame very smoky looking
- Excess oxygen too low.

Flame black and oily-looking with no burning near throat.
- Air register closed.

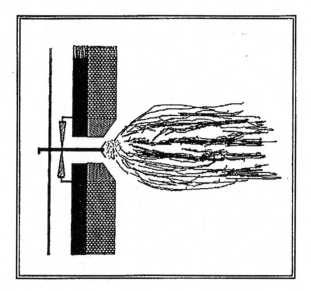

Figure 9.19—Flame is narrow and does not fill the throat. Possible thin stream of oil present in the middle of the flame.

Problem:
- Low oil pressure delta-p or return oil line plugged.
- Air register open too far.

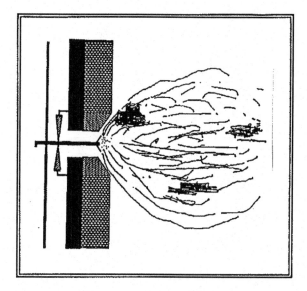

Figure 9.20—Flame pattern with intermittent oil slugs coming out of the tip.

Problem:
- Atomizer problem, tip partially plugged, tip worn or other burner tip problem.

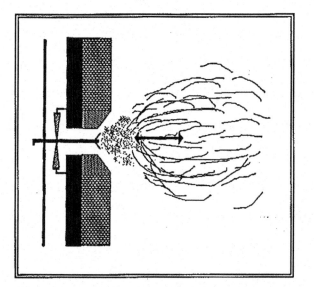

Figure 9.21—Flame blow off. May be continuous or pulsating on and off the tip.

Problem:
- Air register open too far.
- Oil gun positioned too far in toward the furnace.

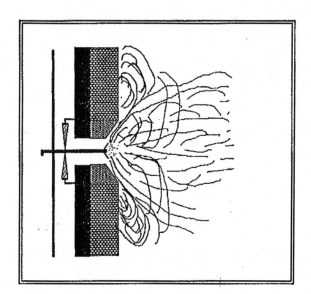

Figure 9.22—Flame clears throat, but rolls back, impinges and rolls up the furnace wall.

Problem:
- Air register closed too far.

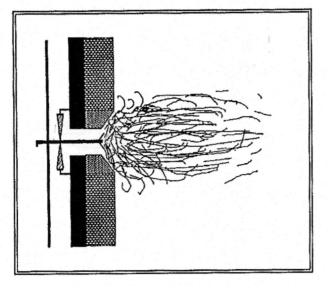

Figure 9.23—Flame impinges on burner throat.

Problem:
- Oil gun not fully extended into firing position.
- Wrong tip or worn tip orifice.

COMBUSTION OF OIL [FIGURE 9.24]

A. The oil is atomized into fine spherical particles to develop the largest possible surface area for vapor formation. The vapor actually is what burns, not the oil.

B. Next the intense heat in the combustion zone ignites the vapor and oxygen supports combustion.

C. The vapor and carbon continues to burn reducing the size of the char as it moves away from the burner. Oxygen infuses in to the envelope around the burning particle and carbon dioxide, carbon monoxide and water are given off by the combustion process.

D. All fuel should be burned in the furnace with no impingement on furnace walls.

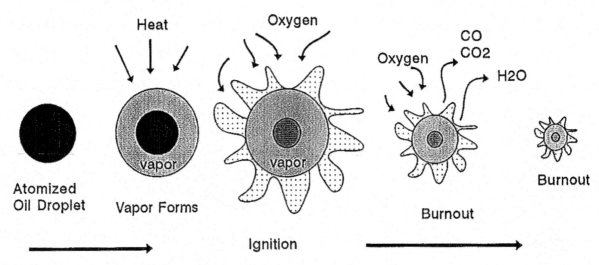

Figure 9.24—Combustion of oil.

Flame appears to be less bright and dense, somewhat transparent.
- Oil pressure delta-p too high. Too much return flow.
- Worn tip returning too much oil flow.

THE BASICS OF A TUNEUP

Keep in mind, the basic criteria for good combustion:

- Time
- Temperature
- Turbulence
- Sufficient air

Figure 9.25 Criteria for good combustion

Problem	Probable Cause	Corrective Action
Heavy White Plume		
• Consistent	High excess oxygen	Lower excess oxygen down to a level slightly above smoke limit.
	High sulfur fuel	Have fuel sample analyzed.
• Transient	Rapid decrease in load	Check cross limiting controls. Make load changes more gradual.
Black smoking		
• At low excess oxygen	Insufficient air	Increase air. Check operation of combustion controls when in automatic mode. Sluggish control response on load swings will cause occasional smoking. Automatic (load control) operation will usually require a slightly higher oxygen setting.
• At high excess O_2 and low wind box-furnace pressure, flames look about the same throughout the furnace; slow and lazy indicating low excess air.	O_2 instrument out of calibration	Check calibration of O2 instruments.
• At high excess O_2 and high windbox-furnace pressure, flames look about the same throughout the furnace.	Oil temperature problems Oil pressure problems	Check oil heater and oil temperature at burner. Check oil supply-return pressure. Check pressure gage calibration.
• At high excess O_2 and high windbox-furnace pressure, flames are not the same throughout the furnace.	Localized combustion problem.	Check flame patterns 1. Oil temperature 2. Oil supply-return pressure 3. Oil gun position If these are ok, then check air register settings and operation, reposition if necessary. Problem may be burner tip related, inspect and change if necessary.
• Excess O_2 cycling with constant forced-draft-fan and fuel flow indications.	Localized air heater pluggage	Check air heater pressure drop.
• Excess O_2, Forced-draft-fan and fuel flow cycling	Control problem	Check combustion control system.

Table 9.1—Troubleshooting chart for problems with smoke.

Note: These tests and adjustments should only be conducted with a through understanding of the test objectives and following a systematic, organized plan.

Instruments

The minimum limits of excess air should be approached cautiously with flue gas analyzers which continuously provide an accurate measurement of average conditions for the burner being adjusted. When the maximum smoke spot number, for oil or the maximum carbon monoxide level for natural gas is reached it should be noted along with the corresponding burner settings. Flame instability is sometimes a limiting factor on reducing excess air. In stoker fired boilers, the onset of clinkering or overheating of the grate sometimes precedes the formation of smoke or carbon monoxide.

SAFETY

- Extremely low excess oxygen operation can result in catastrophic results.

- Know at all times the impact of the modification on fuel flow, air flow and the control system.

- Observe boiler instrumentation, stack and flame conditions while making any changes engineering personnel or the boiler manufacturer.

- Consult the boiler operation and maintenance manual supplied with the unit for details on the combustion control system or methods of varying burner excess air.

FINDING THE SMOKE AND CO THRESHOLD

Once your boiler is in good working order, the next major step in improving efficiency and reducing emissions is to establish the lowest level of excess oxygen at which the unit can operate safely and meet clean air laws.

Maximum Desirable Smoke Spot Number	
Fuel Grade	Maximum Desirable SSN
No. 2	Less than 1
No. 4	2
No. 5	3
No. 6	4

Figure 9.26—Desirable smoke spot numbers for various fuels

Smoke Spot Number	EFFECT OF SMOKE ON SOOT BUILD UP	
	Rating	Sooting Produced
1	Excellent	Extremely light if at all
2	Good	Slight sooting which will not increase stack temperature appreciably.
3	Fair	May be some sooting but will rarely require cleaning more than once a year.
4	Poor	Borderline condition. Some units will require cleaning more than once a year.
5	Very Poor	Sooting occurs rapidly and heavily

Figure 9.27—The effect of smoke spot number on soot build up rate

Since most boilers operate over a reasonably broad range, tests must be run at several firing rates to determine the minimum excess-oxygen level for each. Only then can the combustion-control system be tuned for optimum fuel economy.

At each firing rate investigated, excess oxygen in the flue gas should be varied from 1-2% above the normal operating point down to where the boiler just starts to smoke, or to where CO emissions vary between 150 to 250 Parts per million (PPM). The level of 400 PPM is the legal limit established in many states and by insurance companies. This condition is referred to as the smoke or CO threshold, or simply as the minimum oxygen point.

The smoke threshold generally applies to coal and oil firing, because smoking usually occurs before CO emissions reach significant levels. The CO level pertains to gaseous fuels. The smoke threshold for solid and liquid fuels represents the lowest possible excess-oxygen level at which acceptable stack conditions can be maintained.

The Smoke Spot Number (SSN) is a scale of smoke density which can be related to the soot accumulation in a boiler Figure 9.26 shows the desirable SSN for various fuels. Figure 9.27 shows the relationship of SSN to the rate at which deposits generally accumulate. The SSN is directly related to excess-oxygen levels and burner performance.

MINIMUM EXCESS OXYGEN

A proven method for determining the minimum amount of excess oxygen required for combustion involves developing curves similar to the smoke/oxygen and CO/oxygen curves shown in Figures 9.28 and 9.29. Based on test measurements, these curves show how boiler smoke and CO levels change as excess oxygen is varied.

Each of these figures depicts two distinct curves, illustrating the extremes in smoke and CO behavior that may be encountered. One curve exhibits a very gradual increase in CO or smoke as the minimum excess oxygen condition is reached. The other has a gradual slope at relatively high oxygen levels and a steep slope near the minimum oxygen point. For cases represented by this second curve, unpredictably high levels of smoke and CO, or potentially unstable conditions, can occur with very small changes in excess air.

Caution is required when reducing air flow near the smoke point or CO threshold. Care-

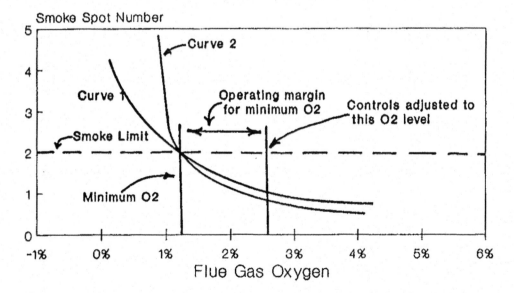

Figure 9.28—Characteristic curve identifies minimum excess air and tuneup control settings for oil fired boilers. Curve 1 gradual smoke/O_2 relationship. Curve 2 is a steep smoke/O_2 relationship. The type of curve is dependent on burner operating characteristics and varies with the firing rate and particular type burner.

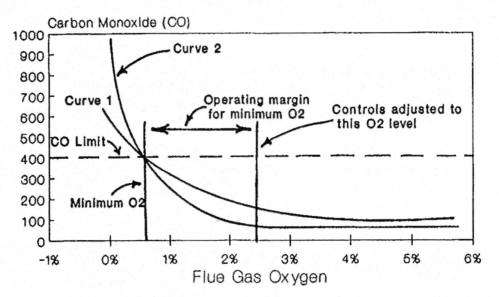

Figure 9.29—Characteristic curve identifies minimum excess air and tuneup control settings for oil fired boilers. Curve 1 gradual carbon monoxide/O_2 relationship. Curve 2 is a steep carbon monoxide/O_2 relationship. The type of curve is dependent on burner operating characteristics and varies with the firing rate and particular type burner.

fully monitor instruments and controls, flame appearance and stack conditions simultaneously. Decrease the level of excess oxygen in very small increments until you find out what kind of a curve is developing. It can be steep and possibly unstable or gradual. Some boilers have a gradual characteristic at one firing rate and a steep characteristic at another.

CONTROL SYSTEM LINEARITY

Before a tuneup of the boiler control system is started a check for system linearity should be done. This requires a measurement of either steam output or fuel input be compared to control system position. Often systems have not been linearized and this causes trouble when trying to run several boilers together. Figure 9.30 shows the situation you could run into. Imagine trying to drive a car on a fast and crowded highway with a vehicle speed control response similar to either the quick response or slow response curve. Such a condition would make driving very difficult and dangerous, the same goes for boilers.

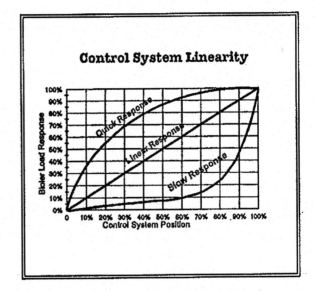

Figure 9.30—Control system Linearity

STEP-BY-STEP PROCEDURE FOR ADJUSTING BOILER CONTROLS FOR LOW-EXCESS OXYGEN

1. Establish the desired firing rate and switch combustion controls from automatic to manual operation. Make sure all safety interlocks are still functioning.

2. Record boiler and stack data (pressure temperature etc.), and observe flame conditions after the boiler operation stabilizes at the particular firing rate selected. If you find that the amount of excess oxygen in the flue gas is at the lower end of the range of typical minimum values and the CO and the smoke are at acceptable levels, the boiler is already operating at a near optimum air to fuel ratio. This may not be so at other firing rates. It may still remaining portion of this procedure, to practical.

3. Increase air flow to the furnace until readings of excess oxygen at the stack increase by 1-2%. Again, be sure to take readings after boiler operation stabilizes and note any changes in flame conditions.

4. Return air flow to normal level and begin to slowly reduce it further, in small increments. Watch the stack for any signs of smoke and constantly observe the flame and stack. Record stack excess-oxygen reading, smoke spot number, the concentration of CO in the flue gas and the stack temperature after each change.

Do not reduce air flow by throttling the burner air registers, because this alters the fuel air mixing characteristics and complicates the tests. Also, if you run tests at low firing rates, which is not generally recommended, keep a close watch on the windbox/furnace differential. If it drops too low, a fuel trip may be activated by the burner safeguard system.

5. Continue to reduce the airflow step- wise until you reach one of theses limits:
   - Unacceptable flame conditions such as flame impingement on furnace walls or burner parts, excessive flame carryover, or flame instability.
   - High level of CO in the flue gas.
   - Smoking at the stack. Do not confuse smoke with water vapor, sulfur or dust plumes which are usually white or gray

in appearance and remember to observe local air pollution ordinances.
   - Incomplete burning of solid fuels. Recognize this by high carbon carryover to dust collectors or increased amounts of combustibles in the ash.
   - Equipment-related limitations such as low windbox/furnace pressure differential, built-in air flow limits, etc.

6. Develop O_2/smoke or O_2/CO characteristic curves, similar to those shown in Figures 9.28 and 9.29 using the excess oxygen and CO or smoke-spot number data obtained at each air-flow setting.

7. Find the minimum excess-oxygen level for the boiler from the curves prepared in step 6, but do not adjust the burner controls to this value. Though this may be the point of maximum efficiency, as well as minimal NO_x emissions, it usually is impractical to operate the boiler controls at this setting, because of the tendency to smoke or to increase CO to dangerously high levels as load changes.

Compare this minimum value of excess oxygen to the expected value provided by the boiler manufacturer. If the minimum level you found is substantially higher then the manufacturer's repairs or parts replacement probably can improve fuel and air mixing, thereby allowing operation with less air.

8. Establish the excess oxygen (buffer zone) margin above the minimum value, required for fuel variations, load changes, and atmospheric conditions. Add this to the minimum value and reset burner controls to operate automatically at the higher level-the lowest practical setting at the particular firing rate.

9. Repeat steps 1-8 for each firing rate being tested. For some control systems, it is not possible to establish the optimum excess-oxygen level at each firing rate. The reason is that control adjustments at one firing rate

may also affect conditions at other firing rates. In such cases, choose the settings that give the best performance over a wide range of firing rates. A trial-and-error approach, one involving repeated tests, may be necessary.

Many experts agree that it generally is best not to make any adjustments to your control system in the lower control range of your boiler without being very careful. Air flow requirements at low-fire conditions usually are dictated by flame ignition characteristics and stability rather than by efficiency. Air/fuel ratios at low loads and at or near light off conditions are very sensitive and any changes may jeopardize safe light-off characteristics. If boiler load requirements force a boiler to operate at low loads much of the time, check with the boiler manufacturers service group or a qualified combustion consultant before establishing excess-oxygen levels.

10. Verify that the new settings can accommodate the sudden load changes that may occur in daily operation without adverse affects. Do this by increasing and decreasing the load rapidly while observing the flame and stack. If you detect undesirable conditions, reset the combustion controls to provide a slightly higher level of excess oxygen at the affected firing rates. Next verify these new settings in a similar fashion. Then make sure that the final control settings are recorded at steady-state operating conditions for future reference.

Repeat these checks at frequent intervals until it becomes obvious that the boiler is not having problems that cause it to exceed smoke or CO limits or that control, burner or fuel system problems are not causing unsafe conditions to develop. It is easy to hide such problems my making high excess oxygen adjustments. Trying to optimize performance will cause these problems to reemerge.

When an alternative fuel is burned, perform these same tests and adjustments for the second fuel. It is not always possible to achieve optimum

excess oxygen levels for both fuels at all firing rates. Based on information gained from the tune-up procedure, a judgment can be made as to the best conditions which are practical.

Evaluation of the New Low O_2 Settings

If energy gains are to be realized, the new low O_2 settings must be realistic and they must be maintained. Pay extra attention to furnace and flame patterns for the first month or two following implementation of the new adjustments. Thoroughly inspect the boiler during the next shutdown. To assure high boiler efficiency, periodically make performance evaluations and compare with the results obtained during the test program.

REVIEW OF THE FINE TUNING PROCESS

It is sometimes possible during the optimization program to lower the CO or smoke limits, to achieve even lower excess air levels achieving greater efficiency gains. If the burner and fuel systems are not functioning properly your best efforts at lowering excess air may be wasted. The approach to this procedure is to insure that everything is in conformance with the manufacturers recommendations and then conduct organized "trial-and-error" (Table 9.2) adjustments in such a way that meaningful comparisons can be made.

Items that may result in lower minimum excess O_2 levels include:

- Burner register settings
- Oil gun tip position
- Diffuser position
- Fuel oil temperature
- Fuel oil atomizing pressure
- Coal spreader adjustments
- Coal particle size

The effect of these adjustments on minimum O_2 are variable from boiler to boiler and difficult to predict.

The principal method used for improving boiler efficiency involves operating the boiler at

Step-by-step Boiler Adjustment Procedure for Low Excess Air Operation.
1. Put the control system in manual control and bring the boiler to the test firing rate.
2. After stabilizing, observe flame conditions and take a complete set of readings.
3. Raise excess O_2 1-2%, allowing time to stabilize and take readings.
4. Reduce excess O_2 in small steps while observing stack and flame conditions. Allow the unit to stabilize following each change and record data.
5. Continue to reduce excess air until a minimum excess O_2 condition is reached.
6. Plot CO or Opacity versus O_2.
7. Compare the minimum excess O_2 value to the expected value provided by the boiler manufacturer. High excess O_2 levels should be investigated.
8. Establish the margin in excess O_2 above the minimum and reset the burner controls to maintain this level. This is the operating "Buffer Zone" which is based on an estimation of the amount of repeatability in the control system and the affects of other influences like temperatures and pressures.
9. Repeat steps 1-8 for each firing rate to be considered. Some compromise in optimum O_2 settings may be necessary since control adjustments at one firing rate may affect conditions at other firing rates if there is no means to characterize the air/fuel ratio.
10. After these adjustments have been completed, verify the operation of these settings by making rapid load pick-ups and drops. If undesirable conditions are encountered, reset controls.

Table 9.2 Boiler Tuneup Procedure.

the lowest practical excess O_2 level with an adequate margin for variations caused by fuel property changes, changes in ambient conditions, and the repeatability and response characteristics of the combustion control system.

The important elements of boiler performance optimization:

- Scientific approach

- Proper analytical test equipment and procedures

- Keep good records

- A total system knowledge of the boiler and the following:

— Boiler safety systems
— Casualty procedures
— Combustion theory
— Burner design and operation
— Control system theory and operation
— Air handling systems
— Fuel systems

How often should a boiler be tuned up? The simple answer to this question is "It depends." The object of a "tune-up" is to insure that a boiler is running efficiently and safely on a permanent basis, so another factor enters the situation:

*Perhaps this is the most important part of tuning up
a boiler—you have to do it until it's right.
A proper tuneup requires follow-up checks
and readjustments until it is proven satisfactory.
A quick brushoff often won't work and could
actually be dangerous.*

The important elements of boiler performance optimization:

- Be Scientific
- Identify dollar losses connected with performance so that reasonable judgments can be made about investments of time and resources.
- Use good instruments
- Do regular spot checks
- Set allowable limits for performance
- Keep records
  - Ideal performance
  - Tuned-up performance
  - Performance history
  - Track excess air performance [Figure 9.31]
  - Track CO, combustibles and smoke behavior
  - Track exhaust temperature [Figure 9.32]

Exit Gas Temperature
Boiler #7 [Deg F]

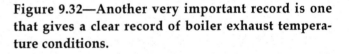

Boiler Operating Load	Ideal	Clean	Latest Data
30%	340	360	410
40%	340	360	410
50%	340	360	410
60%	340	360	410
70%	340	365	425
80%	340	370	430
90%	340	380	440
100%	340	390	450

Figure 9.32—Another very important record is one that gives a clear record of boiler exhaust temperature conditions.

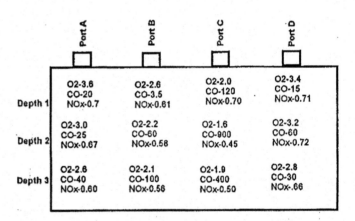

Figure 9.33—On large boilers with multiple burners, test port locations are usually established with the original design. Records of conditions at these locations are necessary to establish oxygen, combustibles and pollution levels. In this case indications are that the burners need adjustment because oxygen and CO are out of range for good and efficient operations.

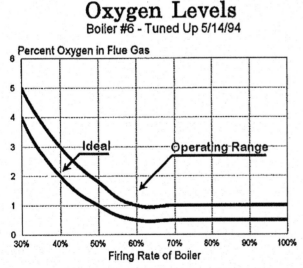

Oxygen Levels
Boiler #6 - Tuned Up 5/14/94

Figure 9.31—One of the records that should be established for a boiler is the percent oxygen [or excess air] in the exhaust

The Effect of Burner Balance on CO

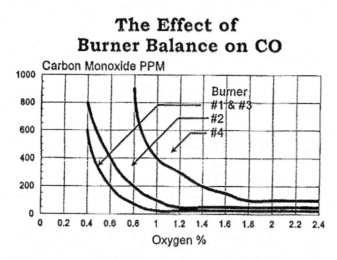

Figure 9.34—This figure shows that differences exist between the performance of individual burners, the reasons should be established and corrected if possible.

The Reasons for Deviations Between Burners

1. Gas. Gas port dimension differences & Piping differences

2. Oil. Atomizer port tolerances, surface finish & erosion. Piping differences.

3. Air. Variations in air flow to burners.

4. Variation in pressure drops across burners. [Air doors, swirl vanes, flow splitters, surface finish & refractory damage.

5. Wind box symmetry & cross flow.

6. Windbox air temperatures and buoyancy forces.

Figure 9.35—Reasons for deviations between performance of otherwise identical burners as illustrated in Figure 9.34

TUNING BOILER CONTROLS

The following section is composed of figures which cover the subject of tuning boiler controls. They are designed for a classroom presentation and make up a simple presentation of a complex subject.

Tuning Boiler Controls
Large, Interactive, Non-linear Control Loops

Important Impacts
- Efficiency
- Ramp Rate
- Turndown
- Low load operations
- Unit availability
- Ability to survive upsets and equipment failures

Figure 9.36—Tuning boiler controls, important impact

Tuning Boiler Controls
Basics
- Start from the bottom up
  — Control drives
  — Valve operators
  — Final control elements

All measurements must be accurate beyond any doubt and backed up by duplicate Instruments

Figure 9.37—Tuning boiler controls, basics

Tuning Boiler Controls
Deadband and Repeatability
Controlling Process Variables

- No amount of upper loop tuning will fix 5% deadband

- Over 3% to 5% deadband is unacceptable

Process variables: feedwater flow, gas flow, air flow, fuel flow, spray valve, etc.

Figure 9.38—Tuning boiler controls, deadband and repeatability

Tuning Boiler Controls
Measurement and Control Response Time

1. Fast measurements improves quality and stability of control loops

2. Fast response times improves quality and stability of control loops

3. Valve stroking rates of 2 to 3 seconds Is required in many cases.

4. If one loop Is upsetting other loops, a compromise must be reached

5. In the final analysis, good engineering judgment Is required for tuning loops

Figure 9.39—Tuning boiler controls, measurement and control, response time

Tuning Boiler Controls
PID
Proportional Gain-Reset-Derivative

• It is desirable to use as little PID as possible

• Stability Is more Important than quality of control

• Excessive reset action is deadly for most control loops

Figure 9.40—Tuning boiler controls, PID proportional Gain-Reset-Derivative

Tuning Boiler Controls
Superheat and Reheat Thermocouples

• Check for good thermocouple contact with the bottom of the thermowell

• At 80% load increase spray valve position 5% as a step function. Observe steam temperature time constant.

• Time constant should be less than 30 seconds.

Figure 9.4—Tuning boiler controls, superheat and reheat thermocouples

Tuning Boiler Controls
Steam Temperature Control

• Fast accurate temperature measurements are desirable

• Slow temperature constants can be compensated for with derivative action at the temperature controller

Figure 9.42—Tuning boiler controls, steam temperature control

Tuning Boiler Controls

Adaptive Tuning
• Final steam temperature controller must be tuned at 5 points.

Figure 9.43—Tuning boiler controls, adaptive tuning

Tuning Boiler Controls

Feed Forward Signals

1. The necessary influence of feed forward signals should be calculated from steady state tests.

2. No more than 80% - 90% of actual influence should be set

Figure 9.44—Tuning boiler controls, feed forward signals (

Tuning Boiler Controls

Feed Forward Signals

Example: superheat spray set point Is 15 F for 100 psi of first stage pressure

— Use only 12 -13 F/100 psi feed-forward signal

— Let feedback control handle last 2 degrees to eliminate Instability

Figure 9.45—Tuning boiler controls, feed forward signals

Tuning Boiler Controls

Fuel Flow

• During the first 30 to 90 seconds after a load change the required energy comes from storage [boiling water in the drum]

Figure 9.46—Tuning boiler controls, fuel flow

Tuning Boiler Controls

Fuel Flow

• Fuel should never lead air supply

• Fuel and air should move together

• Adaptive tuning is most often required for fuel and air to synchronize variations over turn-down range

Figure 9.47—Tuning boiler controls, fuel flow requirements

Tuning Boiler Controls

Drum Level Stability

• Drum level instability can de-stabilize an entire boiler

• Smaller drums may need 3-element control

• Extended range of feedwater and steam flow measurement may also be needed to measure low and high range temporary excursions In boiler operations

Figure 9.48—Tuning boiler controls, drum level stability

Tuning Boiler Controls

Drum Level Stability

Drum level stability is needed for:

1. Good ramp rate

2. Minimum load operations

3. Runback capability (Staying on the line after an upset]

Figure 9.49—Tuning boiler controls, drum level stability

Tuning Boiler Controls

- Adaptive tuning
  Every critical loop should be tuned at 6 different positions

- Tuning requires
  a. Time
  b. Patience
  c. Good engineering judgment

- Tuning   is the art of selecting acceptable compromises

Figure 9.50—Tuning boiler controls, requirements

Tuning Boiler Controls

Runbacks

- Runbacks must be tuned and demonstrated or disconnected.

- A single equipment failure could cause a Domino Trip Effect.

Figure 9.51—Tuning boiler controls, runbacks

Tuning Boiler Controls

Why?

- Improved ramp rate
- Lower excess air
- Higher final steam temperatures
- Economic payback

Figure 9.52—Tuning boiler controls, why?

Reducing Excess Air

1. Oxygen Trimming
   a. Compensates for Many Variations of Buffer Zone Required
   b. Does not Maintain "Optimum" A/F Ratio

2. CO Trimming
   a. Establishes 'Best' A/F Ratio
   b. No Buffer zone Required.

Figure 9.53—Oxygen trim systems vs. carbon monoxide trim systems

Chapter 10

Over 100 Ways to Improve Efficiency

This chapter introduces over 100 ways to reduce the costs of operating boilers and distribution systems. These options are the result of tests and investigations at over a thousand commercial, institutional, industrial and utility plants around the world.

Although more than 100 efficiency improvement options are listed in this chapter, only a few will be appropriate at any one plant. In most cases a understanding of a plant's operation is necessary to identify the causes of wasted energy and to identify the most cost effective remedy.

This chapter has been written with two goals in mind. First it serves as a guide to assist in identifying energy wasting problems for any size boiler. Second, it can serve as a valuable educational tool, showing the logic of how to approach this complex problem.

Section 10-1
High Efficiency Operations

CHECKLIST – 1
High Efficiency Operations

1. Maintain optimum boiler efficiency levels
2. The 2M System: measure and manage.
3. Keep boilers tuned-up.
4. Operate boiler at highest efficiency levels.
5. Lead–Lag boiler configuration
6. Operate boilers at lower steam pressures
7. Avoid low firing rates
8. Maintain low boiler exhaust temperatures
9. Keep accurate statistics

10. Fuel flow meter for each boiler
11. Track performance of small heating units automatically
12. Recover hot condensate
13. Minimize makeup water consumption
14. Isolate unused hot steam piping sections.
15. Reduce venting Steam from the Deareator
16. Boilers converted from Heavy Fuel Oil to Natural Gas
17. Shut off steam coil ahead of air preheater
18. Soot blower operations
19. Shut-down backpressure steam turbine during low steam load periods
20. Utilize fuel additives to improve operating conditions and economy
21. Shut off boilers during weekends and summer months when not needed
22. Consider thermal mass when planning boiler operations
23. Check economizer operation

1. MAINTAIN OPTIMUM
    BOILER EFFICIENCY LEVELS

Figure 10.1—A civilization built on energy.

113

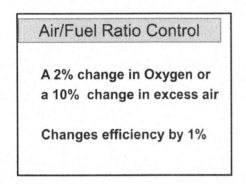

Figure 10.2—Excess air wastes a great deal of energy and money.

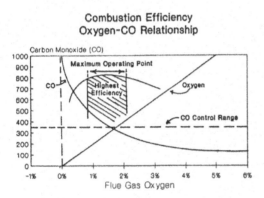

Figure 10.3—Smoke is unburned fuel that fouls the boiler and the environment

Figure 10.4—The key to good efficiency is maintaining low excess air.

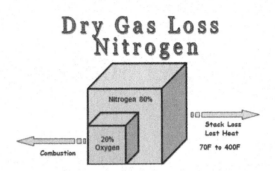

Figure 10.5—Only 20% of the air for combustion is actually used. The remaining 80% is nitrogen which escapes up the stack as an invisible and expensive hot gas.

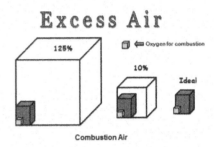

Figure 10.6—These losses climb rapidly as excess air increases.

question is, "Are the switches thrown and the valves managed responsibly?"

Large savings have been recorded just from monitoring energy use, no black boxes were involved. Flow meters and other instruments were installed and monitored. The most remarkable discoveries were made. Pumps and motors ran when they weren't needed, steam to process equipment wasn't turned off when the shift left for the day and boilers were being fired with no load on them, etc. This is a partial listing, but the lesson is clear, without accountability for energy use, without incentives to save utility dollars, waste and indifference will exist.

The key to discovering this type of loss is having instruments installed and monitored. It becomes very clear in a short period, when and where energy is being squandered. Without proper instrumentation and monitoring you will be running blind. The Gas and Electric Utility Companies don't give away energy, they use meters and monitor them closely. It's a good practice to follow when you consider the money spent purchasing their energy.

2.   THE 2M SYSTEM: MEASURE AND MANAGE

There is virtually unlimited access to energy by its users, just throw a switch or open a valve and energy is available for immediate use. The

Advantages/disadvantages

Instruments are often expensive and in themselves do not account for any energy savings, they are however, the only way to find problems. Manpower that may be needed elsewhere must be assigned to collect and interpret the data from field measurements. It's just good management to know what is happening with energy, assigning responsibility for its use and controlling it in a sensible way

3.   KEEP BOILERS TUNED-UP

Description

A good tuneup, using precision test equipment, can detect and correct excess air losses, smoking, unburned fuel losses, sooting, fire side fouling and high stack temperatures.

By using real time instruments to measure flue gas oxygen, carbon monoxide, carbon dioxide, combustibles and temperatures, the current state of boiler operation can be diagnosed. These findings can be used to restore the boiler to its normal efficient operating condition (Figures 10.1-1 to 10.6). This allows the control system and burner to be adjusted and repaired for optimum performance with immediate feedback of results for gas fuels, (Figure 10.7) and oil (Figure 10.8). (Figure 10.9) shows the same information for oil and coal based on smoke limit criteria. When a

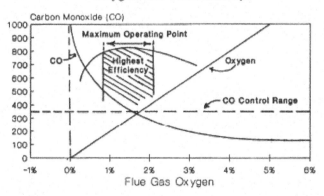

Figure 10.7—The carbon monoxide vs. oxygen relationship showing CO control range and operating range for highest efficiency.

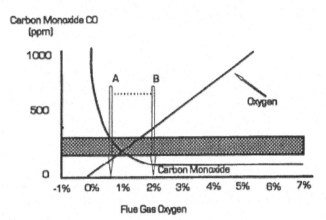

Figure 10.8—Carbon monoxide limit vs. oxygen relationship for adjusting controls for optimum efficiency. Point A is the minimum oxygen control point, B is the chosen set point based on a buffer zone for estimated system errors and influences.

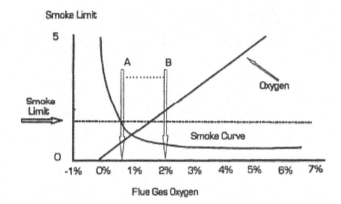

Figure 10.9—Smoke limit (dashed line) vs. oxygen relationship for adjusting controls for optimum efficiency on oil and coal fired plants. In this case point A is minimum oxygen control point. Point B is the chosen set point based on the estimated buffer zone needed to compensate for control system errors and other outside influences affecting combustion.

tune-up is completed there should be a good record of a boiler's excess air and efficiency across the load range for managing boiler operations.

After the initial tune-up, recheck results until assured that the boiler can stay tuned-up for a reasonable amount of time. Tune-ups may be needed more than twice a year. Periodic checks will tell the story; often boiler efficiency will fall by over 5% within 6 months or sooner.

Advantages/Disadvantages

If you intend to manage a boiler plant and control losses, regular tuneups and combustion efficiency checks are necessary. It takes skilled manpower to run these checks and this is an additional expense so judgment should be applied to establish the benefit to cost ratio of such a program.

(Very Important: All energy conservation options should be evaluated from the tuned-up condition to eliminate false estimates of the benefits of retrofit options.)

4.   OPERATE BOILER AT HIGHEST EFFICIENCY LEVELS.

Description

Having specific efficiency information for each boiler, they can be matched to loads for most efficient utilization. Also, by using smaller boilers, which maintain high efficiency during very light load periods, losses associated with larger boilers can be eliminated. The general answer is to use the most efficient boiler combinations possible to match system demands.

Most boilers do not have a straight line efficiency and operate at different efficiencies depending on their firing rate and this must be considered when programming their operation (Figure 10.10).

Boilers can have different efficiency levels at different load points. At the high end the efficiency may fall off because of high exhaust temperatures. At low loads efficiency may go down because of higher excess air needed for proper air-fuel mixing. At mid range the efficiency is very good; so what efficiency do you pick to describe its efficiency?

5.   LEAD – LAG BOILER CONFIGURATION

When setting up automatic operations make most efficient units "Lead" elements and less efficient units "Lag" elements.

6.   OPERATE BOILERS AT LOWER STEAM PRESSURES

Lowering Steam Pressure
1.   Boiler exhaust temperatures are lower

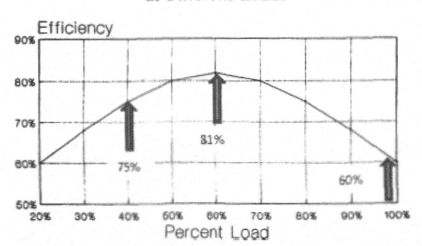

Figure 10.10—Boiler efficiency often varies with load and other operational influences. Long term information comparing fuel burned to steam or hot water generated is needed to understand the true operating efficiency. Quick snap-shot tests are not suitable for obtaining the true efficiency.

2.   Steam system pipe temperatures are lower
3.   Steam leaks are less
4.   Steam trap flash losses are less

Possible Disadvantages

1.   Excessive piping velocities due to higher volume of steam flow

2.   Pressure to emergency pumps and generators may not be high enough

3.   Boiler priming [dangerous slugs of water entering the steam system]

4.   Noisy steam valves [essential element of steam valve design]

5.   Temperature sensitive traps may start to blow through steam

6.   Cavitation noises in feedwater valves

7.   Atomizing steam pressure may be too low

8.   Differential pressure across steam traps may change capacity of some types of traps

9.   Steam traps will have to handle more condensate

10. Pumping capacity of original design may change

11. Calibration of flow meters may change

One way to approach lowering steam pressure is to do it gradually over a period of time and watching for changes in system operation and correcting them if possible..

7.   AVOID LOW FIRING RATES

Some boiler plants are oversized for their current loads, or small boilers are not available to handle light loads of summer or partial plant production schedules. This is caused by the boiler cycling on and off continually because of the low load factor. Each time the boiler starts up or shuts down cold air is blown through it to purge out any possible accumulations of unburned fuel as a safety precaution. This plus normal heat loss to the environment can add up to a significant percent of its operational fuel demand. The shell losses become higher percentage of fuel use and frequent "on" and "off" cold air purge cycles send a lot of valuable heat out the exhaust. Boilers typically have very poor efficiency at low loads so this type of operation can be quite inefficient. Often boilers with measured efficiencies over 80% have been found to be operating at less than 60% efficiency on an annual basis.

There are many solutions to this problem, which usually starts with the original design where large safety factors were employed to insure against inadequate steam supply on the coldest windiest day of the year. However, insulation improvements, energy conservation measures and mild weather can put these over-designed high capacity systems into the energy wasting mode.

The answer to managing boilers efficiently is simply stated: "You have to know the efficiency of your boilers at all firing rates and you have to take into account these standby losses."

Advantages/Disadvantages

Knowing the losses incurred at each level of operation and having a plan to minimize the inefficient levels is a good management practice. Although it may be take extra effort to get the real time performance data required for true operating efficiencies of boiler systems, it must be done if energy dollars are to be saved. In some cases a continuous computerized monitoring of the system may be required.

8.   MAINTAIN LOW BOILER EXHAUST TEMPERATURES

Figure 10.11 and 10.12 show the influence of exhaust temperature on efficiency. Boilers usually run 24 hours a day month after month running up very expensive fuel bills. Keeping a watchful eye on shifts in exhaust temperature can really pay off. Heating oil and heavy fuel oils are especially prone to fouling which retards heat transfer pushing temperatures up. This can happen very quickly, within a few hours and staff must be alert for this.

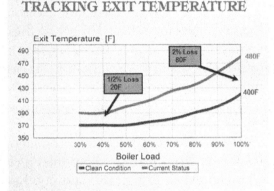

Figure 10.11—Rule of thumb for exhaust temperatures

Figure 10.12—Tracking exit temperatures

9. KEEP ACCURATE STATISTICS

![Keep Accurate Data]

Figure 10.13—How do you keep score? In the field of professional competition like baseball and basketball for instance, statistics and good record keeping are vital to understanding individual and team performance. With the cost of energy having so much influence on the bottom line of business and industry good statistics become very valuable to decision making.

10. FUEL FLOW METER FOR EACH BOILER

1. Detects over-firing
   • high exhaust temperatures
   • dangerous over firing-exceeding safety valve capacity

2. Detects under-firing
   • under utilization requiring additional boilers

3. Detects non-linear response in control system
   • Under-response in control system
   • Over-responding in control system

4. Individual boiler fuel consumption
   • Vital for performance analysis
   • Often required by EPA for emissions calculations

11. TRACK PERFORMANCE OF SMALL HEATING UNITS AUTOMATICALLY

Figure 10.14 and Figure 10.15
A handy way to monitor changes in exhaust temperatures of small boilers and furnaces. A thermometer with a maximum indicating arm continuously indicates the latest (highest) exhaust temperature even when the boiler or furnace is not firing. The green tape shows original "clean" condition and as the exhaust temperature creeps up over time this change is shown by the red maximum indicating needle. In the first case the temperature has risen by 40F for a 1% efficiency loss and in the second figure a 200F change has occurred representing a 5% loss. It would be very time consuming for plant personnel to obtain this information without this special type of thermometer.

12. RECOVER HOT CONDENSATE

Potential Savings
High

The loss of hot condensate wastes energy and money in several ways:

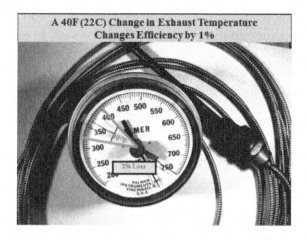

Figure 10.14—Tracking exhaust temperature 1

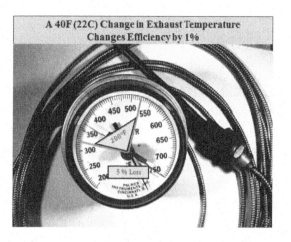

Figure 10.15—Tracking exhaust temperature-2

(a) The condensate should return at 160°F-190°F. If it doesn't return, this heat is lost.

(b) Makeup feedwater must be chemically treated and heated from 50°F to 160°F-190°F to replace the lost condensate.

(c) Condensate is essentially pure water. Having to replace it with "makeup" water containing impurities increases the amount of blowdown required.

(d) Chemical treatment of the makeup feedwater and the cost of the water itself can be avoided.

There is an additional factor here, and that is if the condensate system has leaks or is not functional and steam traps have failed and are blowing steam, all of this high energy steam is also disappearing.

Electric pumps have an uncertain lifetime in environments which occasionally can flood, also centrifugal pumps have difficulty handling condensate near the boiling point. These electrical units can be replaced with pump sets which use steam pressure to pump condensate back to the boiler plant.

Advantages/Disadvantages

Repair and replacement of condensate return systems can be quite expensive. The economic gain or savings available from keeping condensate systems well maintained is also very high.

Managing Energy

Rule of Thumb

Each 11 degree F (6 degree C) change in boiler water temperature changes efficiency by one percent

11 °F Change in Feed water- $\dfrac{11°F}{1,100\ BTU}$ = 1%
1,100 BTU in a pound of steam-

Figure 10.16—Rule of thumb for boiler water temperature.

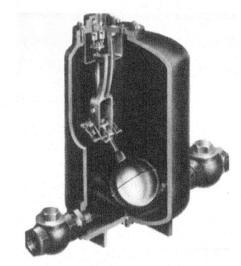

Figure 10.17—A steam powered pumping unit overcomes the reliability problems of electrical powered pumps with a longer life, less expensive installation and lower operating costs.

13.   MINIMIZE MAKEUP WATER CONSUMPTION

The water leaving each steam trap is at the boiling point of 212F in the case of the atmospheric return systems. The temperature is higher in the case of pressurized return systems. In many cases we have the water reaching the main plant at much lower temperatures a case by case situation depending on the design and condition of the individual system. For the sake of example let's say the water is arriving back at the plant has an ideal temperature of 185F. The total heat in a pound of steam is 1150 Btus and the makeup water temperature is 50F. So we have a difference of 135F.

Condensate return temperature 185F
Makeup water temperature 50F
Total heat of atmospheric steam 1150 Btu/lb

Formula 10-1
There is an inherent energy loss of 11.7% with each pound of condensate that doesn't find its way back to the condensate receiver at the plant under these conditions. As the condensate is distilled from pure steam there are additional water consumption and chemical treatment expenses to consider too.

14.   ISOLATE UNUSED HOT STEAM PIPING SECTIONS

If a section of steam piping is not in use it should be isolated from the active steam system as it still represents a significant energy loss whether or not it is transporting steam and condensate. Insulation losses remain the same.

15.   REDUCE VENTING STEAM FROM THE DEAERATOR

The deareator has an open vent to allow oxygen removed from the feedwater to escape from the system so it won't attack the steel piping and boiler components which usually shows up as severe pitting of the metal. However, some plants have been found to be operating with very big steam blows, sometimes hundreds of feet long which is excessive, wasting a lot of steam and energy.

16.   BOILERS CONVERTED FROM HEAVY FUEL OIL TO NATURAL GAS

There are still older boilers around that were designed to burn heavy fuel oil with high sulfur content which required high exhaust temperatures to avoid acid formation on cold metal surfaces in the exhaust system. The use of this high sulfur fuel oil is becoming restricted by modern pollution regulations and being replaced by very low sulfur distillate heating oil and by natural gas with extremely low levels of sulfur. These being the case, much lower exhaust temperatures are possible which the boilers were not originally designed for.

For example you might come across a boiler designed for heavy fuel oil with a 400F exhaust temperature now burning natural gas requiring an exhaust temperature of only 200F. Following the rule of thumb that every 40 degrees you can drop that temperature will improve efficiency by 1%, there is a 5% savings opportunity available. As a bonus, the capacity of the boiler will be improved by the same amount.

17.   SHUT OFF STEAM COIL AHEAD OF AIR PREHEATER

You may sometimes find a heating coil in the air intake to a boiler; this is another hangover from the days when heavy residual, high sulfur oils were burned. It served two purposes, (a) to keep smoke levels down under cold light off conditions and (b) bringing the cold metal temperatures up above the acid dew point temperature quicker shortening the exposure of this section of the boiler to acid formation during initial light off.

These heating coils may still be active or may just be sitting there in the air intakes obstructing air flows. A check with the boiler manufacturer may disclose that these old air preheating systems can be removed saving on steam, maintenance and fan power.

18. SOOT BLOWER OPERATIONS

Cost

Routine operational activity

Potential Savings

Moderate

Description

The rule of thumb applies: every 40 degree decrease in net stack temperature saves about one percent in fuel consumption. Soot is an excellent insulator which can retard heat transfer to a great degree, so soot blowers are used to keep heat exchange surfaces clean to maintain lower stack temperatures and higher efficiencies. Also, there is more of a tendency to form soot with the heavier residual type fuels. On the other hand, soot blowing is an expensive proposition when you consider the loss of energy involved to just keep the tubes clean. It can cost from 50 to 200 thousand dollars a year for a single boiler, so it is important to manage soot blowing properly. Both too little and too much soot-blowing can waste significant amounts of energy. When low excess air levels are maintained, keeping heat exchange surfaces clean becomes a greater challenge because the burners are operating closer to their smoke point. Soot blowers are usually installed on water tube boilers, but they are also available for fire tube type units on special order for applications such as black liquor and other dirty fuel type service. Some plants have even had successful results with a large ship's fog horn to remove deposits with low frequency vibrations. They are less expensive to operate than the conventional units, but their performance has had mixed reviews.

Advantages/Disadvantages

Soot blowers can do an effective job of keeping heat exchange surfaces clean and have been used for years. The cost of steam and compressed air to run soot-blowers can be very high so close control of their use is recommended.

19. SHUT-DOWN BACKPRESSURE STEAM TURBINE DURING LOW STEAM LOAD PERIODS

There may be a risk of over pressurizing the exhaust system of a backpressure turbine being used to let steam down to a lower pressure header or system. In the case where this exhaust steam can not be utilized it may well be being dumped to atmosphere through a pressure relief valve. Occasionally this is evident by a large plume of steam venting out through the roof of a plant. This is especially evident when standby turbines are kept spinning at a low load, the exhaust steam has got to go somewhere.

20. UTILIZE FUEL ADDITIVES TO IMPROVE OPERATING CONDITIONS AND ECONOMY

Class I—Improved storage and handling
- Sludge and gum inhibitors
- Detergents
- Metal deactivators
- Corrosion inhibitors
- Pour-point depressors
- Anti-static
- Anti-icing
- Color stabilizers

Class II—Improved Combustion and pollution reduction
- Smoke
- Particulates
- Carbon monoxide
- Hydrocarbons
- Nitrogen oxides
- Sulfur oxides

Class III—Post-Flame treatment of pollutants, slag and soot
- Soot removal
- Control slag build up
- Corrosion reduction
- Particulate collection   enhancement
- Sulfur oxide scavengers

21. SHUT OFF BOILERS DURING WEEKENDS AND SUMMER MONTHS WHEN NOT NEEDED

For long layup periods consider the wet lay up or dry lay up options suggested by the manufacturer.

22. CONSIDER THERMAL MASS WHEN PLANNING BOILER OPERATIONS

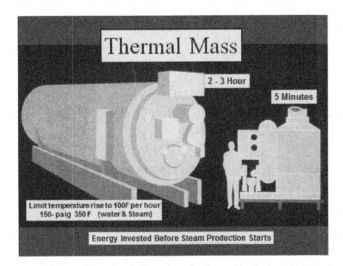

Figure 10.18—The fire Tube type boilers with big drum capacity contain many thousands of pounds of water which must be heated to operational temperatures at a safe rate of about 100F per hour to avoid unusual stresses on the boiler. This takes time on the order of 2 to 3 hours and wastes all the energy invested in this action until the boiler can be brought on line. The smaller flash boilers can be producing steam in 5 to 15 minutes using far less energy and time to get started.

In many applications a large stand-by boiler is kept on hot standby status to insure there is no interruption of steam service in facilities such as hospitals if the active boiler goes down for some reason. Because none of this the fuel us used to provide steam to the facility it is essentially wasted with zero efficiency. When you consider fuel is being used to just keep a boiler hot 24/7 year in and year out, this represents a substantial savings to eliminate this type of energy loss and wear and tear on the boiler. Flash boilers have proven to be quite dependable and relatively efficient.

23. ECONOMIZER OPERATION

An economizer is a simple device that gets hardly any notice and is not expected to have any problems. One problem that could occur with high sulfur fuels is "cold end corrosion" where the tubes are attacked by acids having a dew point in the range of water inlet temperatures to the economizer. The sulfur in the fuel forms SO_2 and SO_3 which combines with water to form sulfuric acids Low feedwater temperatures will cause this problem to accelerate.

The economizer also comes equipped with bypass piping in the event it springs a leak or some other problem. On rare occasion the bypass valve has been left open. This could cause exhaust temperatures to go up as high as 200F and bring efficiency and capacity down by as much as 5%.

Section 10-2
Combustion Systems

CHECKLIST – 2
COMBUSTION SYSTEMS

1. Fuel selection
2. Oxygen trim
3. Carbon monoxide trim
4. Control system linearity
5. Characterizable fuel valve
6. Over fire draft control
7. Cross limiting controls

8. Improve control system accuracy with strong precision parts
9. Stack dampers
10. Slow down exhaust flow
11. Wind deflector for boiler room vents
12. Vent caps on stacks
13. Low excess air burner
14. Oxygen enrichment
15. Oil-water emulsions
16. Replace atmospheric burners with power burners
17. Flame retention head type burners
18. Multi-stage gas valves
19. Reduction of fuel firing rate
20. Replace On/Off controls with modulating controls
21. Convert to air or steam atomizing burners
22. Fuel oil viscosity management
23. Intermittent ignition devices
24. Prevent heat transfer surface fouling
25. Air preheaters Economizers and other heat recovery equipment
26. Reduce radiation losses

1.   FUEL SELECTION

Cost

Low cost, basically analysis and decision making.

High cost, if major equipment change-out is involved.

Potential savings

Large

Description

The bottom line is producing steam for heat and power at the lowest cost possible. At first glance you might think that the "best buy" would be using the fuel with the most Btus per dollar. This type of ranking would show that wood, coal and heavy oil would be the best buy.

There is more to know about the situation than Btus per dollar. There is also the conversion efficiency (Table 10.1) and related problems with transportation and pollution to consider. Know-

ing the cost advantages and problems associated with different fuels can lead to informed decisions.

Table 10.1—Energy conversion efficiencies with different fuels. These numbers are typical but may vary with specific application.

Fuel Oil No. 6	85%	Lignite	80%
Fuel Oil No. 2	84%	Bituminous	87%
Natural Gas	80%	Anthracite	90%
Propane	84%	Methane	81%
Wood	72%	Hydrogen	77.5%

Evaluate Real Cost of Fuels per Million Btus

	Cost Per Million BTUs	Expected Efficiency	Cost of Steam Million BTUs
Wood	$ 1.00	70%	$ 1.43
Coal	$ 2.30	80%	$ 2.88
Natural Gas	$ 10.00	80%	$ 12.50
No. 6 Oil	$ 16.72	85%	$ 19.67
No. 2 Oil	$ 25.56	84%	$ 30.43
Electricity	$ 30.00	97%	$ 30.93

Table 10.2—The price of steam per million Btus can vary to a large degree considering basic fuel cost and conversion efficiency. Unfortunately, this analysis is only the tip of the iceberg and additional study may be needed. Each type of fuel requires different transportation to the plant and some like heavy oil, coal and wood require additional conditioning before they can be burned.

When confronted with the choice of switching from natural gas to residual oil, a more detailed study should be made. Notice that there are many hidden costs in the conversion to heavy oil that should be considered. Figure 10.19 is a plus-point minus-point evaluation example. The (+) in front of an item means that it generally will be a benefit and a (-) indicates that there are drawbacks or additional costs involved.

Different costs will be appropriate for each of these items at individual plants. The important thing is to be aware of the fact that a complete analysis is needed. A simple overlooked fact is

---
Conversion To Oil

+   Higher Combustion Efficiency

-    Cost to preheat oil

-    Cost to atomize oil

-    Soot blowing

-    Tank farm heating

-    Pumping costs

-    Surface temperature rise between soot-blowing

-    Long term stack temperature rise

-    Investment cost to maintain inventory

-    Environmental costs

-    Maintenance of boilers, tanks, atomizing systems, soot blowers, piping, pumps, burners and fuel oil heaters.

Figure 10.19—Plus point minus point evaluation table for conversion from natural gas to residual fuel oil.

that you pay for natural gas after it is used and fuel oil must be purchased and stored for many months or years before it is used. This ties up capital which can be used for other purposes. Another real problem is the cost of environmental compliance in some regions as well as the uncertainty of future environmental requirements for storage tanks and underground piping.

Although the use of electricity to generate steam would seem to be an unwise choice, the whole picture should be considered. Distribution systems can waste a very large percentage of heat, especially if they are lightly loaded. It may be more cost effective to use electric boilers at those locations with only intermittent or light steam use, especially if long or inefficient steam and condensate runs are involved.

Advantages/Disadvantages

A complete analysis could be time consuming, especially if the plant would have to be reconfigured to burn another type of fuel, but on the other hand going through the trouble could be quite profitable.

2.   OXYGEN TRIM

Cost
Moderate to low

Potential Savings
Moderate

Description
There are many factors that can influence and introduce errors into the air/fuel ratio such as fuel property changes, air temperature, control system response, fuel pressure and burner performance (Table 10.3). An oxygen trim system can automatically and continuously compensate for the variables in the combustion process, insuring that the boiler is programmed to operate at near optimum efficiency.

Sources of Errors in Control Systems

System Errors

Error

Remedied With Oxygen Trim Control

5% - Fuel Gravity, Temperature, BTU Content

10% - Fuel System Pressure Variations & wear

15% - Combustion Air Pressure, Temperature, Humidity

5% - Fan Performance, Dampers, Ducting

15% - Exhaust System, Boiler Fouling, Stack Effect

10% - 25% - Control System Accuracy and Repeatability

Table 10.3—The sources of errors in control systems with a rough estimate of how large they could become without an oxygen trim system to automatically compensate for some of the problems. An oxygen trim system can compensate for many system errors, allowing a lower excess air level to be maintained on an automatic basis. However, many boiler plants have not taken full advantage of this potential and are maintaining a higher than necessary set point for excess oxygen. The O_2 trim system can also be used as a tool to diagnose and control system problems.

Advantages/Disadvantages
Without oxygen trim, high levels of excess oxygen are necessary to compensate for many sources of errors in the combustion process; this

Oxygen Trim

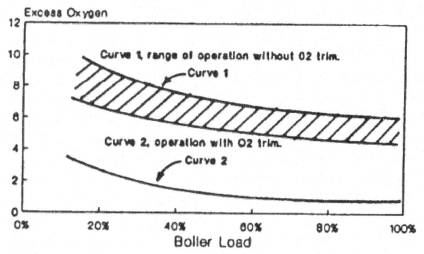

Figure 10.20—An oxygen trim system compensates for many control system errors.

excess oxygen can waste a great deal of energy. Oxygen trim systems are set to a level of excess oxygen established by tests using a carbon monoxide or smoke limit. Some latitude must be given in making this operational setting, so it does not end up at the most efficient fuel/air ratio because it includes this buffer zone too. Also, the oxygen trim system must be maintained in good operating condition and calibrated, adding to the plant maintenance workload. Figure 10.20 shows how an oxygen trim system improves excess air levels.

3.   CARBON MONOXIDE TRIM

Cost

   High

Potential Savings

   Moderate

Description

   A major problem with the oxygen trim system is that it cannot seek out the most efficient operating point and it must operate at fixed oxygen set points. For this reason a fuel wasting buffer zone is programmed into the operation of oxygen trim systems. A carbon monoxide measurement and trimming system is designed to seek out the

most efficient operating level on a continuous basis, insuring the most efficient operation possible (Figure 10.21).

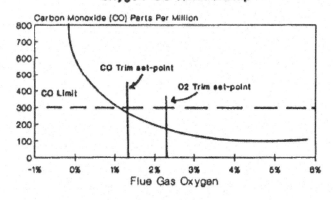

Figure 10.21—A carbon monoxide trim system continuously seeks the optimum air/fuel ratio. In contrast, the O₂ trim system must be set to a fixed value and does not sense poor combustion, combustibles or formation of carbon monoxide.

Advantages/Disadvantages

   Carbon monoxide trim is more expensive and requires more maintenance than other control system components. In the past CO trim has had a poor reliability record in the harsh combustion flue gas environment, but under the right conditions it can save a great deal of money. These are sensitive instruments that require special attention in maintenance and calibration. The CO trim system delivers its best performance and payback when all other parts of the combustion process are working perfectly.

4.   CONTROL SYSTEM LINEARITY

Cost

Low

Potential Savings

High

Description

When control systems have not been designed or calibrated for linearity they do not respond properly especially when programmed to work in coordination with other boilers. If you could compare a boiler's control system to an automobile's accelerator, linearity error is like just resting your foot on the gas pedal and having it accelerate to full speed with very little foot motion. In another case, having your car not respond to large change in position of the gas pedal and only getting a response on the last sixteenth of an inch of travel (Figure 10.22).

Other conditions poor control system linearity may cause, are that a boiler may only operate between certain restricted loads, like from 40% to 65% or from 80% to 100%. This is often a hidden error which makes smooth boiler control almost impossible and it takes a good analysis to uncover.

This is one of the many good reasons for equipping boilers with good instrumentation, including fuel meters. This condition will cause a boiler to cycle on and off or it may cause a boiler to operate below rated capacity making it necessary for additional boilers to carry the load. Both conditions waste fuel and make it very difficult to do a good tuneup.

When a boiler is first put into service, control system response should be checked to insure that its operation is linear. Many boilers have this hidden problem which interferes with efficient operation and makes life very frustrating for operators.

Advantages/Disadvantages

It takes time and additional expense to insure that a boiler's firing system is linearized and operating in its rated operating range. It's one of those things taken for granted when equipment is new. Unless this problem is resolved, it may be difficult to optimize the performance of a boiler to the fullest extent.

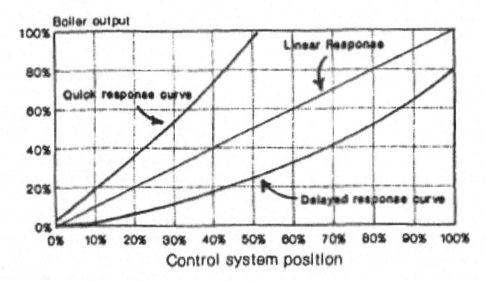

Figure 10.22—Control system linearity. Non-linear response can cause control problems and inefficient operation

5. CHARACTERIZABLE FUEL VALVE

Cost
Low

Potential Savings
2% to 12%

Description
Characterizable fuel valves (Figure 10.23 and 10.24) have a series of adjustments that are used to match the air/fuel ratios across the load range. Without this type of valve, a precise tuneup is almost impossible. It also serves as a valuable tool to correct mechanical problems that can develop in the burner and control systems. One valve is needed for each fuel. The reason a characterizable fuel valve can be so useful becomes apparent when the fuel delivery curve and air delivery curves are compared (Figure 10.25). Often they do not match and many control systems have no way to adjust one to the other in a precise way without sacrificing excess air and lost energy.

Advantages/Disadvantages
This is a low cost option and one of the best investments you can make for boiler optimization and troubleshooting. There should be a characterizable fuel valve for each fuel fired if the control system has no other means to balance fuel flow with air flow.

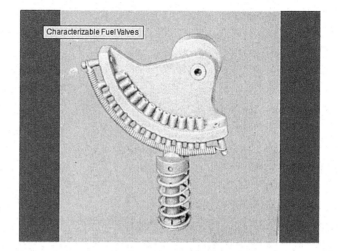

Figure 10.23—Characterizable fuel flow control valve

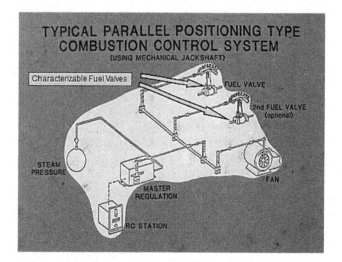

Figure 10.24—Characterizable Fuel Oil Valves installed for dual fuel operations

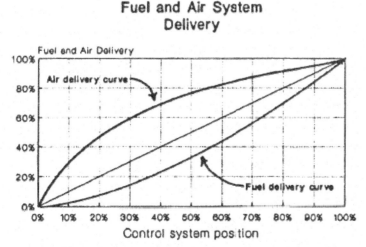

Figure 10.25—Different delivery rates from air and fuel systems make characterizable valves a necessity.

6.   OVER FIRE DRAFT CONTROL

Cost
Low

Potential Savings
Moderate (2% to 10%)

Description
One of the first and simplest devices used to control excess air was with over fire draft control. Because of the change in draft on cold and hot days and the lack of precise draft control experienced with barometric dampers, over fire draft control systems are used to maintain a constant negative pressure in the fire box. This establishes predictable conditions for air fuel ratio control. Also, when the boiler shuts down, the over fire draft control system damper closes, preventing heat from escaping up the stack (Figure 10.26).

Advantages/Disadvantages
The over fire draft control is a simple, low cost ($1,500) device and much cheaper than oxygen trim.

7.   CROSS LIMITING CONTROLS

Cost
Moderate

Potential Savings
Moderate

Description
Fuel delivery responds almost immediately to control system positioning whereas it takes longer for the air to respond, so on upward swings you can get a black smoking condition until the air catches up with the fuel.

This smoking on load changes is a waste of fuel and the soot deposit it leaves behind fouls the heat exchange surfaces causing higher stack temperatures. This is one of the reasons that pneumatic and electronic controls are used. They can be designed for this cross limiting feature which prevents smoking. If a boiler is smoking on load swings, then it is changing load too fast, and cross limiting is needed.

Cross limiting systems simply do not allow the fuel valve positioning signal to be greater than the actual measured air flow requirement, thus eliminating smoking on load changes.

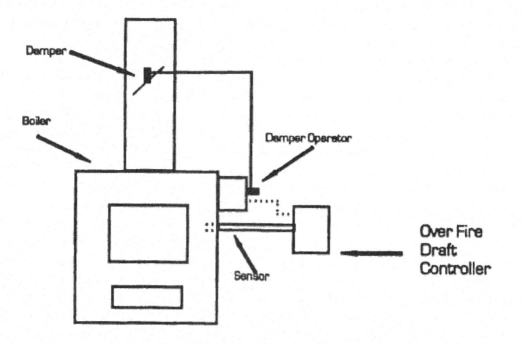

Figure 10.26—Over Fire Draft Control system

Advantages/disadvantages

It may be expensive to install a completely new control system just to have cross limiting. However, if this condition is not corrected, the price in unburned combustibles and higher stack temperatures will surely be taking an undesirable share of your fuel budget.

8. IMPROVE CONTROL SYSTEM ACCURACY WITH STRONG PRECISION PARTS

Cost

Low

Potential Savings

Moderate (2% to 12%)

Description

If your control system is flimsy and does not have strongly built precision parts, the chances are that you can't get a good tuneup or repeatability. If control system linkage has any play at all, precise excess-air levels cannot be maintained. For example, only 1/100 of an inch play in control system linkage can cause the need for a 5% excess air buffer zone.

If low excess air levels are to be maintained and smoking and soot formation prevented, the linkage should be precise with no play. Sometimes pins are lost and threaded bolts are substituted instead and over time, these sharp threads cause elongated holes and a tremendous amount of play. Also, non-precision parts are used to substitute for lost or damaged parts. Control linkages also bind, warp, shear pins and otherwise lose their precision over the years and the only solution is to restore it to a precision system.

Advantages/Disadvantages

You may have a burner and control system that can't be adjusted to maintain low excess air levels and high efficiency. If so, some custom rebuilding may be necessary. It's a cheap enough solution but you may have to be innovative and use aircraft grade precision parts as shown in Figure 10.27.

9. STACK DAMPERS

Cost

Low

Potential Savings

Moderate (5% to 20%)

Figure 10.27

Description

When a boiler isn't firing, a great deal of energy can be lost up the stack due to the chimney effect when the lighter hot air formed in the boiler internals and in the hot stack pull cold air in through the boiler. This situation is most common on atmospheric burner units, but may be present on some older equipment too.

A damper automatically closes off the stack when the boiler isn't firing, holding the heat in the boiler (Figure 10.28).

Advantages/Disadvantages

Stack dampers have full safety approval and should be installed to curb this unnecessary loss.

10.   SLOW DOWN EXHAUST FLOW

Cost

Low

10.   Savings

Moderate to low

Description

In many cases the size of stacks and exhaust systems is excessive allowing unrestricted and often excessive negative draft "draw" to exist. Oversized exhaust systems offer little resistance to the varying conditions of stack "draw" and cause excess air to be pulled into the burner as well as aid the unrestricted escape of large vol-

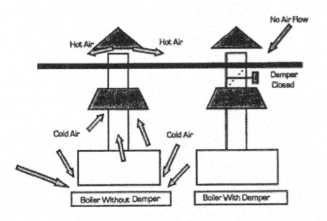

Figure 10.28—Stack dampers used to block heat losses.

umes of high temperature exhaust gasses. Years ago, slots were cut in the stack and metal dampers inserted, but these often were unsatisfactory. A modern adjustable design has been developed, a special exhaust gas buffer that has shown good results (Figure 10.29).

Advantages/disadvantages

This inexpensive device can slow down the escape of hot gases up the stack. It can also be adjusted to a specific location.

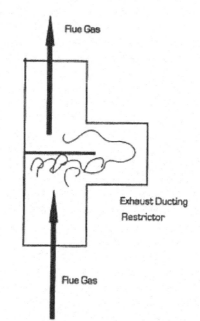

Figure 10.29—A fixed damper buffers high velocity escape of flue gas without limiting cross sectional area of stack

11.   WIND DEFLECTOR FOR BOILER ROOM VENTS

Cost

Small, basically a design consideration

10.   Savings

Moderate

10.   Description

Boiler rooms are required to have a certain square footage of free ventilation access per boiler horsepower. In some locations wind driven air pressure can cause draft control problems. One has only to try to walk into a strong wind to grasp the magnitude of the force involved. When the wind is driven into the outside vents boiler room pressure can rise, when on the sheltered side of a building a partial vacuum can form. This condition can cause burner adjustment problems. The simple solution is to be aware of the potential problem and protect the boiler room from this wind effect or set the excess air high enough to stay out of the smoking zone. This may seem like a small thing, but when low excess air levels are to be achieved a condition like this can become a big challenge.

12.   VENT CAPS ON STACKS

Cost
Low

Savings
Low to Moderate

Description
Something as simple as a vent cap on the stack is not an obvious candidate for fuel savings. A strong wind and a bent or damaged stack cover can produce some wild draft effects. These draft effects will change the air/fuel ratio and cause large swings towards excess air conditions and possibly cause smoking.

With the lack of good draft control, excess air adjustments must be made to compensate for this effect which can be excessive causing wasted fuel. It has always been a mystery as to just how vent covers get bent and pushed over to one side, but they have been known to deflect air into the boiler or furnace in a good wind and cause serious problems and can even back smoke into a building.

13.   LOW EXCESS AIR BURNER

Cost
Moderate to High

Potential Savings
Moderate to High

Description
It doesn't matter how good your controls are, or how fancy the trim system, the burner will ultimately control the level of excess air that can be achieved. Older burners may not have been designed with low excess air in mind or they may have worn out to a point where low excess air levels are no longer possible. The solution is to install a burner which has been designed for low excess air operation, or to repair or refurbish the one you have so that it is capable of dependable low excess air operation.

Advantages/Disadvantages
High cost and redesign of boilers to accommodate these new burners may be a problem. However, low excess air operation may not be possible without burner conversion.

14.   OXYGEN ENRICHMENT

Cost
Varies

Savings
Moderate to small

Description
Substitution of oxygen for combustion air, which contains only 20.9% O_2 reduces the volume of heat absorbing nitrogen flowing through the combustion process and therefore reduces flue gas loss. Energy savings must be measured against the cost of oxygen and the cost involved in additional safety precautions.

The usual oxygen enriching practice in industrial heating is to increase the oxygen in the combustion air from its normal 21% to 25% or 30%. It is preferable to premix it with the combustion air, but lances are sometimes used. Economic justification of the use of oxygen is often marginal unless the user has oxygen facilities with spare capacity.

Advantages/Disadvantages
Good energy savings are possible but safety will be a paramount factor in mixing oxygen with fuel oil.

15. OIL-WATER EMULSIONS

Cost
Moderate

Potential Savings
Low

Description
This is a process where very small amounts of water, usually less than 6% of the fuel, of micron sized droplets are evenly mixed with fuel to cause additional atomization action in the combustion zone. (Figure 10.30)

The benefit from oil/water emulsions is limited, but is of great value in those plants where serious burner problems are causing smoke and excessive particulate emissions. When exposed to the intense heat in the fire box, the micron sized droplets of water expand rapidly causing the oil droplets to break into smaller particles, enhancing atomization. This reduces the smoke level allowing lower excess air and the elimination of the sooting of the heat exchange surfaces.

Advantages/disadvantages
This is an effective way to reduce particulate generation and change the smoke threshold so less excess air can be used. The equipment is expensive and the water used in this process escapes up the stack carrying away a high level of energy.

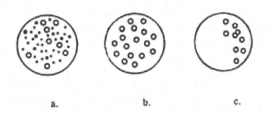

a.   b.   c.

Figure 10.30—Oil-water emulsions, micron sized water droplets are mixed with fuel. This figure shows typical atomized oil droplet with water mixed in fuel:
(a) water droplets are too small,
(b) water droplets uniform and the proper size
(c) non-uniform mixing of water droplets.

16. REPLACE ATMOSPHERIC BURNERS WITH POWER BURNERS

Cost
Moderate

Potential Savings
Moderate

Description
An atmospheric type of burner has practically no excess air control and loses a lot of heat up the stack during the off cycle if a damper is not installed. On the other hand, excess air and stack losses can be more easily controlled with a power burner which has electrically positioned control linkage, motor driven fans and automatic dampers.

Advantages/Disadvantages
Atmospheric type burners have no excess air control. Excess air levels have often been measured at extremely high levels. In addition, because they use air that is pulled into the combustion zone by the chimney effect there is no positive way to prevent cold air from being sucked through the boiler when it is not firing, which wastes a lot of energy up the stack. A power burner has two clear advantages, it can be adjusted for excess air control and its dampers close during the off cycle, keeping heat in the boiler.

17. FLAME RETENTION HEAD TYPE BURNERS

Cost
Moderate

Savings
Low (with new units)

Description
Without good air-fuel mixing, oil fired burners require higher amounts of excess air. Flame retention head burners, however, provide better mixing of oil and combustion air, thereby reduc-

ing excess-air requirements (Figure 10.31). The application of this option depends to a great degree on existing excess-air levels. This option should be considered when replacing old equipment or in the design phase of mechanical systems. U. S Department of Energy tests found an improvement in seasonal efficiency from 5.9 to 12.7 percent when high speed flame retention head burners were used on standard heating units.

Advantages/Disadvantages

This is a small boiler and furnace option. Flame retention head burners should be specified for any new purchases. They may be available as a retrofit option if the manufacturer has redesigned older models. It is a low cost item that can reduce excess air.

18.   MULTI-STAGE GAS VALVES

Cost
Low

Saving
Moderate

Description

Gas fired furnaces and boilers having only one fuel delivery rate can be a disadvantage which wastes fuel because the only firing rate is full fire and off. Having the option of intermediate firing rates (high, low and off) can lower stack

temperature and eliminate to some of the standby losses.

Advantages and Disadvantages

This option requires additional pressure regulating valves and possibly burner modifications to produce stepped firing rates.

19.   REDUCTION OF FUEL FIRING RATE

Cost
Low to Moderate

Savings
Low to Moderate

Description

Conditions change, many burners have been found to be grossly oversized for their applications. Successful energy conservation actions, changes in load requirements and other factors may be producing far less demand for heat. If fuel firing rate can be reduced, then stack losses will decrease along with standby losses.

Advantages/Disadvantages

The cost of installing new burners and the cost of engineering and other fees for redesign of an existing system may not be justified. A study of the economics of the issue will show if it will pay. It makes no sense to drive a lot of heat into a boiler with high exhaust gas temperatures and

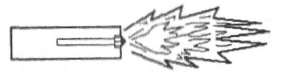

Conventional Burner Flame

Retention Head Burner Flame

Figure 10.31—Flame retention head burner.

then let it sit idle for long periods; this is wasteful. A longer firing time with lower losses makes more sense. Caution may be needed at low firing rates where the exhaust gas temperature is depressed below the acid dew point which can cause damage to the boiler, breechings and stack.

20. REPLACE ON/OFF CONTROLS WITH MODULATING CONTROLS.

Cost
Moderate

Potential Savings
Moderate

Description
Replace on/off type of burner control with controls that can modulate and match load conditions, reducing higher stack temperatures and standby losses. On/off equipment must be set to match the highest expected demand and when the demand has been satisfied the equipment must shut off. Exhaust temperatures are lowered by matching the firing rate to actual demand. Also, the losses associated with purging the boiler before and after each firing cycle could be eliminated by having fewer on/off cycles. Table 10.4 shows typical efficiencies and the enhancements of this type of load matching.

Control Type	Efficiency at % Load			
	25%	50%	75%	100%
On/Off	70.3	74.4	75.6	76.3
On/Off With flue damper	73.3	75.3	76	76.3
High/Low/Off	76.9	76.5	76.4	76.3
Modulating	76.9	77.7	77.2	76.3

Table 10.4—Control system Performance comparison.

Advantages/Disadvantages
On/Off Controls are simpler, with a lower first cost and can be tuned up easily. However, they waste energy with higher stack temperatures and standby losses. The more complex and more expensive modulating controls can overcome this problem.

21. CONVERT TO AIR OR STEAM ATOMIZING BURNERS.

Cost
Moderate

Potential Savings
Moderate 2%-8%

Description
Steam or air atomization allows the firing of a wide range of fuels and high turndown approaching 20:1 instead of 4:1, and good efficiency. Air and steam atomization produces an aerosol in which fine droplets are supported by an expanding cone of air/steam and there is less sensitivity to oil viscosity changes. It also allows more flexibility in shaping the flame to conform to furnace conditions.

Advantages/Disadvantages
The cost of installing and operating a compressed air or steam atomizing system may be a disadvantage. The steam used for atomization is lost up the stack and its energy must be considered. Offsetting this is improved performance turn down ratio and fuel versatility, especially if fuel quality tends to vary.

22. FUEL OIL VISCOSITY MANAGEMENT

Cost
Low

Savings
Moderate

Description
The design of the atomization system is based on certain fuel oil properties, primarily

viscosity. Viscosity is the relative ease or difficulty with which oil flows or is pumped. Viscosity affects the quality of atomization and the smoke point which determines the minimum excess air possible. Because of this, the viscosity of fuel oil must be closely controlled at all times. Since oil is obtained on the open market, its source of origin can be uncertain. Therefore, the properties of one batch will not likely be the same as those of another batch. Oil viscosity is controlled by temperature, the oil temperature necessary at the atomizer depends on the particular fuel. Some heavy fuels are very fluid and require little or no heating. A viscosity of 180 to 200 Saybolt Seconds Universal (SSU) often gives best results for atomization and if the oil at ordinary temperature is of higher viscosity, it must be heated to a temperature which will reduce viscosity to this point.

The temperature control of the fuel oil heater is a key element in controlling atomization and smoke point for excess-air control. Fuel batches should be sampled to determine the temperature set point. If the heater does not maintain an accurate set point, close excess-air tolerances may not be possible. Automatic viscosity controllers will feed back information on viscosity to the heater and maintain a proper temperature.

Advantages/Disadvantages

In general, the lowest fuel temperature permitting good furnace conditions and low excess-air is most desirable. Too high a temperature may cause carbonization of fuel oil heaters and atomizer tips, sparking in the furnace and a tendency for the flame to become unsteady and blow off the tip.

Burner performance controls the minimum possible excess-air levels which can be held consistently. The burner depends for its performance on oil being at the proper viscosity. Poor viscosity (temperature) control can prevent high efficiency. A properly working and maintained fuel oil heater and good viscosity control either through regular tests on oil batches or with an automatic controller is the key to low excess-air.

23. INTERMITTENT IGNITION DEVICES

Cost
Low

Savings
Moderate

Description
It doesn't make any sense to keep a flame lit in a furnace or boiler if heat isn't needed; it wastes energy. Especially with gas fired equipment, there are many smaller sized boilers and furnaces with continuous pilot flames. On a 24-hour-a-day basis, year after year, these small flames can use a lot of Btus. Intermittent ignition technology has been proven to be reliable and safe, so these older continuous pilot lights should be replaced with Intermittent Ignition Devices.

Advantages/Disadvantages
Intermittent ignition devices are proven, reliable equipment. They have been designed to retrofit on most boilers and furnaces and are not expensive. Their payback period is relatively short

24. PREVENT HEAT TRANSFER SURFACE FOULING

Cost
Low

Savings
Moderate

Description
With the increased use of natural gas, combustion source fouling has become much less of a problem, however with distillate fuels, heavy fuel oils and coals heat exchange surface fouling is still a constant threat which requires diligence to keep under control. Inadequate air supply to the burner or localized poor air fuel mixing can cause small smoke sources that over time foul heat exchange surfaces raising exhaust temperatures. This can take just a few minutes or occur

over months; the stack temperature will tell the story. The possible causes are many and varied and may take a detailed investigation once the higher exhaust temperatures become apparent, especially when firing at Maximum Rated Capacity.

Advantages/Disadvantages

Keeping heat exchange surfaces clean prevents wasted fuel, increased pollution and unnecessary carbon emissions. It is a matter of being able to spot this condition when it is developing.

25. INSTALL AIR PREHEAT OR OTHER HEAT RECOVERY EQUIPMENT

Cost

Varies

Savings

Moderate

Description

This option comes under the general heading of what to do when other options don't appear to be workable or cost effective. There are two widely used choices to improve efficiency effectively; put excess heat into an economizer or into a combustion air preheater. Let's say the fuel is the problem and frequent cleaning is not possible, a combustion air pre heater might be the solution. The rule of thumb for this option is that every 40F you bring the exhaust temperature down raises efficiency by 1% and every 11F the feedwater is raised improves efficiency by 1%. So in some cases you may find this to be the most workable solution to investigate. In some cases it might be cost effective to put the energy from high exhaust losses through an older inactive boiler as a hot water source.

Advantages/Disadvantages

This comes under the heading of initiatives, a fall back position if fuel becomes very expensive or scarce or if regulatory constraints block other conventional options like installing new boilers.

26. REDUCE RADIATION LOSSES

Cost

Low

Savings

Moderate

Description

Hot surfaces can waste a lot of energy by means of radiation. Radiation is a very effective heat transfer medium. Unlike convection and conduction radiation losses may not be obvious because it does not heat up the local area but may transfer its energy to surfaces quite a distance away and because of the much greater surface areas affected at a distance this way the actual energy transfer can be greatly under estimated. Some older boilers have large metal doors that radiate heat like this, flame on one side and a much cooler plant on the other side with rather forceful blowers cooling things down with people moving around in their shirt sleeves on very cold days; its radiation just like the sun. The real problem can often be picked up with infrared scans or surface thermometers.

Advantages/Disadvantages

This is not an obvious problem and a condition commonly taken for granted in boiler plants; their hot! It may be awkward and challenging to correct this situation or even convince some people this represents a significant loss of energy especially at low firing rates as this condition becomes a much larger percentage in the losses column. Good insulating materials are available with experts that know how to apply them.

Section 10-3
Boiler Improvements

CHECKLIST - 3
Boiler Improvements

1. Replace binding or worn control system linkages
2. Conduct infrared scans of boiler shell to identify poor refractory and insulation

3. Oxygen sensors
4. Repair a leaking seals on combustion air pre-heaters
5. Install turbulators
6. Chemically clean internal boiler scale to improve heat transfer
7. Clean fire-side deposits to improve heat transfer
8. Check for hot gas short circuits indicating defective baffles

1. REPLACE BINDING OR WORN CONTROL SYSTEM LINKAGES.

Boilers seem to operate for years on a 24/7 basis with no problems while all this time parts are wearing and vibrating with continuous use. Nothing seems to go wrong. Boilers are not like engines that begin to vibrate and make strange noises when they begin to have problems. It's a good idea to check the control linkages for opening clearances and play, just a little bit of play can start costing money. A good check is to test for flue gas oxygen levels at selected load points, this and a hand over hand examination of parts will tell the story. Cycle the controls with the boiler off and feel for unusual vibrations and noises. You might be surprised at what can be found even disconnected linkages and burners not securely bolted to the furnace.

2. CONDUCT INFRARED SCANS OF BOILER AND STEAM LINES.

There are some good instruments available for conducting infrared scans of boiler and steam lines to identify poor refractory and insulation. They can identify hot spots and energy losses quickly and easily identify unusual energy losses. Figure 10.32 shows the graphic results of such a scan.

3. EXCESS AIR INDICATORS

Oxygen sensors provide information for excess air indications on a continuous basis which can be quite helpful for tracking boiler efficiency trends. This is a prime indicator of operating efficiency and also as a trouble indicator for the burner and combustion control systems.

Boiler Surface Survey
for
Temperatures & Radiation Losses
Btu /square ft./hr

Boiler # 143

Front

300	295	310	320	296	310
259	700	635	338	322	415
277	650	598	420	317	352
305	356	420	360	306	325

Back

220	297	296	275	268	225
222	310	350	280	264	230
225	345	960	852	500	460
220	355	720	455	462	350

Front Average	383	Btu/hr
Total	9,186	Btu/hr
Back Average	381	Btu/hr
Total	9,141	Btu/hr

Left Side

	320		456		425		525		350	

Figure 10.32—Shows a sample of a graphic scan of a boiler surface which has been displayed on a spreadsheet grid for easy summation of results. This scan was done with an infrared instrument that directly measured Btus radiated from the boiler surface in Btus per square ft per hour; a handy number to use in a grid like this. Using this and surface temperature information, trouble spots can easily be identified.

4. REPAIR LEAKING SEALS ON COMBUSTION AIR PREHEATERS

The combustion air preheater (Figure 10.33) lowers exhaust temperatures by pre heating incoming combustion air. The unit slowly rotates through the hot exhaust ductwork gaining heat which is then released into the incoming combustion air. After extended use the seals need fixing because of wear. Leaking seals can introduce cold air into exhaust gases causing dew point corrosion problems besides putting an additional load on the combustion air fans.

5. TURBULATORS

Cost
Low

Potential Savings
Low

Figure 10.33—Large Air Heater (Lungstrom) installation at an electric station

Description
Turbulators are a very effective way to reduce stack temperatures and increase efficiency of fire tube boilers (Figure 10.34). In a firetube boiler the hot gasses must pass through relatively long narrow tubes where heat is given up to the side walls causing the gasses to contract and slow down forming layers of colder gas along the heat exchange surfaces. This condition prevents good heat transfer, so turbulators are used break up this insulating film. Turbulators are thin strips of metal that have been bent to break up the flow of hot gasses in the tubes and are designed for minimum obstruction of the passages. They are usually installed in the last pass. By using longer turbulators in the upper tubes of the last pass, hotter gasses are forced through lower tubes eliminating to some degree the stratification of hot gasses at the end of the boiler.

Advantages/Disadvantages
Turbulators may cause condensation of flue gasses when a boiler is first lit off. Generally, they are a good inexpensive way to lower flue gas temperature. They are not recommended for high particulate loading options such as coal and residual fuel oil.

6. CHEMICALLY CLEAN INTERNAL BOILER SCALE TO IMPROVE HEAT TRANSFER

Impurities in boiler feedwater tend to concentrate in the boiler. All boiler water contains dissolved solids. When the feedwater is heated, it evaporates and goes off as distilled steam leaving impurities behind. As more and more water is distilled from the boiler, more feedwater is added to replace it. As a result the amount of solids dissolved in the boiler water gradually increases. After a while these highly soluble solids in solution cause problems like scale accumulation on the tube metal retarding heat transfer and a number of other problems. (For further details see Chapter 15 of this book.)

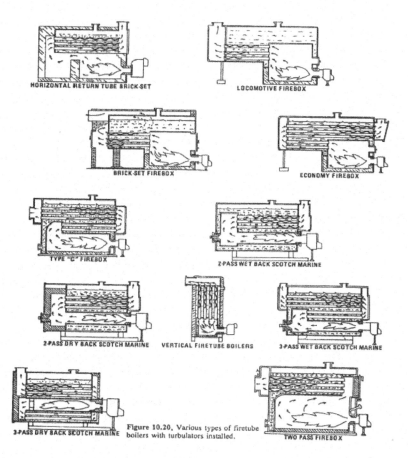

Figure 10.20, Various types of firetube boilers with turbulators installed.

Figure 10.34
Various types of firetube boilers with turbulators installed

7.   CLEAN FIRE-SIDE DEPOSITS TO IMPROVE HEAT TRANSFER

Fireside deposits are quite varied depending on the fuel source, especially Coals and Heavy Fuel Oils (HFO). Slags get so bad sometimes they have to be removed with dynamite or shotgun blasts. In lesser cases, high pressure water wash can be used. Also, fuel additives are used to help with this process.

The key here is rising exhaust temperatures; every 40°F the temperature rises costs 1% in boiler efficiency so it is an basically an economic judgment when to curtail operations and do a cleanup. This could have an important impact on the economy of operations and profits.

8.   CHECK FOR HOT GAS SHORT CIRCUITS INDICATING DEFECTIVE BAFFLES

Boilers are designed to absorb heat from combustion gases with great efficiency. These hot gases are guided on their way through the boiler by baffles that on occasion break down and let hot gasses pass without giving up their heat.

Hot gas short circuits can be detected by doing a temperature traverse across the outlet breechings. Wherever the gases are leaking through the baffles they are causing a hot flow lanes which can be detected by taking temperature measurements. So it is a matter of systematically probing the cross section of the exhaust section searching for hot spots. Specific repairs can be made after shutting down the boiler and inspecting the path indicated for the gases causing the hot spot.

Section 10-4
Metering and Monitoring

CHECKLIST - 4
Metering and Monitoring

1. Improved instrumentation and analysis
2. Monitor and manage with a microprocessor based system.
3. Add a fuel flow meter to the boiler plant for each fuel.
4. Provide a steam flow (or feedwater flow) meter for each boiler.
5. Monitor excess air, carbon monoxide and combustibles either continuously or by regular checks with portable analyzers
6. Install an opacity monitor for oil fired units (smoke is unburned fuel)
7. Monitor flame conditions by visual checks through viewing ports
8. Install fuel cost indicator on each boiler front (i.e.-$/year)
9. Maintain calibration for steam flow meters.
10.   Insure make-up water meter is working and accurate
11.   Insure condensate line thermometers are in place and accurate
12.   Keep instrumentation and chart recorders in good repair.

1.   IMPROVED INSTRUMENTATION

Cost
    Low

Potential Savings
    Moderate to 25% (plus)

Description

There is a difference between "operating" and "managing" a boiler. To manage a boiler you need more information than is generally available with standard instrumentation which only gives information about "operating" conditions. Fuel flow, steam flow, temperatures and make up water use all play an important part in knowing what is happening and tracking trends. A boiler with only 5% utilization or with 300% excess air or a stack temperature of 750°F will not be efficient but it will produce steam and for all practical purposes be operating satisfactory. Likewise, a sudden increase in make up water use may indicate a large loss of energy in the distribution system somewhere. Good instrumentation is needed to identify these efficiency related problems. Plants should have the following instruments:

1.   Stack thermometers.
2.   Fuel meters for each boiler.
3.   Make up feedwater meters.
4.   Oxygen analyzers.
5.   Run time recorder.
6.   Energy output-input metering,
7.   Return condensate thermometers.

The ideal scene would to employ a modern computerized monitoring system to continuously check on performance. If this cannot be done, then this minimum instrumentation should be considered. Performance should be tracked to meet fuel reduction objectives. A simple formula can be used if proper instrumentation exists:

$$\text{EFFICIENCY} \% = \frac{\text{Btu OUTPUT} \times 100}{\text{Btu INPUT}}$$

There are portable boiler efficiency analyzers available which can be used to help assess the efficiency of your boilers. These range from simple wet chemical analyzers to electronic flue gas analyzers that can measure excess air, stack emissions and stack (heat loss) efficiency at the touch of a button. These instruments measure stack losses but do not give information about overall plant performance. Because the instrumentation at most plants does not provide continuous or complete information, there are a lot of blank spaces and assumptions about actual plant performance.

Advantages/Disadvantages

Instruments add extra cost to plants and in many cases they have not been installed or have been neglected. They must be kept calibrated and someone must record and analyze the data and then take positive steps to correct energy wasting conditions. This takes time, effort, money and skill. However, there may be a large pay back in knowing what is happening and discovering where the fuel wasting conditions are.

2.   MONITOR AND MANAGE WITH A MICROPROCESSOR BASED SYSTEM

Potential savings: large

There is a definite reduction in manpower requirements for monitoring energy use with modern microprocessor based systems. The mi-croprocessor is a powerful new tool that can take the pulse of your plant, detect problems and calculate efficiency every few seconds.

The computerized system (Figure 10.35) can store a tremendous amount of data and provide clear and simple reports and graphics. With "Intelligent Programs" the computer can train your personnel and help them to diagnose problems very quickly. With a telephone connection, a plant can be monitored by experts thousands of miles away.

The computer is truly the answer to tracking plant performance. Let's face it, a boiler is only a means of getting heat and power so the facility can accomplish its mission. If steam could be brought into a plant like electricity, many managers would be very happy because a great burden and liability would be lifted from their shoulders. Until this is possible, the next best thing to aid trouble free and efficient boiler plant operation is the use of the computer as a tool to help keep the operation of the plant smooth and trouble free as well as identify where energy (dollars) are being wasted.

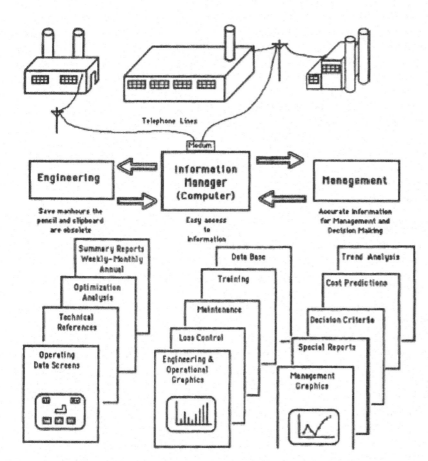

Figure 10.35—Quality plant information is available for engineering, operations and management needs when microprocessor base systems are used.

Many boilers are unattended or are operated by a staff with many other jobs. With a computerized system, the boiler is monitored automatically which can mean a large savings in manpower and energy. (Figures 10.36 and 10.37)

Advantages/Disadvantages

High cost and increased complexity are drawbacks. Properly used and maintained, a microprocessor can save a great deal of money by reducing labor costs, identifying fuel wasting problems, diagnosing plant and system problems and eliminating many service calls.

Some of the features of the microprocessor based system are accuracy of data which can be stored and evaluated providing:

1. Continuous data sampling
2. Immediate alarms
3. Remote surveillance
4. Accurate records
5. Energy and steam system modeling using "what if" scenarios
6. Typical days per month simulation with energy and cash forecasting
7. Production planning tool, economic hourly plant operation
8. Optimal plant operating strategies

3.   PROVIDE A FUEL FLOW METER
      FOR EACH BOILER AND EACH FUEL

Fuel flow meters do not save energy and they are not necessary for producing steam and often are not installed but they are vital for managing an efficient plant. If you know how much fuel a boiler is using then you have a key to its real efficiency, especially when information is collected over a long period of time. Boiler efficiency testing can give you a snapshot of the operating efficiency at specific loads, but a number of significant losses have different causes that are not measured in these tests. For example, if the boiler is oversized for its demands, it can lose a lot of energy when it cycles frequently. Another example is the hot standby boiler that essentially wastes all

of its fuel just being ready to take the load if the on line boilers encounter problems.

4.   PROVIDE A STEAM FLOW (OR FEED-
      WATER FLOW) METER FOR
      EACH BOILER

Flow meters do not save energy and they are not necessary for producing steam but they are vital for managing an efficient plant. Once steam leaves the boiler plant control is lost, it goes somewhere but does anyone have an accurate picture of how it is used with no metering systems downstream. The American Society of Mechanical Engineers (ASME), in their boiler test code indicates that a positive displacement feedwater flow meter serves as a accurate substitute for a steam flow meter. In order to assess boiler performance you must know what the boiler is producing.

5.   MONITOR EXCESS AIR, CARBON
      MONOXIDE AND COMBUSTIBLES

Monitor excess air, carbon monoxide and combustibles either continuously or by regular checks with portable analyzers.

1. Excess air is an accurate indicator of changes in stack losses. The rule of thumb is that every 2% increase in Oxygen in the flue gas represents a 1% loss in efficiency. (2% oxygen is about 10% excess air).

2. Carbon monoxide is incompletely burned carbon which is unburned fuel. Every pound of carbon monoxide has a value of 4,368. Btus which is 20% of the energy in a pound of natural gas.

3. Unburned hydrocarbons are actually unburned fuel too, containing a lot of Btus. Example: with 5% oxygen in the exhaust gas a 2% hydrocarbon reading represents an 8% energy loss.

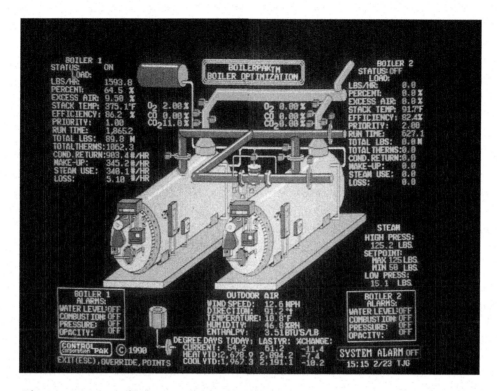

Figure 10.36—typical screen showing vital data for safe efficient operations.

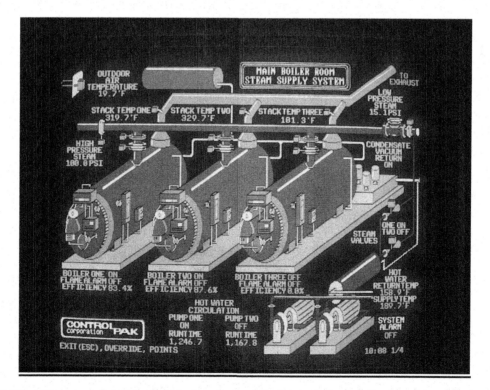

Figure 10.37—These graphic displays are quite handy for training and remote troubleshooting

There is a lot to lose here, someone should be keeping a check on what is going out the stack.

6.   INSTALL AN OPACITY MONITOR FOR OIL AND COAL FIRED UNITS (SMOKE IS UNBURNED FUEL)

Oil and coal plants usually have opacity monitors but if you have an interruptible gas supply contract you need to know if the plant begins to smoke when it has to switch back to oil. Staff is often reduced with gas operations, and the operators are not used to checking to see if the stack is clear. This is particularly important during the fuel switch when people are performing an unfamiliar tasks and are prone to make mistakes.

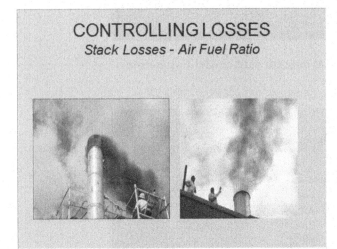

Figure 10.38—**You need to know when you are smoking, its unburned fuel. Many regulatory agencies have maximum opacity limits and levy fines for smoking. An opacity monitor can keep a plant out of trouble. They were once standard equipment on all boilers.**

7.   MONITOR FLAME CONDITIONS BY VISUAL CHECKS THROUGH VIEWING PORTS

When you can't see things they become a mystery. Boiler manufacturers usually design viewing ports into standard boiler construction. However, as time passes, the glass becomes coat-ed with carbon or the glass is broken or someone bricks over the viewing ports during repair activities. These ports are a very good troubleshooting tool, if something goes wrong it can usually be seen during observation of the flame and burner components. Actually it is a good idea to take pictures or make sketches so you have something to refer to as a comparison.

8.   INSTALL FUEL COST INDICATOR ON EACH BOILER FRONT (I.E.-$/YEAR)

Boilers operate on a 24/7 basis consuming great quantities of expensive fuel. Many people do not realize this and do not assign much importance to the amount of money spent on fuel. Company officers work very hard to make a profit which boilers can nullify with their inefficiencies.

One solution to this lack of recognition of the money that boilers can gobble up is to install a meter that indicates the dollars' worth of fuel the boiler has used during the past year. The numbers might be a surprise to everyone and attract help in creating some higher efficiency.

9.   MAINTAIN CALIBRATION FOR STEAM FLOW METERS

Your steam flow meters tell you where the money is going. Would you allow the cash registers to be inaccurate?

10.   INSURE MAKE-UP WATER METER IS WORKING AND ACCURATE

If you spring a steam leak, a water leak or other serious problem with the distribution system, this meter will tell you right away.

11.   INSURE CONDENSATE LINE THERMO-METERS ARE IN PLACE AND ACCURATE

If steam traps have failed or if the piping system has flooded the temperature of the returning condensate will change indicating a problem needs fixing.

12.   KEEP INSTRUMENTATION AND CHART RECORDERS IN GOOD REPAIR.

A lot of things happen in the typical thermal plant that can't be seen, in boilers covered with ibricks and insulation, in piping systems and at remote locations served by the piping systems. Plants don't necessarily run on observable data the instruments tell the story about efficiency. People need correct data to make safe decisions too; there are a lot of dangers connected to thermal plant operations.

Section 10-5
Boiler Water Management

CHECKLIST 5
Boiler Water Management

1.   Improve water treatment
2.   Water pretreatment softener
3.   Boiler water chemistry improvements increase cycles of concentration
4.   Install automated blow down control system
5.   Mechanical scale removal
6.   Chemical scale removal
7.   Boiler blow down heat recovery
8.   Use hot blow down water to keep idle boilers warm

1.   BOILER WATER MANAGEMENT TO CONSERVE ENERGY

When boiler feedwater is heated, it evaporates and goes off as distilled steam leaving impurities behind. As more and more water is distilled in the boiler, more feedwater containing dissolved solids is added to replace it. As a result the amount of these solids dissolved in the boiler water gradually increases. A boiler as small as 100 horsepower, can evaporate more than 10,000 gallons of water in 24 hours producing over 30 pounds of minerals behind every day. This adds up and causes problems by forming foam on the steam-water interface a cause of carry over, scale on tube surfaces and suspended sludge which usually sinks to the bottom of the lower drum. These dissolved solids are kept at a manageable level by draining a portion of the contaminated water from the boiler, called blow-down. This water contains a lot of heat, not to be wasted. The scale when formed retards heat transfer also wasting energy and causing possible boiler damage, so this area is a good target for energy savings activity. See Chapter 15 for more details.

2.   USE WATER SOFTENER

Hardness in water is caused by dissolved calcium and magnesium which precipitates out of the feedwater. As the temperature of the water is raised calcium and magnesium precipitate out of the water and adhere to surfaces such as boiler tubes. The softener is used to remove calcium and magnesium from the make up water. This reduces the potential for scale deposit formation in the boiler allowing the boiler to carry higher cycles of concentration producing both water and energy savings. Another important benefit of this water softening process is the prevention of boiler tube failure caused by overheating of the tube metal caused by these scale deposits

3.   INCREASING CYCLES OF CONCENTRATION

Cost
   Low

Potential Savings
   Moderate

Description
   Cycles of concentration is the ratio of impurities being maintained in the boiler water divided by the impurity level of the boiler feedwater. This

level of concentration is regulated by the percentage of blowdown. If a low level of concentration is maintained, then the blowdown rate could be very high, wasting energy (Figure 10.39). High cycles of concentration require less blowdown and less energy for make up water heating.

Advantages/Disadvantages

The higher the cycles of concentration to be maintained, the closer the scale formation limit is approached so good control is required. If this process is not closely regulated, there is a chance for scale to form in the boiler which could lead to tube failure or costly cleaning. The general practice has been to stay away from higher levels of concentration if good control could not be assured, this can waste money, energy and water (Figure 10.40) so the pros and cons should be carefully evaluated. Feedwater purity is another issue which must be evaluated to really understand the situation. If make-up feedwater has high impurities, then higher cycles of concentration will be approached more quickly whereas very pure feedwater allows very high cycles of concentration, eliminating blowdown losses and make up water heating to a large degree.

4. AUTOMATIC BLOWDOWN CONTROL

Cost
Moderate

Potential Savings
Moderate (2%-3%)

Description

When high temperature, high energy blowdown water is not closely controlled, energy is wasted by dumping too much water out of the boiler at one time (Figure 10.41). This water then must be replaced by cold make up water. Lack of control may also cause the dissolved solids concentration to rise above the scaling level. Automatic control eliminates these problems and saves water, chemicals and energy.

Advantages/Disadvantages

This option requires good reliable equipment which must be kept calibrated.

5. MECHANICALLY SCALE REMOVAL

Cost
Low (essential maintenance)

Figure 10.39—Cycles of Concentration. The purer the feedwater, the less blowdown required and the higher cycles of concentration that can be maintained. For example, if feedwater quality is improved to increase the concentration from 3 to 6 times, the blowdown rate for a 100,000 lb/hr boiler is reduced from 33.3 to 16.7 percent, or about 30,000 lb/hr of hot water drained from the boiler.

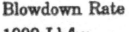
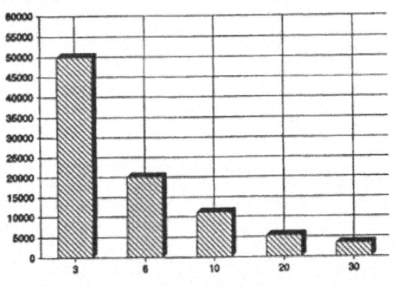

Blowdown Rate
1000 Lb/hr

Cycles of Concentration

Savings
Increasing Cycles of Concentration

Cycles	FW gal/hr	BD gal/hr	Energy $/yr	Water $/yr	Chemicals $/yr
5	3,288	658	67,500	21,900	21,900
10	2,922	292	30,000	9,730	9,730
20	2,637	138	14,200	4,610	4,610
30	2,721	91	9,300	3,020	3,020
40	2,697	67	6,900	2,240	2,240
50	2,684	54	5,500	1,790	1,790

Figure 10.40—Increasing Cycles of Concentration saves on energy, water and chemicals.

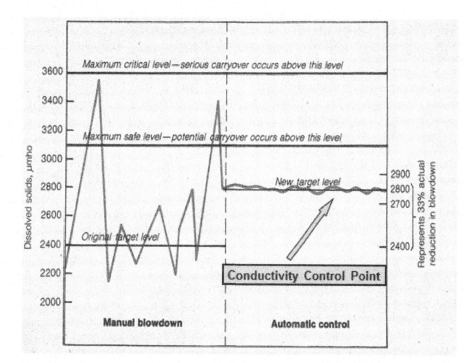

Figure 10.41—Automatic blowdown control enables a 33% reduction in blowdown.

Potential Savings

Moderate (2%-3%)

The buildup of boiler scale constitutes a growth of crystals on waterside heat transfer surfaces and is most severe in those areas of the boiler where maximum heat transfer occurs. (Figure 10.42) shows the relationship between heat transfer efficiency and scale deposit thickness.

Steam generators normally have very high temperature flames and gases heating boiler water and generating steam through the tube walls. The metal tubes in the boiler are kept cool by the boiler water. When scale forms, it acts like an insulation material between the water and the metal. This results in tubes operating at higher temperatures. The greater the thickness of the scale, the greater the insulating effect, and the higher the temperature of the tube metal. At sufficiently high temperatures, in the neighborhood of 1,000°F the tubes begin to lose strength and rupture.

6.   CHEMICAL SCALE REMOVAL

Cost

Moderate (essential maintenance)

Potential Savings

Moderate (2%-3%)

Even a thin layer of scale lowers boiler efficiency. On water-tube boilers the water side scale blocks heat transfer so the tubes may overheat, lose strength and rupture. Scale also creates a breeding ground for under-deposit corrosion which can cause expensive equipment outages and repairs. If you do not acid clean your boiler you may have to replace tubes at some point [expensive repairs].

Once the scale is removed an inspection will reveal the true condition of your boiler, 90% of the time leaks do not occur because of chemical cleaning. In those cases where leaks develop, the scale

was quite possibly plugging an existing hole. At some point these holes would start to leak, possibly causing an expensive shutdown.

The chemical cleaning process is straight forward for a Fire Tube boiler about 10-12 hours is required; cleaning a water tube boiler may take 18 to 24 hours.

7.   BLOWDOWN HEAT RECOVERY

Cost

Low to Moderate

Potential Savings

Moderate (2% to 5%)

Description

Blowdown water temperature is usually over 300°F depending on the pressure and it contains a lot of energy that can be wasted if it is not put back to work in the boiler somehow. About 15% of the blowdown water will flash to low pressure steam so it is a very good source of low pressure steam and is usually used in the deareator/feedwater heater. This steam can be re-

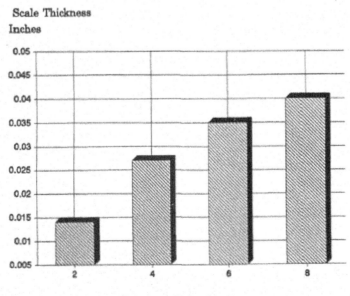

Figure 10.42—Shows how scale build up reduces heat transfer efficiency 2% to 8%. Scaling is most prevalent in the hotter zones in a direct line of sight from the flame.

covered in a flash tank and the rest of the heat in a heat exchanger. If steam is not needed, than a simpler heat exchanger recovery configuration can be used to make-up heat feedwater.

Advantages/Disadvantages

Proven technology. Perhaps the simplest and cost effective approach to recapturing blow down heat.

8.   USE BLOWDOWN TO KEEP
     IDLE BOILERS WARM

Cost

Low

Savings

Low

Description

If you are using steam to protect idle boilers from freezing or need to keep them warm, hot blowdown water can be diverted through idle boilers. Check with the manufacturer

Advantages/Disadvantages

This is a good use of low grade energy. First cost, maintenance and operating costs are involved.

Section 10-6
Capital Investment Projects

CHECKLIST - 6
Capital Investment Projects

1. Replace inefficient boilers beyond economical repair.
2. Install a small boiler for light load operations.
3. Replace oversized boilers to match existing loads.
4. Replace an existing feed-water heater or de-aerator with a new more efficient de-aerator.
5. Install a back pressure control valve to reduce carryover during extreme load conditions
6. Install a steam separator
7. Heat Pumps
8. Cogeneration - Combined Heat and Power
9. Satellite boilers
10. Combined Cycle electrical generation
11. Install backpressure turbines
12. High temperature hot water systems
13. Adjustable speed drives for electric motors.
14. Infrared heaters
15. Submerged Combustion
16. Solar augmented water heating
17. Use flue gas as a source of carbon dioxide
18. Run Around heat recovery systems.
19. Install a hydronic system
20. Pulse combustion systems
21. Install a direct contact water heater.
22. Ceramic wool furnace liner

1.   REPLACE INEFFICIENT
     BOILERS BEYOND ECONOMICAL
     REPAIR.

Boiler systems have many losses connected with their operation that are not obvious at first. The stack loss analysis is a good first look but a longer test period over several months or even a year with accurate measurement of Btus input and Btus output will provide the true story. Unfortunately, this information is not always available and many units perceived to be efficient will be found to be quite inefficient.

2.   INSTALL A SMALL BOILER FOR
     LIGHT LOAD OPERATIONS

Under light loading operations a small boiler is much more efficient than a larger unit because a lot of fuel is expended to keep the boiler hot when it has no real load comparable to its rated capacity especially during off-hours.

3. REPLACE OVERSIZED BOILERS TO MATCH EXISTING LOADS.

There are many reasons for finding oversized boilers. The first question that must be answered is how you would even know it was oversized. This would take some instrumentation to measure steam demand or even length of on cycles verses off cycles with attention to the fact of a certain investment of Btus just to keep the system operating. Without adequate instrumentation and information this condition is hard to detect.

Consider that in the original design the demands for steam were not know so a large safety factor went into the original design. Then an additional safety margin was built into the facility beyond what its highest demands might be. Then add efficiency improvements over the years and other possible load reductions and oversizing becomes a real possibility. This is not normally checked for and more or less guessed at.

4. REPLACE THE EXISTING FEEDWATER HEATER OR DE-AERATOR WITH A NEW MORE EFFICIENT DEAERATOR

Earlier units were not insulated and often insulation has been stripped away for inspection and repairs. If oxygen pitting of pipes and boiler tubing is showing up, then the effectiveness of the deareator is slipping. In any case it should be checked out.

5. INSTALL A BACK PRESSURE CONTROL VALVE TO REDUCE CARRYOVER DURING EXTREME LOAD CONDITIONS

When steam demands exceed plant capacity there is a tendency to reduce boiler pressures caused by lower pressures in the distribution piping. This will cause the water levels in the boilers

to swell up, causing priming or excessive carryover. Priming is the introduction of slugs of water into the steam piping, moving over 100 miles per hour or more. When these high velocity masses of water slam into a sharp turn they can burst piping. In the case of excessive carryover, the boiler water contains a lot dissolved solids which can contaminate steam piping and cause blockages.

This dirty boiler water which gets into the steam system can foul heat exchange surfaces and mess up heat exchange equipment in general.

6. INSTALL A STEAM SEPARATOR TO MAINTAIN CLEAN DRY STEAM

One way to keep boiler water and carryover out of the steam piping is to install a steam separator (Figure 10.43) near the boiler outlets. They have proven to be 99.9% efficient at eliminating priming and carryover, so they have a vital function to perform keeping heat exchange equipment clean and functioning.

7. HEAT PUMPS

Cost
   Moderate

Savings
   Moderate

Description
   Heat pumps have high efficiencies when measured against engines, boilers and turbines (Figure 10.44). In the example below the Heat Pump can bring the apparent efficiency of a boiler up to 128% using waste heat or low pressure steam available in a boiler plant or they can be used to augment old steam systems with air source heat pumps. (Figure 10.45) Heat Pumps can raise efficiencies significantly when combined with low grade waste heat from engines, boilers and turbines.

Waste heat can be put to work at a higher temperature by the use of heat pumps. One system design circulated 150°F water from a waste heat recovery system throughout a plant using

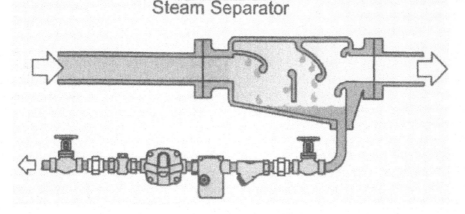

Figure 10.43—Steam separators keeps steam piping system clean and dry.

heat pumps to raise the water temperature for local heating and process demands.

The key to how this operates is in the term: Coefficient of Performance (COP), a measure of the ratio of energy output to energy input. Some heat pump units have a COP of over 5 which means the energy to operate them is only about 20% of the energy produced. (Figure 10.46)

Advantages/Disadvantages

This is one of those options that take imagination and initiative. There appears to be many applications where the efficiency of the Heat Pump has not been applied; for example for heating make up water or to augment old low pressure steam systems. First cost, operation and maintenance expenses are involved which must be worked out. It is a good way to recover low grade waste heat or provide heat to isolated locations without resorting to extensive piping systems and the heat losses they represent.

8.   COEGENERATION— COMBINED HEAT AND POWER

Cost
High

Potential Savings
High

Description
Because of the high price of electricity in some sections of the country, it is very cost effec-

Heat Efficiency

Heat pump CoP 4	128%
Condensing Boiler	95%
Coal fired boiler	83% - 90%
Oil fired boiler	84% - 89%
Gas fired boiler	84% - 86%
Gas Turbine	~50%
Diesel	~40% - 50%

Electrical Efficiency

Combined Cycle Gas	50%
Diesel Generator sets	30% - 40%
Gas Turbine	25% - 40%
Generating Station	32%

Figure 10.44—Heat Pumps can introduce high efficiencies when combined with other technologies.

COP	BTU Out-Kwh	Cost Per MM BTUs	Efficiency
1	3,413	$ 30.00	32%
2	6,826	$ 15.00	64%
3	10,239	$ 10.00	96%
4	13,652	$ 7.50	128%
5	17,065	$ 6.00	160%

Figure 10.45—the Coefficient of performance (COP) of Heat Pumps has the potential to raise the efficiency of some systems over 100%.

tive to use a cogeneration plant to provide steam and power while producing electricity. A utility plant usually wastes about 68% of the input energy and another 5-8% can be lost in transmission and distribution of the power.

Some cogeneration facilities are over 80% efficient. This plus the high cost of electricity puts

Coefficient of Performance = $\dfrac{\text{OUTPUT}}{\text{INPUT}}$

(KW-hr = 3413 BTU)

	Efficiency
CoP 3	= 300%
CoP 4	= 400%
CoP 5	= 500%

Electrical Efficiency   ~ 32%

Figure 10.46—Heat Pumps can raise efficiencies significantly when combined with low grade waste heat from engines, boilers and turbines.

a cogeneration facility at a good economic advantage. Peak shaving is another consideration, a large part of many electric utility bills is for "demand" charges which is a charge based on the peak demand for electricity and not actual energy use. Often the peak demand for electricity occurs at the same time as the demand for other types of energy such as heat and cooling, so if the onsite generation of electricity can reduce electrical demand and provide other forms of energy at the same time, large savings are possible. The key to effective cogeneration is to have both electrical and heat demands occur at the same time. (Figure 10.47)

Advantages/Disadvantages

This option may take an extensive analysis but should not be overlooked as a way to improve the bottom line of your operation. High cost, load matching, utility intertie, red tape and pollution control are a few of the challenges that must be faced with cogeneration. It is a very good way to reduce steam and electrical power costs and may be well worth the trouble.

9.   COMBINED CYCLE ELECTRICAL GENERATION

Cost
    High

Potential Savings
    High

Description
    Because of the high price of electricity in some sections of the country, it is very cost effective to use a Combined Cycle plant to provide steam and power while producing electricity. A utility plant usually wastes about 68% of the input energy and another 5-8% can be lost in transmission and distribution of the power to the plant.

Figure 10.47—Cogeneration is best suited for a location where both the electrical load and waste hear load are consistent like in this case where the electric power requirement and the requirement for steam, a product of the heat from the cogeneration exhaust both exist uninterrupted for a maximum number of hours a year. The economics of cogeneration declines to the degree that either the demand for steam or electricity is not consistent.

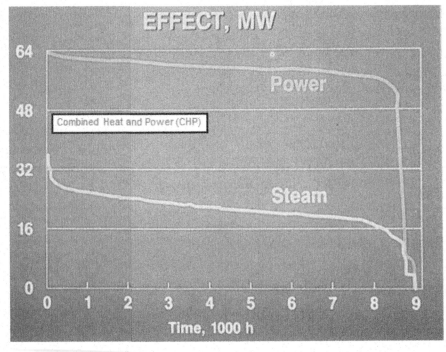

Some Combined Cycle facilities (Figure 10.48) are over 80% efficient. This plus the high cost of electricity puts a Combined Cycle facility at a good economic advantage.

Peak shaving is another consideration, a large part of many electric utility bills is for "demand" charges which is a charge based on the peak demand for electricity and not actual energy use. Often the peak demand for electricity occurs at the same time as the demand for other types of energy such as heat and cooling, so if the onsite generation of electricity can reduce electrical demand and provide other forms of energy at the same time, large savings are possible.

The key to effective cogeneration is to have both electrical and heat demands occur at the same time.

Advantages/Disadvantages

This option may take an extensive analysis but should not be overlooked as a way to improve the bottom line of your operation. High cost, load matching, utility intertie, red tape and pollution control are a few of the challenges that must be taken into account. It is a very good way to reduce steam and electrical power costs and may be well worth the trouble.

10.   INSTALL BACKPRESSURE TURBINES

Cost
High

Potential Savings
High

Description

Steam systems extract the most heat from steam sources when pressure is reduced to match needed temperatures of pressures.(Table 10.4) In a larger facility with sufficient steam flows, turbo-generators can be used instead of PRV valves to do the same function while generating electricity at high efficiency. A electrical utility plant has an efficiency of about 32%; by using turbo-generators this way an efficiency of 80% is possible.

This is a design option requiring a high pressure boiler up front. Backpressure turbines are installed from the high pressure header to the medium pressure and low pressure headers. When steam pressure is reduced this way more heat becomes available and the steam is used more efficiently. This is the same reason why pressure reducing valves are used. The reason for the high efficiency is that the turbo-generators do not con-

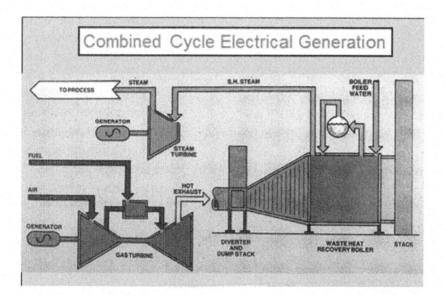

Figure 10.48—Combined Cycle electrical generation.

sume steam but send it on to do additional work in the plant as opposed to an electrical generating plant where the steam is dumped into a condenser to turn it back into water wasting a great deal of heat present in the final steam stages.

This is generally practical on 50 KW units on up and with steam flows of 3,000 pounds of steam per hour significantly reducing local electrical costs.

Table 10.5
Why We Reduce Steam Pressure

Pressure PSIG	Latent Heat Btu/lb	% of Total Heat
125	868	73.5
50	912	77.5
5	961	83.5

When steam pressure is reduced, more heat becomes available and the steam is used more efficiently. This is why pressure reducing valves are used wherever possible.

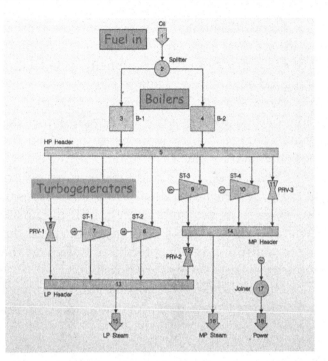

Figure 10.49—The Pressure Reducing Valves (PRV) are the (hour glass shaped symbols) which can be augmented by turbogenerators (cone shaped) which perform the same function with the added benefit of producing electrical power at high efficiency.

11. HIGH TEMPERATURE HOT
     WATER SYSTEMS

Cost
    High

Potential Savings
    High

Description:
1. High temperature hot water has a system temperature from 350°F to 450°F. Medium temperature hot water is between 250°F and 350°F and Low temperature hot water is below 250°F.
2. One cubic foot of water at 350°F contains 50 times as much heat as a cubic foot of saturated steam at 350°F.
3. When a high temperature hot water system cools a cubic foot of water from 350°F to 250°F, it releases approximately 27.6 times more heat than a cubic foot of saturated steam at 350°F.

Advantages/Disadvantages

Losses with a high temperature hot water system are much less than steam because:
1. It is a closed system and leaks are usually less than one half of one percent whereas a steam system can require up to 100% make up feedwater. The actual requirement for makeup water can vary widely.
2. It minimizes losses associated with steam traps, valve stem and packing gland leakage, blowdown and flash losses associated with open condensate receivers.
3. It can pack more usable heat into a given volume than a steam system and needs smaller diameter piping.
4. Return water temperature must be maintained above a certain limit to avoid thermal stress and piping system damage.

A recent study has shown that in a 20 Million Btu/hr system, comparing HTW at 350°F to saturated steam, showed that there is a $300,000 a year savings with HTW. Over a life cycle of 25 years the savings was over $14 million.

12. SATELLITE BOILERS

Cost
   Moderate to High

Potential Savings
   Very High

Description
   Once during the days when coal was King and after Heavy Fuel Oil took its place, large central plants and extensive steam distribution systems were built to accommodate the central fuel supply and need for skilled plant operators for safety and reliable operations on a 24 hour basis month after month year after year. Distribution system losses can be very high and it has been found the best way to eliminate distribution system losses is to eliminate the distribution system. Originally, central plants were designed around the use of coal because it was the only fuel available and now many facilities have miles of old and questionable distribution systems which were built before the day of the reliable and safe satellite boilers. When you see brown grass above buried steam lines in the summer and melted snow over the distribution system in the winter, you are probably looking at heat losses from buried steam and condensate piping. Every foot of piping is a possible source of leaks, energy losses, acid corrosion and other maintenance problems. Rain soaked and degraded insulation, flooded systems, steam plumes along the steam mains in streets and fields and cold or lost condensate are all indicators of major distribution system problems.
   The central steam plant can be a huge liability; one only has to analyze steam production during warm weather or holidays to discover that large losses exist. Reliable and safe control technology and the availability oil and gas supplies now make individual satellite boilers a safe and efficient way to avoid these large distribution system losses.

   In Europe energy engineers have come up with the rule of thumb that it takes 49 times more energy to move a unit of thermal energy from point A to point B than to move the same energy as fuel. This may not be the specific case in all applications, but it is definitely something to consider in the case of satellite boiler projects in old central steam systems.

Advantages/Disadvantages
   The cost of installing satellite boilers and the permitting process can be quite high, but the savings can be very high in both maintenance costs and in energy savings.

13. ADJUSTABLE SPEED DRIVES

Cost
   Moderate

Potential Savings
   Moderate to high.

Description
   When pumps and fans operate at higher rpms than required for a particular load, electrical energy is lost. The pump and fan laws apply which states that the energy used varies with the cube of the rpm. Large savings can be accomplished by reducing the rpm of rotating equipment. Variable speed drives are a new dimension in the cost effective management of plants. For example, if the speed of a pump or fan is reduced by one half, there is a 88% reduction in energy use.

Advantages/Disadvantages
   Many plants have successfully applied adjustable speed drives to boiler operations. Adjustable Speed Drive (ASD) is becoming a proven way to save electrical costs.

14. CONVERT TO INFRARED HEATERS

Cost
   Moderate

Potential Savings

Moderate

Description

Infrared heaters have a high combustion efficiency and an ability to heat selected line of sight areas efficiently, such as hangars, warehouses, gymnasiums and maintenance facilities. They put the heat where it's needed and do not heat up a large volume of air that may uselessly float up to the top of a large open building. People below these units are comfortable in their shirtsleeves in an otherwise cold space.

Advantages/Disadvantages

The infrared heater is usually self contained and mounts in the overhead, out of the way. They may be an intelligent replacement to old steam, hot water and hot air systems.

15.   SUBMERGED COMBUSTION

Cost

Moderate

Potential Savings

High

Description

The combustion process takes place under-water with direct contact heat exchange. Typical use is for heating swimming pools and laundry water (Figure 10.26).

Advantages/Disadvantages

Fuel and air must be pressurized to overcome submersion. Efficiencies approaching 95% are possible.

16.   SOLAR AUGMENTATION FOR WATER HEATING

Cost

Moderate

Savings

Moderate

Description

The rule of thumb of each 11 degrees that you can heat the water going into a boiler will improve efficiency by 1%, applies in this case. In some rare applications, make up feedwater is a good candidate for solar augmentation. Also, domestic hot water heating efficiency can be improved by this option. There are applications where it can pay off such as reducing fuel use with renewable energy for credits towards CO_2 or other air pollution compliance reductions

Advantages/Disadvantages

This application would be most unusual, but possible. Financial analysis and air pollution compliance requirements would be reasons for justification.

17.   USE FLUE GAS AS A SOURCE OF CARBON DIOXIDE

Cost

Moderate

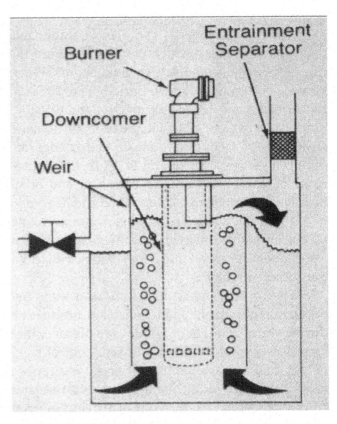

Figure 10.50—Submerged combustion burner unit.

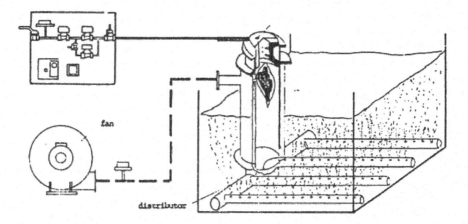

Figure 10.51—Submerged water heating unit capable of efficiencies in the mid 90% range.

Potential Savings
Moderate

Description
In some locations flue gas, with its high CO_2 content, has been used as a source of CO_2 for various purposes.

Advantages/Disadvantages
There would need to be a market for the CO_2 such as oil recovery and the boiler would have to be a reliable source

18. RUN AROUND SYSTEM

19. USE HYDRONIC SYSTEMS FOR EFFICIENT HEATING.

Hydronic systems operate with a higher energy density than steam systems and have several other advantages:

1. It is a closed system and leaks are usually less than one half of one percent whereas a

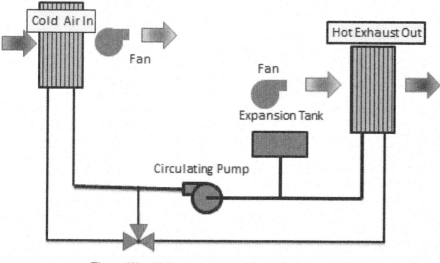

Figure 10.52—When you need to change the air in a space, heat can be conserved using a Run Around system similar to the sketch above which removes heat from exhaust air and heats up incoming air.

steam system can require up to 100% make up feedwater. The actual requirement for makeup water can vary widely.

2. It minimizes losses associated with steam traps, valve stem and packing gland leakage, blowdown and flash losses associated with open condensate receivers.

3. It can pack more usable heat into a given volume than a steam system and needs smaller diameter piping.

4. Water temperatures can be lowered during low heat demands reducing pipe insulation losses. Boiler exhaust temperatures are lowered by this also.

Disadvantages

Circulating pumps must run continuously. Return water temperature must be maintained above a certain limit to avoid thermal stress and piping system noise and damage.

20. PULSE COMBUSTION UNITS (Figure 53)

21. DIRECT CONTACT
HOT WATER HEATER
(Figure 10.54)

Cost
High

Potential Savings
High 15% Range

Description

12%-15% of the waste energy in the exit flue gas is stream formed from the combustion of hydrogen in the fuel and other moisture. For every pound of hydrogen that is burned with the fuel, nine pounds of water is formed which escapes out of the stack as vapor carrying with it a great deal of energy. This flue gas moisture is usually accepted an inevitable energy loss energy loss as it is difficult to recover the latent heat from this vapor, this method successfully recovers this energy. (Figure 10.23).

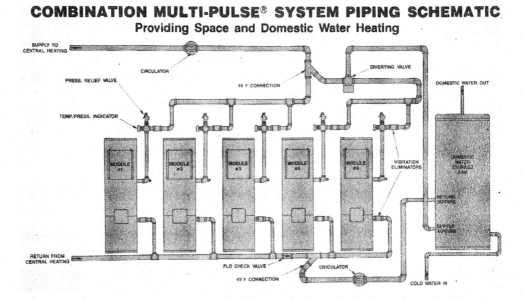

COMBINATION MULTI-PULSE® SYSTEM PIPING SCHEMATIC
Providing Space and Domestic Water Heating

Figure 10.53—Gas fired Pulse Combustion units operate very efficiently because they condense the latent heat energy usually in combustion exhaust producing very high operating efficiencies. They are practical for hot water circulating systems and hot water heaters. They don't need special exhaust systems but do produce water that must be drained away somehow.

Figure10.54—Direct fire hot water heater is very efficient because it reduces stack moisture losses improving efficiency about 15% into the high 90% range. The Manufacturer claims approvals from EPA and other government monitoring agencies indicating that the hot water produced is safe for humans and food processing.

produces inherently lower smoke numbers on light off. This should be considered for low use boilers.

Advantages/Disadvantages

The obvious disadvantage is high first cost if retrofit is required. The use of light refractory improves combustion performance on initial light off and soaks up less heat on/off cycles.

Section 10-7
Steam System Improvements

CHECKLIST 7
Steam Systems Improvements

1. Target steam system losses
2. Insulate hot bare-metal surfaces
3. Steam trap maintenance
4. Repair existing insulation.
5. Optimize condensate recovery

21.   USE SOFT FIREBRICK IN COMBUSTION CHAMBERS

Cost
Moderate (usually a design choice)

Potential Savings
Moderate

Description
If there is a choice in combustion chamber materials for original purchase, for redesign or for repairs, light weight fire brick or ceramic fiber material should be considered. This material heats up faster taking the combustion zone through its cold smoky period more quickly (Figure 10.55). Its use

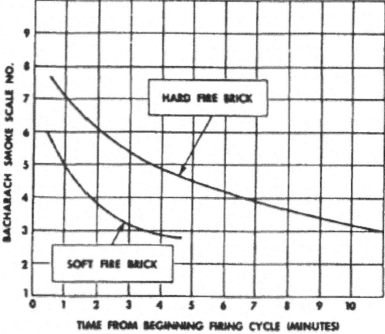

Figure 10.55   The effect the choice of combustion chamber materials has on smoke and soot production on startups.

6. Repair condensate receivers, pumps and piping
7. Repair steam leaks
8. Optimize deaereator vent rate
9. Insulate feedwater System
10. Use high pressure condensate to make low pressure steam
11. Pressure reducing valves
12. Utilize backpressure turbine generation to augment pressure-reducing valves
13. Have a steam system survey done by a professional contractor
14. Install pump traps eliminating electric pumps
15. Repair heat exchanger drains to permit condensate return.
16. Implement a pressurized condensate return system.
17. Separate H.P. and L.P. condensate returns.
18. Isolate un-used sections of steam lines.
19. Insulate condensate return piping systems
20. Flash steam utilization
21. Use hot condensate to keep storage tanks warm (local use)
22. Measure steam loads
23. Insulate oil storage tanks heated with steam.
24. Modulate vacuum pressure in vacuum return systems
25. Steam compression
26. Use steam to drive equipment to reduce electrical demand.

1.   TARGET STEAM SYSTEM LOSSES

      Points of attack:

2.   INSULATE ALL HOT BARE
      METAL SURFACES

Cost
      Moderate

Potential Savings
      Good return on investment

1. Trap flash losses

2. Insulation Losses

3. Condensate system losses

4. External steam leaks

5. Internal steam leaks

6. Steam trap leakage

7. Inactive steam piping.

Figure 10.56—Primary energy losses in steam distribution systems

Description

Boilers and steam systems have surface temperatures from 350°F to 450°F and higher and must have effective insulation to prevent heat from escaping from boiler and piping surfaces (Figure 10.57). Often insulation is removed from valves, piping and boiler parts, for repairs and not replaced. This bare surface is a potential safety hazard as well as a source of lost energy. One danger associated with poor insulation is that fuel and oil leaks have been known to ignite off these hot surfaces and start fires.

Advantages/Disadvantages

Asbestos may be encountered in older systems. Insulating bare steam piping is usually a quick pay back item.

3.   STEAM TRAP MAINTENANCE

Cost
      Low

Potential Savings
      High

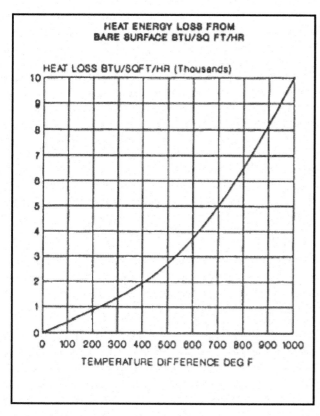

Figure 10.57—Bare pipe heat losses

Description

Steam traps can waste a tremendous amount of energy; even when working properly steam traps are inefficient when you consider the flash losses. Traps may go through millions of cycles a year and develop leaks and many types have high failure rates. Industrial studies have shown that large savings are possible with proper steam trap maintenance programs.

Many condensate systems are "open" which means that they vent to atmosphere. Because of the high pressure on the working side of the trap and the low pressure on the outlet side, over 10% of the condensate going through healthy traps will flash into steam representing a lot of heat that eventually gets lost from the system. Add to this the many problems found in traps, like closing slowly, allowing steam to escape, leaky seats or just not closing at all and it becomes obvious there is a potential for large hidden steam losses, maybe over 50%. Extensive studies have shown that in most plants, a very high percentage of traps have failed one way or the other. The real

question is, "How much of this expensive steam is really being put to productive use and how much is being lost?" When the potential for loss is documented, it often has become obvious that a good maintenance programs are required.

Advantages/Disadvantages

Until the potential savings of good steam trap maintenance is realized by management, big energy dollars will be silently lost in distribution systems.

4. REPAIR EXISTING INSULATION

Cost

Moderate

Potential Savings

Good return on investment

Description

Boilers and steam systems have metal surface temperatures from 350°F to 450°F and above and must have effective insulation to prevent heat from escaping from boiler and piping surfaces. Some older systems have many miles of piping which has been exposed to harsh weather conditions for decades. Infrared instruments are useful in determining these losses and some are calibrated to indicate Btu losses per hour from hot surfaces which is very useful for cost/benefit analysis. Figure 10-...... can be used to assist in estimates.

Advantages/Disadvantages

Asbestos may be encountered in older systems. Insulating bare steam piping is usually a quick pay back item.

5. OPTIMIZE CONDENSATE RECOVERY

The "pure" water from the condensed steam still contains valuable energy after it leaves steam traps and for efficient operation it should be recovered and arrive back the boiler plant where it is reheated into turned back into steam and recycled in the steam system over and over again.

All steam used by the system must pass through steam traps, so the whole system is subject to this loss if atmospheric vents are used. Closed pressurized condensate recovery systems are used in well-designed plants to avoid this problem. If the pure hot water is not brought back to the plant by the condensate recovery system, then the expense of new water, its chemical treatment and heating it back to operating temperature will just add to avoidable plant expenses.

Advantages/Disadvantages

If you are not getting the condensate back from your steam system you are incurring about a 17% loss from that fact alone.

Rule of Thumb
Every 11 deg F [6 deg C] that must be added to boiler feedwater costs 1% in efficiency.

---

6.   REPAIR CONDENSATE RECEIVERS, PUMPS AND PIPING

Water seems to be such a common and cheap commodity and it is not assigned its real value to steam system management.

When conditions are not observable or known, the assumption is often made that they are negligible.

In steam systems, most of the losses are not observable. Unfortunately they are quite large and far from negligible.

7.   REPAIR STEAM LEAKS

External Steam Leaks

External steam leaks are usually visible and the length of the plume can be measured by eye.

For example, if there are 4 visible steam leaks, two 6-foot plumes and two 8-foot plumes, this would work out to nearly $100,000 a year loss plus the cost of water, chemicals and fuel wasted

heating make up water to system operating temperature. One point often overlooked is that boiler inefficiencies are doubled in this case, because the steam in these leaks must be generated twice to use it once.

Internal Steam Leaks

Internal steam leaks are hard to detect, they are inside the piping system, and common sense indicates there are such leaks. Steam traps are installed with bypass valves which are often opened in the event there is a problem with the trap like plugging. The bypass valves are often opened on start up or when water hammer becomes a problem in the system to drain water from the system. Steam driven equipment such as turbines and pumps have drains and warmup systems and these can be left open too. The fact is steam can be slipping through the system without doing any work

8.   OPTIMIZE DEAEREATOR VENT RATE (Figure 10.58)

9.   INSULATE THE FEEDWATER SYSTEM

The water in the feedwater system is at the boiling point which could be 212F or higher depending on the pressure in the system. The deaerator and reserve feedwater tank are typical of such systems. The large surface represents a large loss of heat if not insulated. (Figure 10.59)

10.   USE HIGH-PRESSURE CONDENSATE TO MAKE LOW-PRESSURE STEAM

The high pressure condensate recovery system can be a source for low pressure steam use in the plant or for other uses.

11.   PRESSURE REDUCING VALVES

Potential Savings
Moderate

Figure 10.58—Shown are two steam plumes, the one on the right shows normal venting and the larger plume on the left indicates a lot of energy is being wasted when you consider these systems are active year after year for 24 hours a day.

Figure 10.59—Insulate hot surfaces to preserve system heat.

Description

Installing pressure reducing valves near the point-of-use can reduce flash steam losses. Also, high pressure steam has a higher temperature, lower specific volume and is economical to distribute through steam mains requiring smaller sized piping. As a general rule, steam is most economically distributed at high pressure and if reduced to the lowest pressure that will satisfy the temperature requirements at the point-of-use, the least amount of energy will be lost.

Advantages come with lowering steam pressure; a 600 psig system will have a 30% flash steam rate, at 175 psi this reduces to a 17% flash and at 50 psi its 4%. This represents a considerable amount of energy because every pound of steam flashed contains about 1,100 Btus where the condensate contains only 180 Btus per pound a difference in energy levels of about 6.

Advantages/Disadvantages

The cost of installing and maintaining pressure reducing valves and possible replacement of steam traps will have to be balanced against predicted savings.

12. UTILIZE BACKPRESSURE TURBINE FOR STEAM PRESSURE REDUCTION

Description

The use of turbogenerators as an option for reducing steam pressures in large plants instead of pressure reducing valves is an efficient way to generate electricity. It starts with relatively high pressure steam boilers on the high pressure end of the steam system. The steam pressure is reduced to intermediate and low pressure steam systems through the turbines before it enters the plant's steam distribution system. This is a very efficient way to generate electricity at 80% efficiency rather than efficiencies in the 33% range by electrical generating plants which have no way to utilize their turbine exhausts. The usual design is 50 KW and up with a minimum steam flow of 3,000 pounds per hour.

Advantages/Disadvantages

It is a more complex system to manage but produces electric power much cheaper than commercial power.

13. HAVE A STEAM SYSTEM SURVEY DONE BY A PROFESSIONAL CONTRACTOR

Steam system surveys can be quite informative and useful but requiring special skills, judgment and background. Their usual product is a map of your steam system with all traps identified as to type and manufacturer, quite useful for ordering parts and replacements. The traps are given identification tags and you also get a condition report on each trap with misapplication information and such. It's a good way to identify any problems with your steam system as a start to putting it right and a successful energy conservation program.

Advantages/Disadvantages

It might seem a little expensive or unusual but it is a good way to find out exactly what is going on with your steam distribution system. Good reports and successful energy savings are the usual case concerning these special surveys.

14. INSTALL PRESSURE PUMP RETURNS ELIMINATING ELECTRIC PUMPS

Electrical pumps have a number of drawbacks. Being electrical in a wet environment sometimes gets unhealthy for the electrical end of things. They have centrifugal impellers that can form a partial vacuum at the water inlet with the water near the boiling point. This can cause steam flashing and the loss of suction and effectiveness and of course there is the electrical load and the need to have electricians prowling your steam system's tunnels vaults and confined spaces checking on things and making repairs.

The industry has come up with an alternative pump (Figure 10.60) without these draw-

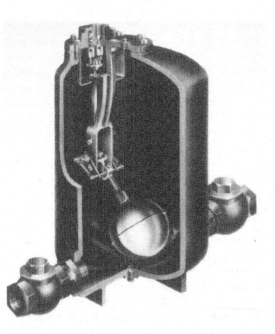

Figure 10.60—Pumping steam trap does not need electrical support and are quite trouble free.

backs which uses the local steam supply to do the pumping without the hazards of an electrical supply with confidence that the valuable condensate will be returned to the boiler plant.

Advantages/Disadvantages
Requires investment but delivers reliability and large savings.

15. REPAIR HEAT EXCHANGER DRAINS TO PERMIT CONDENSATE RETURN

Sometimes things are just not piped up properly and steam condensate is lost to the environment.

16. IMPLEMENT A PRESSURIZED CONDENSATE RETURN SYSTEM

A pressurized condensate return system gets steam and condensate back to the plant where both can be utilized instead of causing steam and water damage in an office building, school or laboratory.

17. SEPARATE H.P. AND L.P. CONDENSATE RETURNS

There are certain economies in separating high pressure and low pressure recovery systems. It will provide a source of two levels of temperature and pressure available for specific applications.

18. ISOLATE UN-USED SECTIONS OF STEAM LINES.

Steam piping loses the same amount of heat whether it is transporting steam or not so if a section of piping is not in use, valve it off or permanently blank it off.

19. INSULATE CONDENSATE RETURN PIPING SYSTEMS

Some insulation programs do not consider it cost effective to insulate condensate recovery piping because of its low temperature and low heat loss. They do not take into consideration the face that for every 11F the condensate loses getting back to the plant costs 1% in overall system efficiency. For example if the condensate temperature drops from 190F to 120F a difference of 70F this represents a 6.3% loss in efficiency; total boiler plant fuel use. So if a plant is burning $2 million worth of fuel a year, this represents about $126,000 a year.

20. FLASH STEAM UTILIZATION

Cost
Moderate

Potential Savings
Moderate

Description
Use a flash steam recovery vessel to collect condensate from high pressure sources. It sepa-

rates steam from condensate and the low pressure flash steam is piped off to some useful purpose. Only as a last resort should flash steam be vented to the atmosphere and lost.

Advantages/Disadvantages

New coils or heat exchangers may be required to utilize the flash steam for heating and other uses. Flash steam possibilities are becoming recognized.

21. USE HOT CONDENSATE TO KEEP STORAGE TANKS WARM (LOCAL USE)

In this case the heat in steam trap drains is used to augment local heating in a process or storage tank.

22. MEASURE STEAM LOADS

Getting an understanding of the steam demands of various components in your system provides you with vital information on managing your steam system and data necessary for decision making for projects under consideration.

23. INSULATE OIL STORAGE TANKS HEATED WITH STEAM

Uninsulated tanks lose a lot of heat, especially tanks out in cold winter weather. If they get too cold there are wax formation and viscosity problems and it becomes very difficult to pump the oil.

24. MODULATE VACUUM PRESSURE IN VACUUM RETURN SYSTEMS

Cost
Original Design (low)

Savings
High

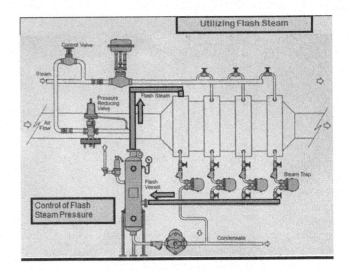

Figure 10.61—Heating unit drains supply flash steam to a heating coil to augment unit capacity. This is an example of utilizing flash steam locally.

Description

There are some older steam systems around that are of the condensing type, where steam is circulated back to a condensing unit in the boiler room. By modulating the steam pressure, the temperature of the distribution system can be matched to the demand for a particular weather condition. For example, on a cold day the steam pressure and temperature can be allowed to rise to meet a higher demand for heat and on warmer days the pressure and steam temperature can be controlled below atmospheric pressure and temperatures, below 212°F, to supply less heat.

25. STEAM COMPRESSION

Cost
Moderate to High

Potential Savings
High

Description

If you have low pressure waste steam or if you need higher pressure steam at some remote location in the distribution system, then steam compression may serve to your advantage. With a source of low pressure steam that would other-

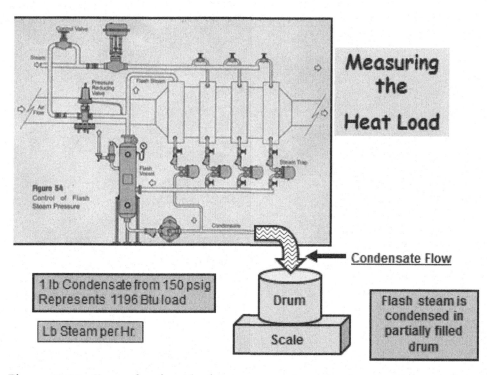

Figure 10.62—Example of method to measure steam loads on various equipment in the steam system; a simple way to find out what is happening in your system.

wise be wasted, it can be salvaged and brought back up to working pressure without large losses using steam compression. In cases where a large distribution system must be maintained at a high pressure to just serve a small high pressure load, the pressure for the system can be reduced and the small high pressure demand can be met with steam compression. Screw compressors and steam eductor sets have been developed for steam compression which are driven by either electric or engine type prime movers.

Advantages/Disadvantages

Using steam compression to satisfy requirements in otherwise low pressure systems makes sense. These systems are cost effective.

26. USE STEAM TO DRIVE EQUIPMENT TO REDUCE ELECTRICAL DEMAND

Cost

Moderate to high

Potential Savings

Moderate to high

Description

Electrically driven pumps and blowers are usually installed because they are simpler and require less maintenance, however when the cost of electricity is considered, steam driven auxiliaries may be more economical. If there is a place to utilize the full volume of exhaust steam from steam driven equipment, then a situation similar to cogeneration exists, where the rejected heat can be utilized improving over all cycle efficiency. Steam driven equipment is especially attractive where electrical demand charges are very high. (Figure 10-63)

Advantages/Disadvantages

High first cost, maintenance requirements and the need for more operator skill work against this option. Energy and cost savings could be very significant, you may have some of these old units around.

Figure 10.63—Steam driven feedwater pump.

Section 10-8
Heat Recovery Projects

CHECKLIST 8
Heat Recovery Projects

1. Install economizers on existing boilers.
2. Spray tower heat recovery system
3. Heat reclaimer
4. Air preheater
5. Heat recovery using plate exchangers
6. Waste heat recovery boilers
7. Blowdown heat recovery
8. Use blowdown to keep idle boilers warm
9. Vent condensers
10. Waste heat for process applications
11. Recover steam and heat from high pressure condensate recovery systems.
12. Install a condensing economizer to heat make-up water.
13. Extract combustion air at the roof level of the plant.

1.   ECONOMIZER

Potential Savings
    ~5%

Description

One of the most effective ways to recover energy from flue-gas is to use an economizer (Figure 10.19) to heat the feedwater going into the boiler. At 100 PSIG the temperature of the steam and water in the boiler is about 334°F. The temperature approach of the flue gas to steam temperature is in the range of 50°F to 150°F, depending on firing rate. This can result in a stack temperature from 385°F to over 480°F providing a good source for heat recovery.

Hot flue gas contains a lot of wasted energy and it is very beneficial to put this energy back into the boiler with heat recovery equipment. The rule of thumb for reducing stack temperature is: for every 40°F the temperature is reduced, a 1% efficiency increase occurs. In this case lowering stack temperature by 200F will result in efficiency increase of 5%.

Preheating the incoming feedwater also improves efficiency. The usual temperature of incoming feedwater is 212 F to 220 F or higher. This water can be heated in the economizer improving efficiency on the water system side. The Rule of Thumb for feedwater entering a boiler, is the efficiency goes up by one percent for each 11F rise in temperature so, if you can heat the water 220°F to 275°F, will result an efficiency increase of 5%.

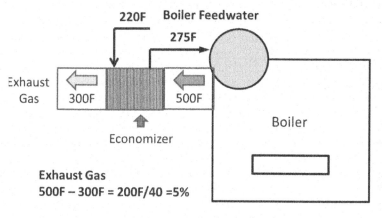

Exhaust Gas
500F – 300F = 200F/40 =5%

Feedwater
275F – 220F = 55F/11 = 5%

Figure 10.64—The Economizer lowers boiler exhaust from 500F to 300F for an efficiency and capacity improvement of 5%. At the same time the boiler feedwater temperature goes up about 57F. Two Rules of Thumb apply here (a) for every 40F the exhaust temperature is brought down the boiler efficiency goes up 1% and (b) for every 11F the boiler feedwater goes up efficiency of the boiler improves 1%. The 5% savings (and capacity increase) can be counted only once however.

Advantages/Disadvantages

Once installed, they are usually trouble free and require little maintenance. If stack temperature or incoming economizer water temperature is too low, an acid formation problem may develop with fuels containing sulfur.

2.   SPRAY TOWER HEAT RECOVERY SYSTEM

Cost

High

Potential Savings

High 15% Range

Description

12%-15% of the waste energy in the exit flue gas is superheated stream is formed from the combustion of hydrogen in the fuel and other moisture. (Table 10.5) For every pound of hydrogen that is burned with the fuel, about nine pounds of water is formed which escapes out of the stack as superheated vapor carrying with it a great deal of energy

The acid dew point temperature is the temperature at which sulfuric acid and other acids form. An efficiency increase of 11% to 15% is possible with most boilers if the flue gasses can be cooled below the dew point.(Figures 10.65 and 10.66) Natural gas, because of its higher hydrogen

content, produces higher efficiency increases with this method than oil.

When flue gas temperatures can be dropped below the dew point, thereby extracting latent heat from the flue gas moisture, efficiencies above 95% can be expected. Examples of this technology are: pulse combustion, advanced fiber burner designs, special coated plate exchangers, glass tube exchangers, Teflon coated tubes and spray tower direct-contact heat exchangers.

Advantages/Disadvantages

This emerging technology is the only way to get efficiencies in the mid 90% range. Special designs are required; most installations have been highly successful. Acid formation does not seem to be a problem in proper designs.

3.   HEAT RECLAIMER

Cost

Moderate/Low

Potential Savings

Moderate

Description

The heat reclaimer (Figure 10.67 and Figure 10.68) is similar to the economizer, except that it is used on smaller hydronic or hot water systems.

Table 10.5—Sources of vaporized water in boiler exhaust. Combustion processes create lots of moisture which turns to superheated steam leaving the stack with a lot of energy; a major loss of efficiency.

**Sources of Vaporized
Water in Boiler Exhaust**
Every pound of hydrogen burned forms 9 pounds of water

Natural gas is 22.5% hydrogen   (2 lb/lb fuel)
Distillate oil is 12.5% hydrogen   (1.25 lb/lb fuel)
Residual oil is 9.5% hydrogen   (0.86 lb/lb fuel)
Bituminous coal is about 5%
    hydrogen but may contain
    5% to 25% fixed moisture by wt.   (0.45 to 2.25 lb/lb fuel)

Other Sources:
1.   Atomizing steam at burners
2.   Soot blowers
3.   Tube leaks
4.   Surface moisture on coal [rain]

Figure 2.65—As the exhaust temperature drops to approximately 135F a 3% efficiency increase occurs. As the exhaust temperature drops to 50F, caused by the cold water spray, another 11% efficiency increase is possible. Actual results depend upon specific conditions.

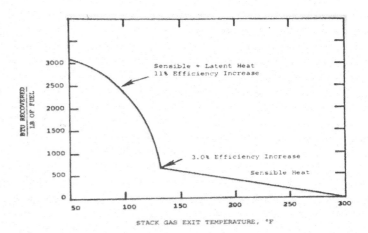

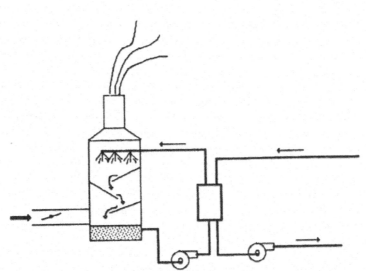

Figure 10.66—Hot boiler exhaust enters the spray tower where the steam vapor from combustion moisture sources condensed back into water liberating about 3,000 Btus per pound of fuel.

It can also be used on boilers for heating process or domestic hot water. A heat exchanger is installed in the stack with its own pump and regulating valve and is basically an additional heat exchange surface for the heating unit. The flue gas temperature can be controlled by regulating the flow of water through the heat reclaimer. Many of the older hydronic boilers have high stack temperatures and high excess air levels, making this option very attractive.

Advantages/Disadvantages

Design and installation costs on the smaller units may be hard to justify, otherwise they are relatively trouble free low cost units. The heat reclaimer has two advantages, (a) it recovers waste heat and (b) it lowers excess air on atmospheric type boilers by cooling exhaust gasses creating less furnace draw in both the on and off condition.

4.   AIR PREHEATER

Cost
Moderate to High

Potential Savings
Moderate

Description
Another means to recover the wasted energy resource in the flue gas is the air preheater, (Figure 10.69) which is used to heat the incoming combustion air. This is an excellent way to capture energy which would otherwise be lost and to put it back to work in the boiler. The rule of thumb which applies to this type of heater is that you can expect a one percent efficiency increase for every 40 degrees that you can decrease the net stack temperature (outlet stack temperature mi-

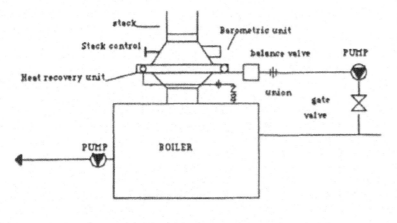

Figure 10.67—Heat reclaimer installed in a hydronic boiler exhaust.

Figure 10.68—Heat reclaimer assembly displayed

nus inlet combustion air temperature). There are several types of air heaters: The tubular type air heater (Figure 10.69) circulates cooler combustion air around hot tubes which are heated by exhaust gases. Several novel but effective designs have emerged for this type of heater; glass tubes have been used for applications where corrosion is a problem. Also, Teflon coated tubes are used if you wish to bring flue gas temperatures below the dew point. Heat pipes are also being introduced as an effective means to capture as much heat as possible. Plate type heat recovery units also offer another excellent way to recover latent heat from the flue gas stream. They are good heat exchange devices and can be constructed of various materials including stainless steel, Teflon coated steel and titanium to name a few.

Advantages and disadvantages

Air heaters are usually very large and the need for large supporting ducting systems to the burners is expensive. They are an excellent way to capture low grade heat. Also, using hot air for combustion raises the furnace temperature and there is a possibility that the refractory may be affected.

5. HEAT RECOVERY USING PLATE EXCHANGERS

Cost
Moderate

Potential Savings
Moderate to High

Description
Plate type heat exchangers (Figures 10.70 and 10.71) can be used to recover waste heat for air preheating as well as feedwater heating. Because the plates can be fabricated from materials such as stainless steel, titanium and Teflon coated steel, units can be designed to bring exhaust gas temperatures well below the acid dew point. This allows the recovery of latent heat from the flue gas, increasing efficiency from 5 to 15 percent. One manufacturer has made a breakthrough with their plate exchanger which is not welded, but torqued together with resilient sections. This type of construction allows the unit to experience thermal expansion without cracking welds or causing other stress related damage.

Advantages/Disadvantages
Materials must withstand acid condensate. Efficiencies approaching 97% possible.

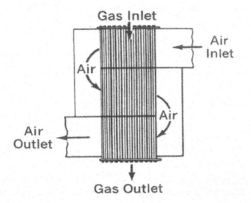

Recuperative Air Preheater

Figure 10.69—Air preheater also brings exhaust temperature down by heating incoming combustion air. The Rule of thumb applies to this method too. Every 40F drop in exhaust temperature increases efficiency and capacity by 1%

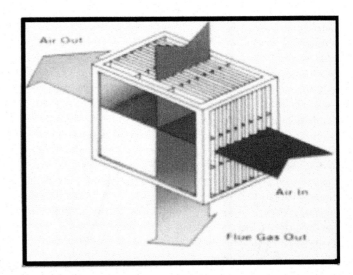

Figure 10.70—Plate type combustion air preheater

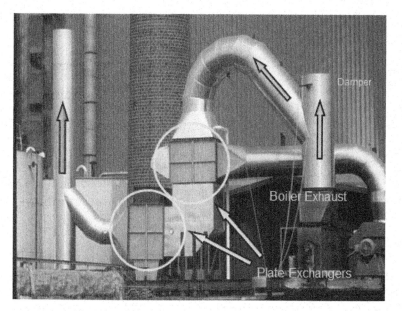

Figure 10.71—Combustion air preheater and feedwater heater unit installed in processing plant boiler exhaust system.

6.   WASTE HEAT RECOVERY BOILERS

Cost
High

Savings
High

Description

There are many processes and applications where heat is wasted that could be captured in a waste heat recovery boiler (Figure 10.72) especially where exhaust temperatures are 500°F and above. This type of system has been with us for a long time, but opportunities still exist to tap this virtually free source of steam and heat. The classic use of waste heat boilers is on reciprocating engines (Figure 10.73 and Figure 10.74) and gas turbines which have exhaust temperatures near the 1,000°F range.

Advantages/Disadvantages

First cost, additional maintenance and operational costs are involved in waste heat boiler planning. They can be an excellent source of free energy.

The "combined cycle" uses high temperature turbine exhaust to generate steam for a second generator utilizing a waste heat recovery boiler. The steam from the turbine exhaust is also used making a very efficient operation.

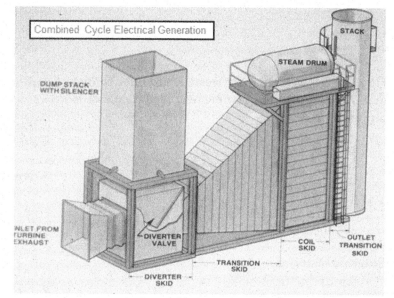

Figure 10.72—Waste Heat boiler that captures the gas turbine exhaust heat to generate steam for a second generator and for other steam applications for Heat and Power.

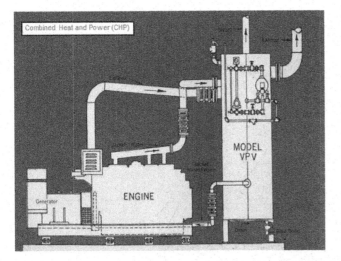

Figure 10.73—This waste heat recovery configuration is suited to hot water applications utilizing jacket water cooling and exhaust gasses for a heat source in addition to the engine-generator set electrical output.

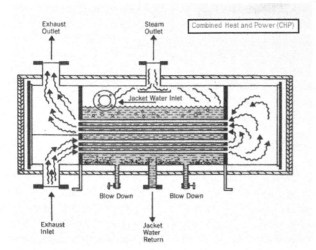

Figure 10.74—An alternate waste heat boiler configuration for applications where overhead clearance is limited.

7.   BLOWDOWN HEAT RECOVERY

Cost
    Low to Moderate

Potential Savings
    Moderate (2% to 5%)

Description
    Blowdown heat recovery system (Figure 10.75). Blowdown water temperature is usually over 300°F depending on Boiler pressure and it contains a lot of energy that can be wasted if it is not put back to work in the boiler somehow. About 15% of the blowdown water will flash to low pressure steam so it is a very good source of low pressure steam and is usually used in the deareator/feedwater heater. This steam can be recovered in a flash tank and the rest of the heat contained in the hot water in a heat exchanger. If steam is not needed, than a simpler heat exchanger recovery configuration can be used to just heat feedwater. (Figure 10.75)

Advantages/Disadvantages
    It's a proven technology.

8.   USE BLOWDOWN TO KEEP IDLE BOILERS WARM

Cost
    Low

Savings
    Low

Description
    If you are using steam to protect idle boilers from freezing or need to keep them warm, hot blowdown water can be diverted through idle boilers. This is a tricky option and boiler water levels and chemical build up must be watched carefully.
    An alternative is to use a heat exchanger with a circulating system or a special pipe in the mud drum for the hot water to circulate through.

Advantages/Disadvantages
    This is a good use of low grade energy. First cost, maintenance and operating costs are involved also.

9.   VENT CONDENSERS

Cost
    Moderate

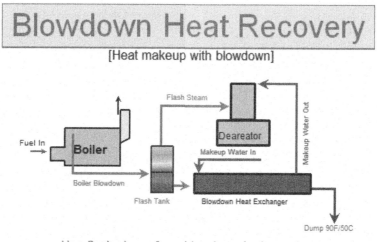

Figure 10.75—Blowdown heat exchanger application. On larger high capacity boilers there is a surface blow system in the steam drum used to discharge contaminated boiler water containing a high level of dissolved solids. This blowdown water contains a lot of heat that when recovered is used to heat the cooler incoming and much less contaminated makeup water.

Potential Savings

Moderate

Description

Wherever steam is being vented to the atmosphere, a vent condenser (Figure 10.76) can be used to capture this high level energy (~1,040 Btu/lb). Not only is the energy in the steam involved, but the energy lost by the boiler and steam system in getting the steam to the point where it is being lost. The boiler may be 70% efficient with a 30% loss just producing the steam that is being lost through a vent. Locations for vent condensers are deareators, feedwater heaters, receivers and condensate system vents. Make up water is usually heated in vent condensers, but domestic hot water and process water can also be heated.(Figure 10.77) A vent condenser can keep this type of loss from large trap leaks to a minimum until repairs can be made.

Advantages/Disadvantages

The design and installation of piping, heat exchangers and pumps are an additional cost and complexity for the steam system. They are proven energy savers.

Figure 10.76—It adds up, a lot of lost energy and pure water when you consider lost steam on a 24/7 basis with an energy value of about 1040 Btu per pound.

Figure 10.77—Vent condenser opportunities. Sometimes there is a lot of pure steam available leaving plants containing a lot of energy.

10. WASTE HEAT FOR PROCESS APPLICATIONS

Cost
Moderate

Potential Savings
High

Description
If waste heat can not be recycled into the boiler process, then it can be transferred for other applications like water heating and warm air for drying processes. The best choice is to return the heat to the boiler itself, to eliminate load matching and other complications.

Advantages/Disadvantages
The complexity of transporting the heat to its point of use may be expensive. When waste heat is returned to the boiler, automatic load matching usually occurs. When using waste heat for process loads, the demand may not always match the boiler loads and special controls may be needed

11. RECOVER STEAM AND HEAT FROM HIGH PRESSURE CONDENSATE RECOVERY SYSTEMS

This is usually a design function but if low pressure steam or hot water is being wasted, it may be a good candidate for heat recovery.

12. INSTALL A CONDENSING ECONOMIZER TO HEAT MAKE-UP WATER

With natural gas fuel a deep economizer Teflon coated tubes, glass, stainless steel or titanium that condenses flue gas water vapor can capture a lot of wasted energy.

13. EXTRACT COMBUSTION AIR AT THE ROOF LEVEL OF THE PLANT

The high locations in boiler rooms can get quite hot. Hot air from this location and from around the exhaust stacks can be drawn into the combustion air system. A percent or so efficiency improvement might be available from such measures.

Section 10-9
Process Heating Systems

CHECKLIST 9
Process Heating Systems

1. Minimize air leakage into the furnace by sealing openings

2. Maintain proper, slightly positive furnace pressure

3. Reduce weight of or eliminate material handling fixtures [reduce thermal mass]

4. Modify the furnace system or use a separate heating system to recover furnace exhaust gas heat. [regenerative heat recovery]

5. Other Check Lists apply to process heating.

1.   MINIMIZE AIR LEAKAGE INTO THE FURNACE BY SEALING OPENINGS

One of your best tools for identifying heat losses are infra red scanning instruments which will identify hot spots that need fixing. High energy losses are produced by high temperature locations.

2.   MAINTAIN PROPER, SLIGHTLY POSITIVE FURNACE PRESSURE

You don't want to be drawing in cold outside air that is heated up then exhausted from the process as wasted heat.

3.   REDUCE WEIGHT OF OR ELIMINATE MATERIAL HANDLING FIXTURES [REDUCE THERMAL MASS]

Invest heat into your product not into heating and cooling of massive support equipment. Use soft ceramic furnace liners instead of fire brick. Keep the massive weights down.

4.   MODIFY THE FURNACE SYSTEM OR USE A SEPARATE HEATING SYSTEM TO RECOVER FURNACE EXHAUST GAS HEAT. [REGENERATIVE HEAT RECOVERY]

Find a place to invest the heat lost from the process like preheating the next batch. Recover part of the furnace exhaust heat for use in lower-temperature processes.

Section 10-10
Building HVAC Projects

CHECKLIST 10
Building HVAC Projects

1. Use heat pumps
2. Condensing furnaces, water heaters & boilers
3. Install temperature setback to operate in off hours.
4. Close shipping/hangar doors when not in use.
5. Reduce make-up air, increase re-circulated air.
6. Convert steam air make-up units to natural gas.
7. Reduce infiltration air.
8. Shut-down exhaust fans during off production periods
9. Install heat recovery ventilators.
10. Install run around heat recovery units between exhaust and intakes

1.   USE HEAT PUMPS WITH SOURCE HEAT FROM WASTE ENERGY OR STEAM FROM BOILER PLANT

Savings
Some of the highest savings possible

Description

Heat pumps have high efficiencies when measured against engines, boilers and turbines. In the example below theh heat pump can bring the apparent efficiency of a boiler up to 128% using waste heat or low pressure steam available from a boiler plant or they can be used to augment old steam systems with air source heat pumps.

Waste heat can be put to work at a higher temperature by the use of heat pumps. One system design circulated 150°F water from a waste heat recovery system throughout a plant using heat pumps to raise the water temperature for local heating and process demands.

The key to how this operates is in the term: Coefficient of Performance (COP), a measure of the ratio of energy output to energy input. Some heat pump units have a COP of over 5 which means the energy to operate them is only about 20% of the energy gained. See Table 10.6 and Figure 10.78.

Advantages/Disadvantages

First cost, operation and maintenance expenses involved. It is a good way to recover low grade waste heat.

Table 10.6—Heat Pumps can raise efficiencies significantly when combined with low grade waste heat from engines, boilers and turbines.

Comparing Efficiencies of Available Heat Sources
Heat Efficiency
Heat pump CoP 4	128%
Condensing Boiler	95%
Coal fired boiler	83% - 90%
Oil fired boiler	84% - 89%
Gas fired boiler	84% – 86%
Gas Turbine	~50%
Diesel	~40% - 50%

Comparing Efficiencies Electrical Sources
Electrical Efficiency
Combined Cycle Gas	50%
Diesel Generator sets	30% - 40%
Gas Turbine	25% - 40%
Generating Station	32%

Table 10.7—Heat Pumps can introduce high efficiencies when combined with other technologies.

Heat Pump Efficiency

Coefficient of Performance = OUTPUT / INPUT

(KW-hr = 3413 BTU)

		Efficiency
CoP	3	= 300%
CoP	4	= 400%
CoP	5	= 500%

Electrical Efficiency   ~ 32%

Table 10.8—The Coefficient of performance (COP) of Heat Pumps has the potential to raise the efficiency of heating systems to over 100%.

COP	BTU Out-Kwh	Cost Per MM BTUs	Efficiency
1	3,413	$ 30.00	32%
2	6,826	$ 15.00	64%
3	10,239	$ 10.00	96%
4	13,652	$ 7.50	128%
5	17,065	$ 6.00	160%

2. CONDENSING FURNACES, WATER HEATERS & BOILERS

Cost
High

Potential Savings
High

Description

A greater percentage of the waste energy in the exit flue gas stream is in the form of superheated steam formed from the combustion of hydrogen in the fuel and other moisture. For every pound of hydrogen that is burned with the fuel, vapor carrying with it a great deal of energy. It is difficult to recover the latent heat from this vapor without dropping the flue gas temperature below the acid dew point temperature (Figure 10.78 and Figure 10.79). The acid dew point temperature is the temperature at which sulfuric acid and other acids form. An efficiency increase of 11% to 15% is possible with most boilers if the flue gasses can be cooled below the dew point. Natural gas, because of its higher hydrogen content, produces higher efficiency increases with this method than oil. When flue gas temperatures can be dropped below the dew point, thereby extracting latent heat from the flue gas moisture, efficiencies above 95% can be expected. Examples of this technology are: pulse combustion, advanced fiber burner designs, special coated plate exchangers, glass tube exchangers, Teflon coated tubes and spray tower direct-contact heat exchangers.

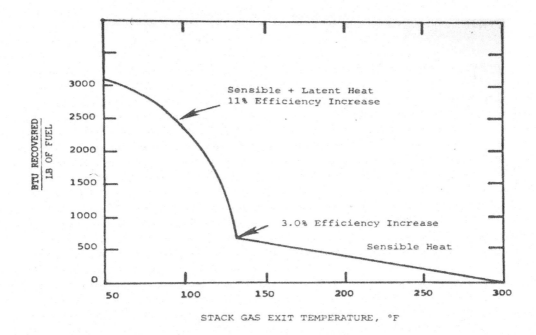

Figure 10.78—Condensing flue gases. As the combustion system exhaust goes down below ~135F the steam in the combustion gases begins to condense out liberating a lot of latent heat into the spray water.

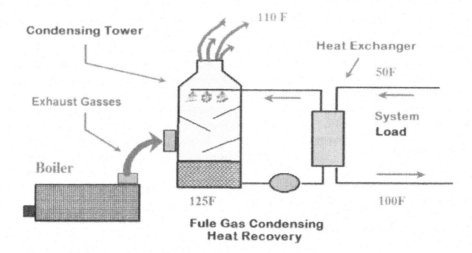

Figure 10.79—Condensing flue gases. As the combustion system exhaust goes down below ~135F the steam in the combustion gases begins to condense out liberating a lot of latent heat into the spray water.

Advantages/Disadvantages

This emerging technology is the only way to get efficiencies in the mid 90% range. Special designs are required, most installations have been highly successful. Acid formation does not seem to be a problem in proper designs

Standard Practices

3. INSTALL TEMPERATURE SETBACK TO OPERATE IN OFF HOURS

4. CLOSE SHIPPING/HANGAR DOORS WHEN NOT IN USE

5. REDUCE MAKE-UP AIR, INCREASE RE-CIRCULATED AIR

6. CONVERT STEAM AIR MAKE-UP UNITS TO NATURAL GAS

7. REDUCE INFILTRATION AIR

8. SHUT-DOWN EXHAUST FANS DURING OFF PRODUCTION PERIODS

9. INSTALL HEAT RECOVERY VENTILATORS

10. INSTALL RUN AROUND HEAT RECOVERY UNITS BETWEEN EXHAUST AND INTAKES
    (See Figure 10.52)

Chapter 11

Boiler Plant Calculations

FINDING EXCESS AIR USING FLUE GAS ANALYSIS OF CARBON DIOXIDE (CO_2) AND OXYGEN (O_2)

The purpose of this section is to demonstrate how to find excess air when either the carbon dioxide (CO_2) or oxygen (O_2) content of the combustion products is known.

The use of a greater amount of combustion air than is necessary constitutes a major waste of energy. Measuring the excess air of a combustion process is necessary to control this loss.

Explanation:

a. In the Combustion Tables find the table for the fuel being fired.

b. In the three left hand columns, find either the carbon dioxide or oxygen content of the combustion products and read the corresponding excess air level.

1. The measured oxygen level in a natural gas boiler is 8%, what is the excess air? See Table 11-1.
   (Ans.: 54%)

2. The measured CO_2 level for a boiler burning natural gas is 9%, what is the excess air? See Table 11-1.
   (Ans.: 28%)

3. The measured CO_2 level for a boiler burning No. 6 fuel oil is 15%, what is the excess air? See Table 11-2.
   (Ans.: 10%)

4. Rule of thumb. Using the oxygen scale, can you develop a rule of thumb relationship for excess air for oxygen levels between 0% and 4%. See Table 11-1.
   (Ans.: Multiply oxygen by 5 to get excess air or divide excess air by 5 to get oxygen)

5. Why is the maximum CO_2 level at 0% excess air higher for no. 6 fuel oil than for Natural Gas? See Tables 11-1 and 11-2.
   (Ans.: There is more carbon in heavier fuels.)

USING CARBON DIOXIDE (CO_2) MEASUREMENTS

1. Find the excess air when firing natural gas for a carbon dioxide (CO_2) level of 10%. See Table 11-1.
   (Ans.: 14.8%)

2. Find the excess air when firing no. 2 fuel oil for a carbon dioxide (CO_2) level of 15%. See Table 11-3.
   (Ans.: 4.6%)

3. Find the excess air when firing no. 6 fuel oil for a carbon dioxide (CO_2) level of 15.5%. See Table 11-1.
   (Ans.: 7.2%)

USING OXYGEN MEASUREMENTS

1. Find the excess air when firing natural gas for an oxygen (O_2) level of 10%. See Table 11-1.
   (Ans. 80 %)

2. Find the excess air when firing no. 6 fuel oil

for an oxygen (O_2) level of 2%. See Table 11-2.
(*Ans.: 9.9%*)

3. Find the excess air when firing bituminous coal for an oxygen (O_2) level of 5%. See Table 11-4.
(*Ans.: 29.9%*)

4. What is the ultimate CO_2 value for the following fuels?

	Table	Page	(Ans.)
a.	Natural Gas 11-1	226	(11.7%)
b.	No. 6 fuel oil 11-2	227	(16.7%)
c.	No 2 fuel oil 11-3	228	(15.7%)
d.	Kerosene 11-5	230	(15.1%)
e.	No. 5 fuel oil 11-6	231	(16.3%)
f.	Anthracite coal 11-7	232	(19.9%)

Example 1.

Fuel Type	No. 6	[See Table 11-2]
Carbon Dioxide	9.1%	
Stack Temperature	560F	[293C]
Air Temperature	60F	[16C]
Net Temperature	500F	[260C]
Efficiency	———	(*Ans.: Loss 20.7% Efficiency 79.3%*)

Example 2.

Fuel Type	No.2	[See Table 11-3]
Oxygen	9%	
Stack Temperature	420F	[216C]
Air Temperature	80F	[27C]
Net Temperature	340F	[171C]

Efficiency   (*Ans.: Loss 16.5% Efficiency 83.5%*)

Example 3.

Fuel Type	Natural Gas	[See Table 11-1]
Carbon Dioxide	8.9%	
Stack Temperature	580F	[304C]
Air Temperature	60F	[160C]
Net Temperature	520F	[271C]

Efficiency   (*Ans.: Loss 20.6% Efficiency 79.4%*)

BOILER STACK GAS LOSSES

The purpose of this section is to learn how to determine combustion efficiency using prepared tables.

Explanation:
Using the Heat Loss Tables:
(a) Look up the fuel type being fired.
(b) Locate the page with the closest exit gas temperature.
(c) Find in the table where the excess air and net stack temperature intersect. This point will be the value of the dry gas heat losses, heat losses due to the formation of moisture (steam) from the hydrogen in the fuel and the losses from the moisture in the fuel.

Determine Heat Loss Efficiency of the following examples using the combustion tables.

Example 4.

Fuel Type	Wood 30% Moisture	
	[See Table 11-8]	
Oxygen	10%	
Stack Temperature	680F	[360C]
Air Temperature	80F	[27C]
Net Temperature	600F	[316C]

Efficiency   *(Ans.: Loss 30% Efficiency 70%)*

CALCULATING FUEL SAVINGS

The purpose of this section is to show how to compute fuel savings based on change in efficiency.

$$\text{Fuel Savings (\%)} = \frac{\text{New Efficiency} - \text{As Found Efficiency}}{\text{New Efficiency}}$$

Example 1.

The as-found efficiency of a Natural Gas fired boiler is 73%, the boiler is tuned up to a new efficiency of 75%. What is the percent fuel savings? _____

(Ans.: 2.67%)

Example 2.

The Maximum Achievable Efficiency for this boiler is about 81.7%. What will the % fuel savings be when this is achieved? ———————

(Ans.: 10.6%)

Example 3.

The Maximum Attainable Efficiency for this boiler is about 86.2%. What will the % fuel savings be when this is achieved? _____

(Ans.: 15.31%)

ESTIMATING LOSSES AND SAVINGS

a. Excess Air & Oxygen Changes
b. Exhaust Temperature
c. Carbon Monoxide & Combustibles
d. Coal Moisture
e. Wood Moisture
f. Bagasse Moisture
g. Condensate ad Feedwater System Temperature Changes
h. Changing rpm of Pumps and Blowers

a. Changing excess air by 10% or oxygen by 2% changes efficiency by 1%.
   Example. Excess air has been improved (reduced) by 15%, what is the estimated efficiency improvement? _____
   (Ans. 1.5%)

b. Each 40F Change in exhaust temperature changes efficiency by 1%.
   Example. Burner problems are causing soot accumulations and the exhaust temperature has risen by 120F. How has this changed efficiency? _____
   (Ans. lower by 3%)

c. 0.5% or 5,000 ppm carbon monoxide changes efficiency by 1%.
   Example. A small boiler is found to have 1/4% CO (2,500 ppm). What is the energy loss? _____
   (Ans. 1/2%) (Note- The legal limit in many areas is 400 ppm CO)

d. A 10% moisture change in coal changes efficiency by 1%.

Example. Heavy rains have soaked the coal supply and surface moisture is over 15%, how will this change the capacity and efficiency of a 200,000 pph (100 ton) boiler? _____

(Ans. efficiency will drop by 1.5%, capacity will drop by 3,000 pph or 1.5 tons/hr)

e. A 10% change in wood/bark moisture changes efficiency by 1.5%
    Example. Moisture has changed from 30% to 60%, how will this change efficiency and output? _____
    (Ans. efficiency and capacity will drop by 4.5%)

f. A 10% change in Bagasse moisture will change efficiency by 3.0%.
    Example. A new system to dry bagasse is being evaluated, it can reduce moisture from 50% to 15%, how will this change efficiency and capacity? _____
    (Ans. Efficiency and Capacity change by 10.5%. Fuel savings if the original efficiency is 65% will be 14%)

g. A 11°F condensate temperature change improves efficiency by 1%.
    Example. Condensate temperature drops from 195F [91C] to 160F [71C] during a cold period. How has this changed system efficiency? _____
    (Ans. 3.2%)

    Example. Condensate temperature suddenly rises from 160F [71C] to 200F [93C], estimate the amount of steam trap leakage if no make up water is being used.
    (Ans. 3.6% of system load)

    Example. High pressure feedwater heater temperature drops from 350F [177C] to 320F [160], how will this change boiler firing rate?
    (Ans. 2.7% more capacity will be needed to heat the feedwater)

    Example. If an economizer raises the feedwater temperature 55F [13C], how much

does this reduce the boiler firing rate?
(Ans. 5%)

h. Power used in pumps and fans varies with the cube of the rpm.
Note: Tests have shown that power reduction may vary with the square of the rpm when a *static head* is involved.

Example. Estimate the power demand reductions produced by the following rpm changes using the "cube" relationship.

Reduction of rpm to   90%   80%   50%
Power Reduction   ___   ___   ___

Ans. (90% Cubed—73% of original power)
      (80% Cubed—51% of original power)
      (50% Cubed—12.5% of original power)

REDUCING EXCESS AIR TO IMPROVE EFFICIENCY

The purpose of this section is to show how changing excess air changes heat losses.

1. A Natural Gas boiler has a stack temperature of 600F [316C]. What will the efficiency increase and fuel savings be if the average excess air is reduced from 90% to 15% (CO_2 from 5.9% to 10.1%)?

As found efficiency	73.8%
Net stack temperature	
600F-80F	520F [271C]
CO_2	5.9%
Efficiency with reduced	
excess air	80.6%
Net stack temperature	
600F-80F	520F [271C]
CO_2	10%
Fuel savings ___%	

(Ans.: 8.4%)

2. What will the fuel savings be with the same excess air reduction with a stack temperature of 300F [149C] instead of 600F [316C]?

As found efficiency	83.3%
Net stack temperature	
300F-80F [149-27C]	220F [104C]
CO_2	5.9%
Efficiency with reduced	
excess air	86.2%
Net stack temperature	
300F-80F [149-27C]	220F [104C]
CO_2	10%
Fuel savings	_____%

(Ans.: 3.4%)

3. What conclusion can you draw from these examples? _____
(Ans.: There is 5% greater opportunity for fuel savings at the higher temperature)

4. A tuneup has reduced the excess air for a natural gas fired boiler from 60% to 12% with the net stack temperature remaining at 600F [316C]. What is the change in Stack Loss? See the Table 11-1.
(Ans.: Losses from 25.8% to 20.7%; a 5.1% change)

5. A tune-up has reduced the excess air for a natural gas fired boiler from 60% to 12% with a net stack temperature remaining at 300F [149C]. What is the change in stack loss? See Table 11-9.
(Ans.: losses from 17.7% to 15.2%; a 2.5% change)

6. A tune-up has reduced the oxygen level in the combustion gases of a wood (40% moisture) fired boiler from 14% to 5% with the net stack temperature remaining at 750F [399C]. What is the change in stack loss? See Table 11-10.
(Ans.: Losses from 49.5% to 28.8%; a 20.7% change)

7. A tune-up has reduced the oxygen level in the combustion gases of a wood (40% moisture) fired boiler from 14% to 5% with the net stack temperature remaining at 400F [204C]. What is the change in stack loss? See Table 11-10.
(Ans.: Losses from 31.8% to 20.8%; a 11.0% change)

8. The measured oxygen level in the flue gases of a boiler firing bituminous coal has increased from 2.5% to 14% with the net stack temperature remaining at 600F [316C]. What is the change in stack loss? See Table 11-4.
(Ans.: losses from 16.5% to 36.4%; a 19.9% increase)

9. The measured oxygen level in the flue gases of a boiler firing number 6 fuel oil has increased from 6% to 15% with the net stack temperature remaining at 700F [371C]. What is the change in stack loss? See Table 11-6.
(Ans.: losses from 22.1% to 46.9%; a 24.8% increase)

10. The measured oxygen level in the flue gases of a boiler firing number 2 fuel oil has increased from 2% to 9% with the net stack temperature remaining at 500F [260C]. What is the change in efficiency? See Table 11-13.
(Ans.: losses from 16% to 21.4%; a 5.4% increase)

AIR INFILTRATION

The infiltration of cold air into a boiler or furnace can represent a serious energy loss. This usually occurs in negative draft boilers and furnaces where dampers, doors and inspection ports are inadvertently left open or no longer form a tight seal after years of use. Because these boilers are usually constructed of brick and are often quite old, cracks from thermal expansion and other reasons allow cold air to be pulled into the boiler to mix with the hot combustion gases.

1. The over-fire draft control system for a natural gas fired boiler with a very tall stack has failed and the change in the negative over-fire draft pressure has caused a problem. The oxygen level in the exhaust combustion gases has risen from 5% to 13%, with the net stack temperature remaining at 500F [260C]. What losses have been created by this condition? See Table 11-1.
    (Ans.: losses from 20.2% to 30.5%; efficiency loss 10.3%)

2. In a boiler burning bituminous coal, the damper doors have become warped and can not be shut. This and other problems have caused the carbon dioxide in the flue gas to drop from 12.3% to 6.6%. With a net flue gas exit temperature of 730F [388C], how is this affecting efficiency? See Table 11-4.
    (Ans.: Losses have gone up from 24% to 40.9% with an efficiency change of 16.9%)

REDUCING STACK TEMPERATURE
TO IMPROVE EFFICIENCY

Because it represents a measure of the unrecoverable heat being wasted to the atmosphere, the temperature of the stack gas is an important indicator of boiler efficiency.

Exit gas temperature, should be measured at the boiler outlet, after the economizer or air heater depending on which is last before the gasses exit to boiler. If the measurements are not typical, being cooled by wall losses of the duct or stack or by the infiltration of cold air, the resulting heat efficiency measurements will be erroneous.

1. In a natural gas fired boiler with 6% flue gas oxygen, recent maintenance has resulted in the net stack temperature going down from 600F [316C] to 400F [204]. How has this affected losses? See Table 11-1 and 11-9.
    (Ans.: losses drop from 23.1% to 18.6% a 4.5% improvement)

2. An economizer has been installed in a boiler firing No. 6 fuel oil which has a combustion gas oxygen level of 5% and the average net stack temperature has dropped from 600F [316C] to 420F [204C]. How has this affected losses?   See tables on pages 12 and 24.
    (Ans.: Flue gas losses have dropped from 18.7% to 14.5% with a 4.2% improvement)

3. The purpose of the following example is to demonstrate the loss of fuel dollars caused by the gradual fouling of heat exchange surfaces.

The stack temperature has risen from 480F [249C] to 680F [360C] since the boiler firesides were last cleaned. How much is this costing in terms of the monthly fuel bill?

Fuel No. 2 Fuel Oil
Monthly fuel costs $400,000

Tuned Up Conditions:

Flue gas oxygen	2%	
Stack temperature	480F	[249C]
Air Temperature	80F	[28C]
Net stack temperature	400F	[204C]
Tuned up Efficiency	**86%**	

Present Conditions:

Flue gas oxygen	2%	
Stack temperature	**680F**	[360C]
Air Temperature	80F	[28C]
Net stack temperature	600F	[316C]
Efficiency	**82.1%**	

1. Percent efficiency loss   _____%
   (Ans.: 3.9%)

2. Percent fuel loss   _____%
   (Ans.: 4.75%)

3. Monthly dollar loss $_____
   (Ans. $19,000/Mo)

4. Annual dollar loss $_____
   (Ans.: $228,000/Yr)

5. Would it be cost effective to clean the heat exchange surfaces?

"It would depend on plant conditions.
(a) A production shutdown could be quite expensive.
(b) If the boiler is not vital and could be spared, it would cost effective"

Table 11-1.

NATURAL GAS—450°F to 600°F

EXIT GAS HEAT LOSSES

% EXCESS AIR	% OXYGEN	% CO2	NET STACK TEMPERATURE DEG F EXIT FLUE GAS TEMPERATURE - COMBUSTION AIR TEMPERATURE															
			450	460	470	480	490	500	510	520	530	540	550	560	570	580	590	600
0.0	0.0	11.7	17.0	17.1	17.3	17.4	17.6	17.8	17.9	18.1	18.3	18.4	18.6	18.7	18.9	19.1	19.2	19.4
2.2	0.5	11.4	17.1	17.3	17.5	17.6	17.8	18.0	18.1	18.3	18.5	18.6	18.8	19.0	19.1	19.3	19.5	19.6
4.4	1.0	11.1	17.3	17.5	17.6	17.8	18.0	18.2	18.3	18.5	18.7	18.8	19.0	19.2	19.3	19.5	19.7	19.9
6.8	1.5	10.9	17.5	17.7	17.8	18.0	18.2	18.4	18.5	18.7	18.9	19.1	19.2	19.4	19.6	19.8	19.9	20.1
9.3	2.0	10.6	17.7	17.9	18.1	18.2	18.4	18.6	18.8	18.9	19.1	19.3	19.5	19.7	19.8	20.0	20.2	20.4
12.0	2.5	10.3	17.9	18.1	18.3	18.5	18.6	18.8	19.0	19.2	19.4	19.6	19.7	19.9	20.1	20.3	20.5	20.7
14.8	3.0	10.0	18.1	18.3	18.5	18.7	18.9	19.1	19.3	19.4	19.6	19.8	20.0	20.2	20.4	20.6	20.8	21.0
17.7	3.5	9.8	18.4	18.6	18.8	18.9	19.1	19.3	19.5	19.7	19.9	20.1	20.3	20.5	20.7	20.9	21.1	21.3
20.8	4.0	9.5	18.6	18.8	19.0	19.2	19.4	19.6	19.8	20.0	20.2	20.4	20.6	20.8	21.0	21.2	21.4	21.6
24.1	4.5	9.2	18.9	19.1	19.3	19.5	19.7	19.9	20.1	20.3	20.5	20.7	20.9	21.1	21.3	21.5	21.7	22.0
27.6	5.0	8.9	19.2	19.4	19.6	19.8	20.0	20.2	20.4	20.6	20.8	21.1	21.3	21.5	21.7	21.9	22.1	22.3
31.4	5.5	8.6	19.5	19.7	19.9	20.1	20.3	20.5	20.8	21.0	21.2	21.4	21.6	21.9	22.1	22.3	22.5	22.7
35.4	6.0	8.4	19.8	20.0	20.2	20.4	20.7	20.9	21.1	21.3	21.6	21.8	22.0	22.2	22.5	22.7	22.9	23.1
39.6	6.5	8.1	20.1	20.3	20.6	20.8	21.0	21.3	21.5	21.7	22.0	22.2	22.4	22.7	22.9	23.1	23.4	23.6
44.2	7.0	7.8	20.5	20.7	21.0	21.2	21.4	21.7	21.9	22.2	22.4	22.6	22.9	23.1	23.4	23.6	23.8	24.1
49.0	7.5	7.5	20.9	21.1	21.4	21.6	21.9	22.1	22.4	22.6	22.9	23.1	23.4	23.6	23.9	24.1	24.4	24.6
54.3	8.0	7.2	21.3	21.6	21.8	22.1	22.3	22.6	22.8	23.1	23.4	23.6	23.9	24.1	24.4	24.7	24.9	25.2
60.0	8.5	7.0	21.7	22.0	22.3	22.6	22.8	23.1	23.4	23.6	23.9	24.2	24.4	24.7	25.0	25.2	25.5	25.8
66.1	9.0	6.7	22.2	22.5	22.8	23.1	23.4	23.6	23.9	24.2	24.5	24.8	25.0	25.3	25.6	25.9	26.2	26.4
72.8	9.5	6.4	22.8	23.1	23.4	23.7	23.9	24.2	24.5	24.8	25.1	25.4	25.7	26.0	26.3	26.6	26.9	27.2
80.0	10.0	6.1	23.4	23.7	24.0	24.3	24.6	24.9	25.2	25.5	25.8	26.1	26.4	26.7	27.0	27.3	27.6	27.9
88.0	10.5	5.9	24.0	24.3	24.6	25.0	25.3	25.6	25.9	26.2	26.6	26.9	27.2	27.5	27.8	28.2	28.5	28.8
96.7	11.0	5.6	24.7	25.0	25.4	25.7	26.0	26.4	26.7	27.1	27.4	27.7	28.1	28.4	28.7	29.1	29.4	29.7
106.3	11.5	5.3	25.5	25.8	26.2	26.5	26.9	27.3	27.6	28.0	28.3	28.7	29.0	29.4	29.7	30.1	30.4	30.8
117.0	12.0	5.0	26.4	26.7	27.1	27.5	27.8	28.2	28.6	29.0	29.3	29.7	30.1	30.4	30.8	31.2	31.6	31.9
128.9	12.5	4.7	27.3	27.7	28.1	28.5	28.9	29.3	29.7	30.1	30.5	30.9	31.3	31.6	32.0	32.4	32.8	33.2
142.3	13.0	4.5	28.4	28.8	29.2	29.7	30.1	30.5	30.9	31.3	31.7	32.2	32.6	33.0	33.4	33.8	34.3	34.7
157.4	13.5	4.2	29.6	30.1	30.5	31.0	31.4	31.9	32.3	32.8	33.2	33.6	34.1	34.5	35.0	35.4	35.9	36.3
174.6	14.0	3.9	31.1	31.5	32.0	32.5	33.0	33.4	33.9	34.4	34.9	35.3	35.8	36.3	36.8	37.2	37.7	38.2
194.4	14.5	3.6	32.7	33.2	33.7	34.2	34.7	35.2	35.8	36.3	36.8	37.3	37.8	38.3	38.8	39.3	39.9	40.4
217.4	15.0	3.3	34.6	35.1	35.7	36.2	36.8	37.4	37.9	38.5	39.0	39.6	40.1	40.7	41.2	41.8	42.3	42.9
244.4	15.5	3.1	36.8	37.4	38.0	38.6	39.2	39.8	40.5	41.1	41.7	42.3	42.9	43.5	44.1	44.7	45.3	45.9
276.7	16.0	2.8	39.5	40.2	40.8	41.5	42.2	42.8	43.5	44.2	44.8	45.5	46.2	46.8	47.5	48.2	48.8	49.5

Table 11-2.

FUEL OIL NO. 6—450°F to 600°F

EXIT GAS HEAT LOSSES

% EXCESS AIR	% OXYGEN	% CO2	NET STACK TEMPERATURE DEG F EXIT FLUE GAS TEMPERATURE - COMBUSTION AIR TEMPERATURE															
			450	460	470	480	490	500	510	520	530	540	550	560	570	580	590	600
0.0	0.0	16.7	12.9	13.1	13.2	13.4	13.6	13.8	13.9	14.1	14.3	14.5	14.7	14.8	15.0	15.2	15.4	15.6
2.3	0.5	16.3	13.1	13.2	13.4	13.6	13.8	14.0	14.2	14.3	14.5	14.7	14.9	15.1	15.3	15.4	15.6	15.8
4.7	1.0	15.9	13.3	13.4	13.6	13.8	14.0	14.2	14.4	14.6	14.8	14.9	15.1	15.3	15.5	15.7	15.9	16.1
7.2	1.5	15.5	13.5	13.6	13.8	14.0	14.2	14.4	14.6	14.8	15.0	15.2	15.4	15.6	15.8	16.0	16.1	16.3
9.9	2.0	15.1	13.7	13.9	14.1	14.3	14.5	14.7	14.9	15.1	15.2	15.4	15.6	15.8	16.0	16.2	16.4	16.6
12.7	2.5	14.7	13.9	14.1	14.3	14.5	14.7	14.9	15.1	15.3	15.5	15.7	15.9	16.1	16.3	16.5	16.7	16.9
15.6	3.0	14.3	14.1	14.3	14.6	14.8	15.0	15.2	15.4	15.6	15.8	16.0	16.2	16.4	16.6	16.8	17.0	17.2
18.8	3.5	13.9	14.4	14.6	14.8	15.0	15.2	15.5	15.7	15.9	16.1	16.3	16.5	16.7	16.9	17.2	17.4	17.6
22.1	4.0	13.5	14.7	14.9	15.1	15.3	15.5	15.8	16.0	16.2	16.4	16.6	16.8	17.1	17.3	17.5	17.7	17.9
25.6	4.5	13.1	14.9	15.2	15.4	15.6	15.8	16.1	16.3	16.5	16.7	17.0	17.2	17.4	17.6	17.9	18.1	18.3
29.3	5.0	12.7	15.2	15.5	15.7	15.9	16.2	16.4	16.6	16.9	17.1	17.3	17.6	17.8	18.0	18.3	18.5	18.7
33.3	5.5	12.3	15.6	15.8	16.0	16.3	16.5	16.8	17.0	17.2	17.5	17.7	18.0	18.2	18.4	18.7	18.9	19.2
37.5	6.0	11.9	15.9	16.2	16.4	16.6	16.9	17.1	17.4	17.6	17.9	18.1	18.4	18.6	18.9	19.1	19.4	19.6
42.0	6.5	11.5	16.3	16.5	16.8	17.0	17.3	17.5	17.8	18.1	18.3	18.6	18.8	19.1	19.3	19.6	19.8	20.1
46.8	7.0	11.1	16.7	16.9	17.2	17.5	17.7	18.0	18.2	18.5	18.8	19.0	19.3	19.6	19.8	20.1	20.4	20.6
52.0	7.5	10.7	17.1	17.4	17.6	17.9	18.2	18.5	18.7	19.0	19.3	19.5	19.8	20.1	20.4	20.6	20.9	21.2
57.5	8.0	10.3	17.5	17.8	18.1	18.4	18.7	19.0	19.2	19.5	19.8	20.1	20.4	20.7	20.9	21.2	21.5	21.8
63.6	8.5	9.9	18.0	18.3	18.6	18.9	19.2	19.5	19.8	20.1	20.4	20.7	21.0	21.3	21.6	21.9	22.1	22.4
70.1	9.0	9.5	18.6	18.9	19.2	19.5	19.8	20.1	20.4	20.7	21.0	21.3	21.6	21.9	22.2	22.5	22.8	23.2
77.1	9.5	9.1	19.1	19.5	19.8	20.1	20.4	20.7	21.1	21.4	21.7	22.0	22.3	22.6	23.0	23.3	23.6	23.9
84.8	10.0	8.7	19.8	20.1	20.4	20.8	21.1	21.4	21.8	22.1	22.4	22.8	23.1	23.4	23.8	24.1	24.4	24.8
93.2	10.5	8.4	20.5	20.8	21.2	21.5	21.9	22.2	22.5	22.9	23.2	23.6	23.9	24.3	24.6	25.0	25.3	25.7
102.4	11.0	8.0	21.2	21.6	22.0	22.3	22.7	23.0	23.4	23.8	24.1	24.5	24.9	25.2	25.6	26.0	26.3	26.7
112.6	11.5	7.6	22.1	22.4	22.8	23.2	23.6	24.0	24.4	24.7	25.1	25.5	25.9	26.3	26.7	27.0	27.4	27.8
123.9	12.0	7.2	23.0	23.4	23.8	24.2	24.6	25.0	25.4	25.8	26.2	26.6	27.0	27.4	27.8	28.2	28.6	29.1
136.5	12.5	6.8	24.0	24.5	24.9	25.3	25.7	26.2	26.6	27.0	27.5	27.9	28.3	28.7	29.2	29.6	30.0	30.4
150.7	13.0	6.4	25.2	25.7	26.1	26.6	27.0	27.5	27.9	28.4	28.8	29.3	29.7	30.2	30.6	31.1	31.6	32.0
166.7	13.5	6.0	26.5	27.0	27.5	28.0	28.5	28.9	29.4	29.9	30.4	30.9	31.4	31.8	32.3	32.8	33.3	33.8
184.9	14.0	5.6	28.1	28.6	29.1	29.6	30.1	30.6	31.2	31.7	32.2	32.7	33.2	33.7	34.2	34.8	35.3	35.8
205.8	14.5	5.2	29.8	30.4	30.9	31.5	32.0	32.6	33.1	33.7	34.2	34.8	35.4	35.9	36.5	37.0	37.6	38.1
230.2	15.0	4.8	31.8	32.4	33.0	33.6	34.2	34.8	35.5	36.1	36.7	37.3	37.9	38.5	39.1	39.7	40.3	40.9
258.8	15.5	4.4	34.3	34.9	35.6	36.2	36.9	37.5	38.2	38.8	39.5	40.2	40.8	41.5	42.1	42.8	43.4	44.1
292.9	16.0	4.0	37.2	37.9	38.6	39.3	40.0	40.8	41.5	42.2	42.9	43.6	44.3	45.1	45.8	46.5	47.2	47.9

Table 11-3.

FUEL OIL NO. 2—300°F to 450°F

EXIT GAS HEAT LOSSES

% EXCESS AIR	% OXYGEN	% CO2	NET STACK TEMPERATURE DEG F EXIT FLUE GAS TEMPERATURE - COMBUSTION AIR TEMPERATURE															
			300	310	320	330	340	350	360	370	380	390	400	410	420	430	440	450
0.0	0.0	15.7	11.5	11.7	11.9	12.1	12.2	12.4	12.6	12.8	12.9	13.1	13.3	13.5	13.7	13.8	14.0	14.2
2.3	0.5	15.3	11.6	11.8	12.0	12.2	12.4	12.6	12.7	12.9	13.1	13.3	13.5	13.7	13.8	14.0	14.2	14.4
4.6	1.0	15.0	11.8	12.0	12.1	12.3	12.5	12.7	12.9	13.1	13.3	13.5	13.6	13.8	14.0	14.2	14.4	14.6
7.1	1.5	14.6	11.9	12.1	12.3	12.5	12.7	12.9	13.1	13.3	13.4	13.6	13.8	14.0	14.2	14.4	14.6	14.8
9.8	2.0	14.2	12.1	12.2	12.4	12.6	12.8	13.0	13.2	13.4	13.6	13.8	14.0	14.2	14.4	14.6	14.8	15.0
12.5	2.5	13.8	12.2	12.4	12.6	12.8	13.0	13.2	13.4	13.6	13.8	14.0	14.2	14.4	14.6	14.8	15.0	15.2
15.5	3.0	13.5	12.4	12.6	12.8	13.0	13.2	13.4	13.6	13.8	14.0	14.2	14.4	14.6	14.8	15.1	15.3	15.5
18.5	3.5	13.1	12.5	12.7	13.0	13.2	13.4	13.6	13.8	14.0	14.2	14.4	14.7	14.9	15.1	15.3	15.5	15.7
21.8	4.0	12.7	12.7	12.9	13.1	13.4	13.6	13.8	14.0	14.2	14.5	14.7	14.9	15.1	15.3	15.6	15.8	16.0
25.3	4.5	12.3	12.9	13.1	13.4	13.6	13.8	14.0	14.2	14.5	14.7	14.9	15.1	15.4	15.6	15.8	16.0	16.3
29.0	5.0	12.0	13.1	13.3	13.6	13.8	14.0	14.3	14.5	14.7	15.0	15.2	15.4	15.6	15.9	16.1	16.3	16.6
32.9	5.5	11.6	13.3	13.6	13.8	14.0	14.3	14.5	14.7	15.0	15.2	15.5	15.7	15.9	16.2	16.4	16.7	16.9
37.0	6.0	11.2	13.5	13.8	14.0	14.3	14.5	14.8	15.0	15.3	15.5	15.8	16.0	16.3	16.5	16.7	17.0	17.2
41.5	6.5	10.8	13.8	14.0	14.3	14.6	14.8	15.1	15.3	15.6	15.8	16.1	16.3	16.6	16.8	17.1	17.4	17.6
46.2	7.0	10.5	14.1	14.3	14.6	14.8	15.1	15.4	15.6	15.9	16.2	16.4	16.7	16.9	17.2	17.5	17.7	18.0
51.4	7.5	10.1	14.3	14.6	14.9	15.2	15.4	15.7	16.0	16.2	16.5	16.8	17.1	17.3	17.6	17.9	18.1	18.4
56.9	8.0	9.7	14.6	14.9	15.2	15.5	15.8	16.1	16.3	16.6	16.9	17.2	17.5	17.7	18.0	18.3	18.6	18.9
62.8	8.5	9.3	15.0	15.3	15.6	15.8	16.1	16.4	16.7	17.0	17.3	17.6	17.9	18.2	18.5	18.8	19.1	19.4
69.2	9.0	9.0	15.3	15.6	15.9	16.2	16.5	16.8	17.2	17.5	17.8	18.1	18.4	18.7	19.0	19.3	19.6	19.9
76.2	9.5	8.6	15.7	16.0	16.3	16.7	17.0	17.3	17.6	17.9	18.3	18.6	18.9	19.2	19.5	19.8	20.2	20.5
83.8	10.0	8.2	16.1	16.5	16.8	17.1	17.5	17.8	18.1	18.5	18.8	19.1	19.4	19.8	20.1	20.4	20.8	21.1
92.1	10.5	7.9	16.6	16.9	17.3	17.6	18.0	18.3	18.7	19.0	19.4	19.7	20.1	20.4	20.8	21.1	21.5	21.8
101.2	11.0	7.5	17.1	17.5	17.8	18.2	18.6	18.9	19.3	19.6	20.0	20.4	20.7	21.1	21.5	21.8	22.2	22.6
111.3	11.5	7.1	17.7	18.0	18.4	18.8	19.2	19.6	20.0	20.3	20.7	21.1	21.5	21.9	22.2	22.6	23.0	23.4
122.5	12.0	6.7	18.3	18.7	19.1	19.5	19.9	20.3	20.7	21.1	21.5	21.9	22.3	22.7	23.1	23.5	23.9	24.3
134.9	12.5	6.4	19.0	19.4	19.8	20.3	20.7	21.1	21.5	22.0	22.4	22.8	23.2	23.7	24.1	24.5	24.9	25.4
148.9	13.0	6.0	19.8	20.2	20.7	21.1	21.6	22.0	22.5	22.9	23.4	23.8	24.3	24.7	25.2	25.6	26.1	26.5
164.7	13.5	5.6	20.6	21.1	21.6	22.1	22.6	23.1	23.5	24.0	24.5	25.0	25.5	25.9	26.4	26.9	27.4	27.9
182.7	14.0	5.2	21.7	22.2	22.7	23.2	23.7	24.2	24.8	25.3	25.8	26.3	26.8	27.3	27.8	28.4	28.9	29.4
203.4	14.5	4.9	22.8	23.4	23.9	24.5	25.0	25.6	26.2	26.7	27.3	27.8	28.4	28.9	29.5	30.0	30.6	31.2
227.5	15.0	4.5	24.2	24.8	25.4	26.0	26.6	27.2	27.8	28.4	29.0	29.6	30.2	30.8	31.4	32.0	32.6	33.2
255.8	15.5	4.1	25.8	26.5	27.1	27.8	28.4	29.1	29.7	30.4	31.0	31.7	32.3	33.0	33.7	34.3	35.0	35.6
289.5	16.0	3.7	27.7	28.5	29.2	29.9	30.6	31.3	32.0	32.8	33.5	34.2	34.9	35.6	36.4	37.1	37.8	38.5

Table 11-4.

BITUMINOUS COAL—600°F to 750°F

% EXCESS AIR	% OXYGEN	% CO2	NET STACK TEMPERATURE DEG F EXIT FLUE GAS TEMPERATURE - COMBUSTION AIR TEMPERATURE															
			600	610	620	630	640	650	660	670	680	690	700	710	720	730	740	750
0.0	0.0	18.5	15.0	15.2	15.4	15.6	15.8	16.0	16.2	16.4	16.6	16.8	16.9	17.1	17.3	17.5	17.7	17.9
2.3	0.5	18.1	15.3	15.5	15.7	15.9	16.1	16.3	16.5	16.7	16.9	17.1	17.2	17.4	17.6	17.8	18.0	18.2
4.8	1.0	17.6	15.6	15.8	16.0	16.2	16.4	16.6	16.8	17.0	17.2	17.4	17.6	17.8	18.0	18.2	18.4	18.6
7.4	1.5	17.2	15.9	16.1	16.3	16.5	16.7	16.9	17.1	17.3	17.5	17.7	17.9	18.1	18.3	18.5	18.7	18.9
10.1	2.0	16.7	16.2	16.4	16.6	16.8	17.0	17.2	17.4	17.6	17.8	18.0	18.3	18.5	18.7	18.9	19.1	19.3
13.0	2.5	16.3	16.5	16.7	16.9	17.1	17.3	17.5	17.8	18.0	18.2	18.4	18.6	18.8	19.1	19.3	19.5	19.7
16.0	3.0	15.9	16.8	17.0	17.3	17.5	17.7	17.9	18.1	18.4	18.6	18.8	19.0	19.2	19.5	19.7	19.9	20.1
19.2	3.5	15.4	17.2	17.4	17.6	17.8	18.1	18.3	18.5	18.8	19.0	19.2	19.4	19.7	19.9	20.1	20.3	20.6
22.6	4.0	15.0	17.5	17.8	18.0	18.2	18.5	18.7	18.9	19.2	19.4	19.6	19.9	20.1	20.3	20.6	20.8	21.0
26.1	4.5	14.5	17.9	18.2	18.4	18.7	18.9	19.1	19.4	19.6	19.9	20.1	20.3	20.6	20.8	21.1	21.3	21.5
29.9	5.0	14.1	18.4	18.6	18.9	19.1	19.4	19.6	19.8	20.1	20.3	20.6	20.8	21.1	21.3	21.6	21.8	22.1
34.0	5.5	13.7	18.8	19.1	19.3	19.6	19.8	20.1	20.3	20.6	20.9	21.1	21.4	21.6	21.9	22.1	22.4	22.6
38.3	6.0	13.2	19.3	19.6	19.8	20.1	20.4	20.6	20.9	21.1	21.4	21.7	21.9	22.2	22.5	22.7	23.0	23.2
42.9	6.5	12.8	19.8	20.1	20.4	20.6	20.9	21.2	21.4	21.7	22.0	22.3	22.5	22.8	23.1	23.3	23.6	23.9
47.8	7.0	12.3	20.4	20.7	20.9	21.2	21.5	21.8	22.1	22.3	22.6	22.9	23.2	23.5	23.7	24.0	24.3	24.6
53.1	7.5	11.9	21.0	21.3	21.5	21.8	22.1	22.4	22.7	23.0	23.3	23.6	23.9	24.2	24.4	24.7	25.0	25.3
58.8	8.0	11.5	21.6	21.9	22.2	22.5	22.8	23.1	23.4	23.7	24.0	24.3	24.6	24.9	25.2	25.5	25.8	26.1
65.0	8.5	11.0	22.3	22.6	22.9	23.2	23.5	23.9	24.2	24.5	24.8	25.1	25.4	25.7	26.0	26.4	26.7	27.0
71.6	9.0	10.6	23.0	23.4	23.7	24.0	24.3	24.7	25.0	25.3	25.6	26.0	26.3	26.6	26.9	27.3	27.6	27.9
78.8	9.5	10.1	23.9	24.2	24.5	24.9	25.2	25.5	25.9	26.2	26.6	26.9	27.2	27.6	27.9	28.2	28.6	28.9
86.7	10.0	9.7	24.7	25.1	25.4	25.8	26.2	26.5	26.9	27.2	27.6	27.9	28.3	28.6	29.0	29.3	29.7	30.0
95.3	10.5	9.3	25.7	26.1	26.4	26.8	27.2	27.6	27.9	28.3	28.7	29.0	29.4	29.8	30.1	30.5	30.9	31.2
104.7	11.0	8.8	26.8	27.2	27.6	27.9	28.3	28.7	29.1	29.5	29.9	30.3	30.6	31.0	31.4	31.8	32.2	32.6
115.1	11.5	8.4	28.0	28.4	28.8	29.2	29.6	30.0	30.4	30.8	31.2	31.6	32.0	32.4	32.8	33.2	33.7	34.1
126.6	12.0	7.9	29.3	29.7	30.1	30.6	31.0	31.4	31.8	32.3	32.7	33.1	33.6	34.0	34.4	34.8	35.3	35.7
139.5	12.5	7.5	30.7	31.2	31.6	32.1	32.5	33.0	33.5	33.9	34.4	34.8	35.3	35.7	36.2	36.6	37.1	37.5
154.0	13.0	7.0	32.4	32.9	33.3	33.8	34.3	34.8	35.3	35.7	36.2	36.7	37.2	37.7	38.2	38.6	39.1	39.6
170.3	13.5	6.6	34.3	34.8	35.3	35.8	36.3	36.8	37.3	37.8	38.3	38.9	39.4	39.9	40.4	40.9	41.4	41.9
188.9	14.0	6.2	36.4	36.9	37.5	38.0	38.6	39.1	39.7	40.2	40.8	41.3	41.9	42.4	43.0	43.5	44.1	44.6
210.3	14.5	5.7	38.9	39.4	40.0	40.6	41.2	41.8	42.4	43.0	43.6	44.1	44.7	45.3	45.9	46.5	47.1	47.7
235.1	15.0	5.3	41.7	42.4	43.0	43.6	44.3	44.9	45.5	46.2	46.8	47.5	48.1	48.7	49.4	50.0	50.6	51.3
264.4	15.5	4.8	45.1	45.8	46.5	47.2	47.9	48.6	49.3	50.0	50.7	51.4	52.1	52.7	53.4	54.1	54.8	55.5
299.2	16.0	4.4	49.2	50.0	50.7	51.5	52.3	53.0	53.8	54.5	55.3	56.1	56.8	57.6	58.3	59.1	59.9	60.6

Table 11-5.

KEROSENE—450°F to 600°F

EXIT GAS HEAT LOSSES

% EXCESS AIR	% OXYGEN	% CO2	NET STACK TEMPERATURE DEG F EXIT FLUE GAS TEMPERATURE - COMBUSTION AIR TEMPERATURE															
			450	460	470	480	490	500	510	520	530	540	550	560	570	580	590	600
0.0	0.0	15.1	14.3	14.5	14.7	14.8	15.0	15.2	15.4	15.5	15.7	15.9	16.0	16.2	16.4	16.6	16.7	16.9
2.2	0.5	14.8	14.5	14.7	14.8	15.0	15.2	15.4	15.6	15.7	15.9	16.1	16.3	16.4	16.6	16.8	17.0	17.1
4.6	1.0	14.4	14.7	14.9	15.0	15.2	15.4	15.6	15.8	15.9	16.1	16.3	16.5	16.7	16.8	17.0	17.2	17.4
7.1	1.5	14.0	14.9	15.1	15.3	15.4	15.6	15.8	16.0	16.2	16.4	16.5	16.7	16.9	17.1	17.3	17.5	17.6
9.7	2.0	13.7	15.1	15.3	15.5	15.7	15.8	16.0	16.2	16.4	16.6	16.8	17.0	17.2	17.4	17.5	17.7	17.9
12.5	2.5	13.3	15.3	15.5	15.7	15.9	16.1	16.3	16.5	16.7	16.9	17.1	17.2	17.4	17.6	17.8	18.0	18.2
15.4	3.0	13.0	15.5	15.7	15.9	16.1	16.3	16.5	16.7	16.9	17.1	17.3	17.5	17.7	17.9	18.1	18.3	18.5
18.4	3.5	12.6	15.8	16.0	16.2	16.4	16.6	16.8	17.0	17.2	17.4	17.6	17.8	18.0	18.2	18.4	18.6	18.9
21.7	4.0	12.2	16.0	16.3	16.5	16.7	16.9	17.1	17.3	17.5	17.7	17.9	18.1	18.4	18.6	18.8	19.0	19.2
25.1	4.5	11.9	16.3	16.5	16.8	17.0	17.2	17.4	17.6	17.8	18.0	18.3	18.5	18.7	18.9	19.1	19.3	19.6
28.8	5.0	11.5	16.6	16.8	17.1	17.3	17.5	17.7	17.9	18.2	18.4	18.6	18.8	19.1	19.3	19.5	19.7	20.0
32.6	5.5	11.2	16.9	17.2	17.4	17.6	17.8	18.1	18.3	18.5	18.8	19.0	19.2	19.4	19.7	19.9	20.1	20.4
36.8	6.0	10.8	17.3	17.5	17.7	18.0	18.2	18.4	18.7	18.9	19.1	19.4	19.6	19.9	20.1	20.3	20.6	20.8
41.2	6.5	10.4	17.6	17.9	18.1	18.3	18.6	18.8	19.1	19.3	19.6	19.8	20.1	20.3	20.5	20.8	21.0	21.3
45.9	7.0	10.1	18.0	18.2	18.5	18.7	19.0	19.3	19.5	19.8	20.0	20.3	20.5	20.8	21.0	21.3	21.5	21.8
51.0	7.5	9.7	18.4	18.7	18.9	19.2	19.4	19.7	20.0	20.2	20.5	20.8	21.0	21.3	21.5	21.8	22.1	22.3
56.5	8.0	9.4	18.8	19.1	19.4	19.6	19.9	20.2	20.5	20.7	21.0	21.3	21.6	21.8	22.1	22.4	22.6	22.9
62.4	8.5	9.0	19.3	19.6	19.9	20.2	20.4	20.7	21.0	21.3	21.6	21.8	22.1	22.4	22.7	23.0	23.3	23.5
68.8	9.0	8.6	19.8	20.1	20.4	20.7	21.0	21.3	21.6	21.9	22.2	22.5	22.8	23.1	23.3	23.6	23.9	24.2
75.7	9.5	8.3	20.4	20.7	21.0	21.3	21.6	21.9	22.2	22.5	22.8	23.1	23.4	23.7	24.1	24.4	24.7	25.0
83.2	10.0	7.9	21.0	21.3	21.6	21.9	22.3	22.6	22.9	23.2	23.5	23.9	24.2	24.5	24.8	25.1	25.5	25.8
91.5	10.5	7.6	21.7	22.0	22.3	22.7	23.0	23.3	23.7	24.0	24.3	24.7	25.0	25.3	25.7	26.0	26.3	26.7
100.6	11.0	7.2	22.4	22.7	23.1	23.4	23.8	24.1	24.5	24.8	25.2	25.5	25.9	26.2	26.6	26.9	27.3	27.7
110.6	11.5	6.8	23.2	23.6	23.9	24.3	24.7	25.0	25.4	25.8	26.1	26.5	26.9	27.3	27.6	28.0	28.4	28.7
121.7	12.0	6.5	24.1	24.5	24.9	25.3	25.7	26.0	26.4	26.8	27.2	27.6	28.0	28.4	28.8	29.2	29.5	29.9
134.0	12.5	6.1	25.1	25.5	25.9	26.3	26.7	27.2	27.6	28.0	28.4	28.8	29.2	29.6	30.0	30.5	30.9	31.3
147.9	13.0	5.8	26.2	26.7	27.1	27.5	28.0	28.4	28.9	29.3	29.7	30.2	30.6	31.0	31.5	31.9	32.3	32.8
163.6	13.5	5.4	27.5	28.0	28.4	28.9	29.4	29.8	30.3	30.8	31.2	31.7	32.2	32.6	33.1	33.6	34.0	34.5
181.5	14.0	5.0	29.0	29.5	30.0	30.5	31.0	31.5	32.0	32.5	33.0	33.5	34.0	34.5	35.0	35.5	36.0	36.5
202.1	14.5	4.7	30.7	31.2	31.7	32.3	32.8	33.4	33.9	34.4	35.0	35.5	36.0	36.6	37.1	37.6	38.2	38.7
226.0	15.0	4.3	32.7	33.2	33.8	34.4	35.0	35.5	36.1	36.7	37.3	37.9	38.4	39.0	39.6	40.2	40.8	41.3
254.1	15.5	4.0	35.0	35.6	36.2	36.9	37.5	38.1	38.8	39.4	40.0	40.7	41.3	41.9	42.6	43.2	43.8	44.4
287.6	16.0	3.6	37.8	38.5	39.2	39.9	40.6	41.2	41.9	42.6	43.3	44.0	44.7	45.4	46.1	46.8	47.5	48.2

Table 11-6.

FUEL OIL #5—750°F to 900°F

EXIT GAS HEAT LOSSES

% EXCESS AIR	% OXYGEN	% CO2	750	760	770	780	790	800	810	820	830	840	850	860	870	880	890	900
0.0	0.0	16.3	18.5	18.7	18.9	19.0	19.2	19.4	19.6	19.7	19.9	20.1	20.3	20.4	20.6	20.8	21.0	21.2
2.3	0.5	15.9	18.8	19.0	19.2	19.4	19.5	19.7	19.9	20.1	20.3	20.4	20.6	20.8	21.0	21.2	21.3	21.5
4.7	1.0	15.5	19.1	19.3	19.5	19.7	19.9	20.1	20.2	20.4	20.6	20.8	21.0	21.2	21.3	21.5	21.7	21.9
7.2	1.5	15.1	19.5	19.7	19.8	20.0	20.2	20.4	20.6	20.8	21.0	21.2	21.4	21.5	21.7	21.9	22.1	22.3
9.8	2.0	14.7	19.8	20.0	20.2	20.4	20.6	20.8	21.0	21.2	21.4	21.6	21.8	21.9	22.1	22.3	22.5	22.7
12.6	2.5	14.4	20.2	20.4	20.6	20.8	21.0	21.2	21.4	21.6	21.8	22.0	22.2	22.4	22.6	22.8	23.0	23.2
15.6	3.0	14.0	20.6	20.8	21.0	21.2	21.4	21.6	21.8	22.0	22.2	22.4	22.6	22.8	23.0	23.2	23.4	23.6
18.7	3.5	13.6	21.0	21.2	21.4	21.6	21.8	22.0	22.3	22.5	22.7	22.9	23.1	23.3	23.5	23.7	23.9	24.1
22.0	4.0	13.2	21.4	21.7	21.9	22.1	22.3	22.5	22.7	22.9	23.2	23.4	23.6	23.8	24.0	24.2	24.5	24.7
25.5	4.5	12.8	21.9	22.1	22.3	22.6	22.8	23.0	23.2	23.5	23.7	23.9	24.1	24.3	24.6	24.8	25.0	25.2
29.2	5.0	12.4	22.4	22.6	22.9	23.1	23.3	23.5	23.8	24.0	24.2	24.5	24.7	24.9	25.1	25.4	25.6	25.8
33.1	5.5	12.0	22.9	23.2	23.4	23.6	23.9	24.1	24.3	24.6	24.8	25.0	25.3	25.5	25.7	26.0	26.2	26.5
37.3	6.0	11.6	23.5	23.7	24.0	24.2	24.5	24.7	24.9	25.2	25.4	25.7	25.9	26.2	26.4	26.6	26.9	27.1
41.8	6.5	11.3	24.1	24.3	24.6	24.8	25.1	25.3	25.6	25.8	26.1	26.3	26.6	26.8	27.1	27.3	27.6	27.8
46.6	7.0	10.9	24.7	25.0	25.3	25.5	25.8	26.0	26.3	26.5	26.8	27.1	27.3	27.6	27.8	28.1	28.4	28.6
51.7	7.5	10.5	25.4	25.7	26.0	26.2	26.5	26.8	27.0	27.3	27.6	27.8	28.1	28.4	28.6	28.9	29.2	29.5
57.3	8.0	10.1	26.2	26.4	26.7	27.0	27.3	27.6	27.8	28.1	28.4	28.7	29.0	29.2	29.5	29.8	30.1	30.3
63.3	8.5	9.7	27.0	27.3	27.6	27.8	28.1	28.4	28.7	29.0	29.3	29.6	29.9	30.2	30.4	30.7	31.0	31.3
69.7	9.0	9.3	27.8	28.1	28.4	28.7	29.0	29.3	29.7	30.0	30.3	30.6	30.9	31.2	31.5	31.8	32.1	32.4
76.7	9.5	8.9	28.8	29.1	29.4	29.7	30.0	30.4	30.7	31.0	31.3	31.6	31.9	32.2	32.6	32.9	33.2	33.5
84.4	10.0	8.5	29.8	30.2	30.5	30.8	31.1	31.5	31.8	32.1	32.4	32.8	33.1	33.4	33.8	34.1	34.4	34.7
92.8	10.5	8.2	31.0	31.3	31.6	32.0	32.3	32.7	33.0	33.4	33.7	34.0	34.4	34.7	35.1	35.4	35.7	36.1
102.0	11.0	7.8	32.2	32.6	32.9	33.3	33.6	34.0	34.4	34.7	35.1	35.4	35.8	36.2	36.5	36.9	37.2	37.6
112.1	11.5	7.4	33.6	34.0	34.3	34.7	35.1	35.5	35.8	36.2	36.6	37.0	37.4	37.7	38.1	38.5	38.9	39.2
123.4	12.0	7.0	35.1	35.5	35.9	36.3	36.7	37.1	37.5	37.9	38.3	38.7	39.1	39.5	39.9	40.3	40.7	41.1
135.9	12.5	6.6	36.8	37.2	37.7	38.1	38.5	38.9	39.3	39.8	40.2	40.6	41.0	41.4	41.9	42.3	42.7	43.1
150.0	13.0	6.2	38.7	39.2	39.6	40.1	40.5	41.0	41.4	41.9	42.3	42.8	43.2	43.6	44.1	44.5	45.0	45.4
165.9	13.5	5.8	40.9	41.4	41.9	42.3	42.8	43.3	43.8	44.2	44.7	45.2	45.7	46.1	46.6	47.1	47.6	48.0
184.0	14.0	5.4	43.4	43.9	44.4	44.9	45.4	45.9	46.5	47.0	47.5	48.0	48.5	49.0	49.5	50.0	50.5	51.0
204.9	14.5	5.0	46.3	46.8	47.4	47.9	48.5	49.0	49.6	50.1	50.7	51.2	51.7	52.3	52.8	53.4	53.9	54.5
229.1	15.0	4.7	49.6	50.2	50.8	51.4	52.0	52.6	53.2	53.8	54.4	55.0	55.5	56.1	56.7	57.3	57.9	58.5
257.6	15.5	4.3	53.6	54.2	54.9	55.5	56.2	56.8	57.5	58.1	58.7	59.4	60.0	60.7	61.3	62.0	62.6	63.3
291.5	16.0	3.9	58.3	59.1	59.8	60.5	61.2	61.9	62.6	63.3	64.0	64.7	65.4	66.1	66.8	67.5	68.3	69.0

Table 11-7.

ANTHRACITE COAL—300°F to 450°F

EXIT GAS HEAT LOSSES

% EXCESS AIR	% OXYGEN	% CO2	NET STACK TEMPERATURE DEG F EXIT FLUE GAS TEMPERATURE - COMBUSTION AIR TEMPERATURE															
			300	310	320	330	340	350	360	370	380	390	400	410	420	430	440	450
0.0	0.0	19.9	8.2	8.4	8.6	8.8	9.0	9.2	9.4	9.6	9.8	10.0	10.2	10.4	10.6	10.8	11.0	11.2
2.4	0.5	19.4	8.3	8.5	8.7	8.9	9.1	9.3	9.5	9.7	9.9	10.1	10.3	10.6	10.8	11.0	11.2	11.4
4.9	1.0	19.0	8.5	8.7	8.9	9.1	9.3	9.5	9.7	9.9	10.1	10.3	10.5	10.7	11.0	11.2	11.4	11.6
7.5	1.5	18.5	8.6	8.8	9.0	9.2	9.5	9.7	9.9	10.1	10.3	10.5	10.7	10.9	11.2	11.4	11.6	11.8
10.3	2.0	18.0	8.8	9.0	9.2	9.4	9.6	9.8	10.1	10.3	10.5	10.7	10.9	11.2	11.4	11.6	11.8	12.0
13.2	2.5	17.5	8.9	9.1	9.4	9.6	9.8	10.0	10.3	10.5	10.7	10.9	11.2	11.4	11.6	11.8	12.1	12.3
16.3	3.0	17.1	9.1	9.3	9.6	9.8	10.0	10.2	10.5	10.7	10.9	11.2	11.4	11.6	11.9	12.1	12.3	12.5
19.5	3.5	16.6	9.3	9.5	9.8	10.0	10.2	10.5	10.7	10.9	11.2	11.4	11.6	11.9	12.1	12.3	12.6	12.8
23.0	4.0	16.1	9.5	9.7	10.0	10.2	10.4	10.7	10.9	11.2	11.4	11.7	11.9	12.1	12.4	12.6	12.9	13.1
26.6	4.5	15.6	9.7	9.9	10.2	10.4	10.7	10.9	11.2	11.4	11.7	11.9	12.2	12.4	12.7	12.9	13.2	13.4
30.5	5.0	15.2	9.9	10.2	10.4	10.7	10.9	11.2	11.4	11.7	12.0	12.2	12.5	12.7	13.0	13.2	13.5	13.8
34.6	5.5	14.7	10.1	10.4	10.7	10.9	11.2	11.5	11.7	12.0	12.3	12.5	12.8	13.0	13.3	13.6	13.8	14.1
39.0	6.0	14.2	10.4	10.7	10.9	11.2	11.5	11.8	12.0	12.3	12.6	12.8	13.1	13.4	13.7	13.9	14.2	14.5
43.6	6.5	13.7	10.7	10.9	11.2	11.5	11.8	12.1	12.3	12.6	12.9	13.2	13.5	13.8	14.0	14.3	14.6	14.9
48.7	7.0	13.3	10.9	11.2	11.5	11.8	12.1	12.4	12.7	13.0	13.3	13.6	13.9	14.1	14.4	14.7	15.0	15.3
54.0	7.5	12.8	11.2	11.5	11.8	12.2	12.5	12.8	13.1	13.4	13.7	14.0	14.3	14.6	14.9	15.2	15.5	15.8
59.8	8.0	12.3	11.6	11.9	12.2	12.5	12.8	13.1	13.5	13.8	14.1	14.4	14.7	15.0	15.3	15.6	16.0	16.3
66.1	8.5	11.8	11.9	12.3	12.6	12.9	13.2	13.6	13.9	14.2	14.5	14.9	15.2	15.5	15.8	16.2	16.5	16.8
72.8	9.0	11.4	12.3	12.7	13.0	13.3	13.7	14.0	14.3	14.7	15.0	15.4	15.7	16.0	16.4	16.7	17.0	17.4
80.2	9.5	10.9	12.7	13.1	13.4	13.8	14.1	14.5	14.8	15.2	15.6	15.9	16.3	16.6	17.0	17.3	17.7	18.0
88.2	10.0	10.4	13.2	13.6	13.9	14.3	14.7	15.0	15.4	15.8	16.1	16.5	16.9	17.2	17.6	18.0	18.3	18.7
96.9	10.5	10.0	13.7	14.1	14.5	14.9	15.2	15.6	16.0	16.4	16.8	17.2	17.5	17.9	18.3	18.7	19.1	19.5
106.5	11.0	9.5	14.3	14.7	15.1	15.5	15.9	16.3	16.7	17.1	17.5	17.9	18.3	18.7	19.1	19.5	19.9	20.3
117.1	11.5	9.0	14.9	15.3	15.7	16.1	16.6	17.0	17.4	17.8	18.2	18.7	19.1	19.5	19.9	20.4	20.8	21.2
128.8	12.0	8.5	15.5	16.0	16.4	16.9	17.3	17.8	18.2	18.7	19.1	19.5	20.0	20.4	20.9	21.3	21.8	22.2
141.9	12.5	8.1	16.3	16.8	17.2	17.7	18.2	18.7	19.1	19.6	20.1	20.5	21.0	21.5	21.9	22.4	22.9	23.4
156.6	13.0	7.6	17.2	17.7	18.2	18.7	19.2	19.6	20.1	20.6	21.1	21.6	22.1	22.6	23.1	23.6	24.1	24.6
173.2	13.5	7.1	18.1	18.7	19.2	19.7	20.2	20.8	21.3	21.8	22.4	22.9	23.4	24.0	24.5	25.0	25.6	26.1
192.1	14.0	6.6	19.2	19.8	20.4	20.9	21.5	22.1	22.6	23.2	23.8	24.3	24.9	25.5	26.0	26.6	27.2	27.7
213.9	14.5	6.2	20.5	21.1	21.7	22.3	22.9	23.6	24.2	24.8	25.4	26.0	26.6	27.2	27.8	28.4	29.0	29.7
239.2	15.0	5.7	22.0	22.7	23.3	24.0	24.6	25.3	26.0	26.6	27.3	27.9	28.6	29.3	29.9	30.6	31.2	31.9
268.9	15.5	5.2	23.8	24.5	25.2	25.9	26.6	27.3	28.1	28.8	29.5	30.2	30.9	31.7	32.4	33.1	33.8	34.5
304.3	16.0	4.7	25.9	26.7	27.4	28.2	29.0	29.8	30.6	31.4	32.2	33.0	33.7	34.5	35.3	36.1	36.9	37.7

Table 11-8.

WOOD 30% MOISTURE—450°F to 600°F

EXIT GAS HEAT LOSSES

% EXCESS AIR	% OXYGEN	% CO2	NET STACK TEMPERATURE DEG F — EXIT FLUE GAS TEMPERATURE - COMBUSTION AIR TEMPERATURE															
			600	610	620	630	640	650	660	670	680	690	700	710	720	730	740	750
0.0	0.0	20.0	21.1	21.3	21.5	21.6	21.8	22.0	22.2	22.3	22.5	22.7	22.9	23.0	23.2	23.4	23.6	23.8
2.4	0.5	19.5	21.3	21.5	21.7	21.9	22.1	22.2	22.4	22.6	22.8	23.0	23.1	23.3	23.5	23.7	23.9	24.1
4.9	1.0	19.0	21.6	21.8	22.0	22.1	22.3	22.5	22.7	22.9	23.1	23.3	23.4	23.6	23.8	24.0	24.2	24.4
7.5	1.5	18.6	21.9	22.0	22.2	22.4	22.6	22.8	23.0	23.2	23.4	23.6	23.8	23.9	24.1	24.3	24.5	24.7
10.3	2.0	18.1	22.1	22.3	22.5	22.7	22.9	23.1	23.3	23.5	23.7	23.9	24.1	24.3	24.5	24.7	24.9	25.0
13.2	2.5	17.6	22.4	22.6	22.8	23.0	23.2	23.4	23.6	23.8	24.0	24.2	24.4	24.6	24.8	25.0	25.2	25.4
16.3	3.0	17.1	22.7	22.9	23.1	23.3	23.6	23.8	24.0	24.2	24.4	24.6	24.8	25.0	25.2	25.4	25.6	25.8
19.5	3.5	16.7	23.1	23.3	23.5	23.7	23.9	24.1	24.3	24.5	24.7	25.0	25.2	25.4	25.6	25.8	26.0	26.2
23.0	4.0	16.2	23.4	23.6	23.8	24.1	24.3	24.5	24.7	24.9	25.1	25.4	25.6	25.8	26.0	26.2	26.4	26.6
26.6	4.5	15.7	23.8	24.0	24.2	24.4	24.7	24.9	25.1	25.3	25.6	25.8	26.0	26.2	26.4	26.7	26.9	27.1
30.5	5.0	15.2	24.2	24.4	24.6	24.9	25.1	25.3	25.5	25.8	26.0	26.2	26.5	26.7	26.9	27.1	27.4	27.6
34.6	5.5	14.8	24.6	24.8	25.1	25.3	25.5	25.8	26.0	26.2	26.5	26.7	26.9	27.2	27.4	27.6	27.9	28.1
39.0	6.0	14.3	25.0	25.3	25.5	25.8	26.0	26.2	26.5	26.7	27.0	27.2	27.5	27.7	27.9	28.2	28.4	28.7
43.7	6.5	13.8	25.5	25.8	26.0	26.3	26.5	26.8	27.0	27.3	27.5	27.8	28.0	28.3	28.5	28.8	29.0	29.3
48.7	7.0	13.3	26.0	26.3	26.5	26.8	27.0	27.3	27.6	27.8	28.1	28.3	28.6	28.9	29.1	29.4	29.6	29.9
54.1	7.5	12.9	26.6	26.8	27.1	27.4	27.6	27.9	28.2	28.4	28.7	29.0	29.2	29.5	29.8	30.0	30.3	30.6
59.9	8.0	12.4	27.1	27.4	27.7	28.0	28.3	28.5	28.8	29.1	29.4	29.6	29.9	30.2	30.5	30.8	31.0	31.3
66.2	8.5	11.9	27.8	28.1	28.4	28.6	28.9	29.2	29.5	29.8	30.1	30.4	30.7	31.0	31.2	31.5	31.8	32.1
72.9	9.0	11.4	28.5	28.8	29.1	29.4	29.7	30.0	30.3	30.6	30.9	31.2	31.5	31.8	32.1	32.4	32.7	33.0
80.3	9.5	11.0	29.2	29.5	29.8	30.2	30.5	30.8	31.1	31.4	31.7	32.0	32.3	32.7	33.0	33.3	33.6	33.9
88.3	10.0	10.5	30.0	30.4	30.7	31.0	31.3	31.7	32.0	32.3	32.6	33.0	33.3	33.6	33.9	34.3	34.6	34.9
97.0	10.5	10.0	30.9	31.3	31.6	32.0	32.3	32.6	33.0	33.3	33.7	34.0	34.3	34.7	35.0	35.4	35.7	36.0
106.6	11.0	9.5	31.9	32.3	32.6	33.0	33.3	33.7	34.1	34.4	34.8	35.1	35.5	35.8	36.2	36.6	36.9	37.3
117.2	11.5	9.0	33.0	33.4	33.7	34.1	34.5	34.9	35.3	35.6	36.0	36.4	36.8	37.1	37.5	37.9	38.3	38.6
129.0	12.0	8.6	34.2	34.6	35.0	35.4	35.8	36.2	36.6	37.0	37.4	37.8	38.2	38.6	39.0	39.3	39.7	40.1
142.1	12.5	8.1	35.6	36.0	36.4	36.8	37.2	37.6	38.1	38.5	38.9	39.3	39.7	40.2	40.6	41.0	41.4	41.8
156.8	13.0	7.6	37.1	37.5	38.0	38.4	38.8	39.3	39.7	40.2	40.6	41.1	41.5	42.0	42.4	42.8	43.3	43.7
173.4	13.5	7.1	38.8	39.3	39.7	40.2	40.7	41.2	41.6	42.1	42.6	43.0	43.5	44.0	44.5	44.9	45.4	45.9
192.4	14.0	6.7	40.8	41.3	41.8	42.3	42.8	43.3	43.8	44.3	44.8	45.3	45.8	46.3	46.8	47.3	47.8	48.3
214.2	14.5	6.2	43.0	43.6	44.1	44.7	45.2	45.7	46.3	46.8	47.4	47.9	48.5	49.0	49.5	50.1	50.6	51.2
239.4	15.0	5.7	45.7	46.3	46.8	47.4	48.0	48.6	49.2	49.8	50.4	51.0	51.5	52.1	52.7	53.3	53.9	54.5
269.2	15.5	5.2	48.8	49.4	50.1	50.7	51.4	52.0	52.6	53.3	53.9	54.6	55.2	55.8	56.5	57.1	57.8	58.4
304.7	16.0	4.8	52.6	53.3	54.0	54.7	55.4	56.1	56.8	57.5	58.2	58.9	59.6	60.3	61.0	61.7	62.4	63.1

Table 11-9.

NATURAL GAS—300°F to 450°F

EXIT GAS HEAT LOSSES

% EXCESS AIR	% OXYGEN	% CO2	NET STACK TEMPERATURE DEG F EXIT FLUE GAS TEMPERATURE - COMBUSTION AIR TEMPERATURE															
			300	310	320	330	340	350	360	370	380	390	400	410	420	430	440	450
0.0	0.0	11.7	14.5	14.7	14.8	15.0	15.2	15.3	15.5	15.7	15.8	16.0	16.1	16.3	16.5	16.6	16.8	17.0
2.2	0.5	11.4	14.6	14.8	15.0	15.1	15.3	15.5	15.6	15.8	16.0	16.1	16.3	16.5	16.6	16.8	17.0	17.1
4.4	1.0	11.1	14.8	14.9	15.1	15.3	15.4	15.6	15.8	15.9	16.1	16.3	16.5	16.6	16.8	17.0	17.1	17.3
6.8	1.5	10.9	14.9	15.1	15.2	15.4	15.6	15.8	15.9	16.1	16.3	16.4	16.6	16.8	17.0	17.1	17.3	17.5
9.3	2.0	10.6	15.0	15.2	15.4	15.5	15.7	15.9	16.1	16.3	16.4	16.6	16.8	17.0	17.2	17.3	17.5	17.7
12.0	2.5	10.3	15.2	15.3	15.5	15.7	15.9	16.1	16.3	16.4	16.6	16.8	17.0	17.2	17.4	17.5	17.7	17.9
14.8	3.0	10.0	15.3	15.5	15.7	15.9	16.1	16.2	16.4	16.6	16.8	17.0	17.2	17.4	17.6	17.8	17.9	18.1
17.7	3.5	9.8	15.5	15.7	15.8	16.0	16.2	16.4	16.6	16.8	17.0	17.2	17.4	17.6	17.8	18.0	18.2	18.4
20.8	4.0	9.5	15.6	15.8	16.0	16.2	16.4	16.6	16.8	17.0	17.2	17.4	17.6	17.8	18.0	18.2	18.4	18.6
24.1	4.5	9.2	15.8	16.0	16.2	16.4	16.6	16.8	17.0	17.2	17.4	17.6	17.9	18.1	18.3	18.5	18.7	18.9
27.6	5.0	8.9	16.0	16.2	16.4	16.6	16.8	17.0	17.3	17.5	17.7	17.9	18.1	18.3	18.5	18.7	18.9	19.2
31.4	5.5	8.6	16.2	16.4	16.6	16.8	17.1	17.3	17.5	17.7	17.9	18.1	18.4	18.6	18.8	19.0	19.2	19.5
35.4	6.0	8.4	16.4	16.6	16.8	17.1	17.3	17.5	17.7	18.0	18.2	18.4	18.6	18.9	19.1	19.3	19.5	19.8
39.6	6.5	8.1	16.6	16.9	17.1	17.3	17.6	17.8	18.0	18.3	18.5	18.7	19.0	19.2	19.4	19.6	19.9	20.1
44.2	7.0	7.8	16.9	17.1	17.3	17.6	17.8	18.1	18.3	18.6	18.8	19.0	19.3	19.5	19.8	20.0	20.2	20.5
49.0	7.5	7.5	17.1	17.4	17.6	17.9	18.1	18.4	18.6	18.9	19.1	19.4	19.6	19.9	20.1	20.4	20.6	20.9
54.3	8.0	7.2	17.4	17.7	17.9	18.2	18.4	18.7	19.0	19.2	19.5	19.7	20.0	20.3	20.5	20.8	21.0	21.3
60.0	8.5	7.0	17.7	18.0	18.3	18.5	18.8	19.1	19.3	19.6	19.9	20.1	20.4	20.7	20.9	21.2	21.5	21.7
66.1	9.0	6.7	18.0	18.3	18.6	18.9	19.2	19.4	19.7	20.0	20.3	20.6	20.8	21.1	21.4	21.7	22.0	22.2
72.8	9.5	6.4	18.4	18.7	19.0	19.3	19.6	19.9	20.2	20.4	20.7	21.0	21.3	21.6	21.9	22.2	22.5	22.8
80.0	10.0	6.1	18.8	19.1	19.4	19.7	20.0	20.3	20.6	20.9	21.2	21.5	21.8	22.1	22.4	22.8	23.1	23.4
88.0	10.5	5.9	19.2	19.5	19.9	20.2	20.5	20.8	21.1	21.5	21.8	22.1	22.4	22.7	23.0	23.4	23.7	24.0
96.7	11.0	5.6	19.7	20.0	20.4	20.7	21.0	21.4	21.7	22.0	22.4	22.7	23.0	23.4	23.7	24.0	24.4	24.7
106.3	11.5	5.3	20.2	20.6	20.9	21.3	21.6	22.0	22.3	22.7	23.0	23.4	23.7	24.1	24.4	24.8	25.1	25.5
117.0	12.0	5.0	20.8	21.2	21.5	21.9	22.3	22.6	23.0	23.4	23.8	24.1	24.5	24.9	25.2	25.6	26.0	26.4
128.9	12.5	4.7	21.4	21.8	22.2	22.6	23.0	23.4	23.8	24.2	24.6	25.0	25.4	25.8	26.1	26.5	26.9	27.3
142.3	13.0	4.5	22.2	22.6	23.0	23.4	23.8	24.2	24.7	25.1	25.5	25.9	26.3	26.7	27.2	27.6	28.0	28.4
157.4	13.5	4.2	23.0	23.4	23.9	24.3	24.8	25.2	25.6	26.1	26.5	27.0	27.4	27.9	28.3	28.8	29.2	29.6
174.6	14.0	3.9	23.9	24.4	24.9	25.3	25.8	26.3	26.8	27.3	27.7	28.2	28.7	29.2	29.6	30.1	30.6	31.1
194.4	14.5	3.6	25.0	25.5	26.0	26.5	27.1	27.6	28.1	28.6	29.1	29.6	30.1	30.6	31.1	31.7	32.2	32.7
217.4	15.0	3.3	26.3	26.8	27.4	27.9	28.5	29.0	29.6	30.2	30.7	31.3	31.8	32.4	32.9	33.5	34.0	34.6
244.4	15.5	3.1	27.8	28.4	29.0	29.6	30.2	30.8	31.4	32.0	32.6	33.2	33.8	34.4	35.0	35.6	36.2	36.8
276.7	16.0	2.8	29.6	30.2	30.9	31.6	32.2	32.9	33.5	34.2	34.9	35.5	36.2	36.9	37.5	38.2	38.9	39.5

Table 11-10.

WOOD 40% MOISTURE—600°F to 750°F

WOOD 40% MOISTURE

EXIT GAS HEAT LOSSES

% EXCESS AIR	% OXYGEN	% CO2	NET STACK TEMPERATURE DEG F EXIT FLUE GAS TEMPERATURE - COMBUSTION AIR TEMPERATURE															
			600	610	620	630	640	650	660	670	680	690	700	710	720	730	740	750
0.0	0.0	20.0	22.3	22.5	22.6	22.8	23.0	23.2	23.3	23.5	23.7	23.9	24.1	24.2	24.4	24.6	24.8	24.9
2.4	0.5	19.5	22.5	22.7	22.9	23.1	23.2	23.4	23.6	23.8	24.0	24.1	24.3	24.5	24.7	24.9	25.1	25.2
4.9	1.0	19.0	22.8	23.0	23.1	23.3	23.5	23.7	23.9	24.1	24.3	24.4	24.6	24.8	25.0	25.2	25.4	25.6
7.5	1.5	18.6	23.0	23.2	23.4	23.6	23.8	24.0	24.2	24.4	24.6	24.7	24.9	25.1	25.3	25.5	25.7	25.9
10.3	2.0	18.1	23.3	23.5	23.7	23.9	24.1	24.3	24.5	24.7	24.9	25.1	25.3	25.5	25.6	25.8	26.0	26.2
13.2	2.5	17.6	23.6	23.8	24.0	24.2	24.4	24.6	24.8	25.0	25.2	25.4	25.6	25.8	26.0	26.2	26.4	26.6
16.3	3.0	17.1	23.9	24.1	24.3	24.5	24.7	24.9	25.1	25.3	25.6	25.8	26.0	26.2	26.4	26.6	26.8	27.0
19.5	3.5	16.7	24.2	24.5	24.7	24.9	25.1	25.3	25.5	25.7	25.9	26.1	26.3	26.6	26.8	27.0	27.2	27.4
23.0	4.0	16.2	24.6	24.8	25.0	25.2	25.5	25.7	25.9	26.1	26.3	26.5	26.7	27.0	27.2	27.4	27.6	27.8
26.6	4.5	15.7	25.0	25.2	25.4	25.6	25.8	26.1	26.3	26.5	26.7	27.0	27.2	27.4	27.6	27.8	28.1	28.3
30.5	5.0	15.2	25.3	25.6	25.8	26.0	26.3	26.5	26.7	26.9	27.2	27.4	27.6	27.9	28.1	28.3	28.5	28.8
34.6	5.5	14.8	25.8	26.0	26.2	26.5	26.7	26.9	27.2	27.4	27.6	27.9	28.1	28.4	28.6	28.8	29.1	29.3
39.0	6.0	14.3	26.2	26.5	26.7	26.9	27.2	27.4	27.7	27.9	28.2	28.4	28.6	28.9	29.1	29.4	29.6	29.9
43.7	6.5	13.8	26.7	26.9	27.2	27.4	27.7	27.9	28.2	28.4	28.7	28.9	29.2	29.4	29.7	29.9	30.2	30.4
48.7	7.0	13.3	27.2	27.5	27.7	28.0	28.2	28.5	28.7	29.0	29.3	29.5	29.8	30.0	30.3	30.6	30.8	31.1
54.1	7.5	12.9	27.7	28.0	28.3	28.5	28.8	29.1	29.3	29.6	29.9	30.2	30.4	30.7	31.0	31.2	31.5	31.8
59.9	8.0	12.4	28.3	28.6	28.9	29.2	29.4	29.7	30.0	30.3	30.6	30.8	31.1	31.4	31.7	31.9	32.2	32.5
66.2	8.5	11.9	29.0	29.3	29.5	29.8	30.1	30.4	30.7	31.0	31.3	31.6	31.8	32.1	32.4	32.7	33.0	33.3
72.9	9.0	11.4	29.7	30.0	30.3	30.6	30.9	31.2	31.5	31.8	32.1	32.4	32.7	33.0	33.3	33.6	33.9	34.2
80.3	9.5	11.0	30.4	30.7	31.0	31.3	31.6	32.0	32.3	32.6	32.9	33.2	33.5	33.8	34.2	34.5	34.8	35.1
88.3	10.0	10.5	31.2	31.5	31.9	32.2	32.5	32.8	33.2	33.5	33.8	34.2	34.5	34.8	35.1	35.5	35.8	36.1
97.0	10.5	10.0	32.1	32.5	32.8	33.1	33.5	33.8	34.2	34.5	34.8	35.2	35.5	35.9	36.2	36.5	36.9	37.2
106.6	11.0	9.5	33.1	33.4	33.8	34.2	34.5	34.9	35.2	35.6	36.0	36.3	36.7	37.0	37.4	37.7	38.1	38.5
117.2	11.5	9.0	34.2	34.6	34.9	35.3	35.7	36.1	36.4	36.8	37.2	37.6	37.9	38.3	38.7	39.1	39.4	39.8
129.0	12.0	8.6	35.4	35.8	36.2	36.6	37.0	37.4	37.8	38.2	38.6	38.9	39.3	39.7	40.1	40.5	40.9	41.3
142.1	12.5	8.1	36.7	37.2	37.6	38.0	38.4	38.8	39.2	39.7	40.1	40.5	40.9	41.3	41.8	42.2	42.6	43.0
156.8	13.0	7.6	38.3	38.7	39.1	39.6	40.0	40.5	40.9	41.4	41.8	42.2	42.7	43.1	43.6	44.0	44.5	44.9
173.4	13.5	7.1	40.0	40.4	40.9	41.4	41.9	42.3	42.8	43.3	43.8	44.2	44.7	45.2	45.6	46.1	46.6	47.1
192.4	14.0	6.7	41.9	42.4	43.0	43.5	44.0	44.5	45.0	45.5	46.0	46.5	47.0	47.5	48.0	48.5	49.0	49.5
214.2	14.5	6.2	44.2	44.8	45.3	45.8	46.4	46.9	47.5	48.0	48.6	49.1	49.6	50.2	50.7	51.3	51.8	52.4
239.4	15.0	5.7	46.9	47.4	48.0	48.6	49.2	49.8	50.4	51.0	51.6	52.1	52.7	53.3	53.9	54.5	55.1	55.7
269.2	15.5	5.2	50.0	50.6	51.3	51.9	52.5	53.2	53.8	54.5	55.1	55.7	56.4	57.0	57.7	58.3	58.9	59.6
304.7	16.0	4.8	53.7	54.4	55.1	55.8	56.5	57.2	57.9	58.7	59.4	60.1	60.8	61.5	62.2	62.9	63.6	64.3

Table 11-11.

WOOD 40% MOISTURE—300°F to 450°F

WOOD 40% MOISTURE

EXIT GAS HEAT LOSSES

% EXCESS AIR	% OXYGEN	% CO2	NET STACK TEMPERATURE DEG F																
			EXIT FLUE GAS TEMPERATURE - COMBUSTION AIR TEMPERATURE															
			300	310	320	330	340	350	360	370	380	390	400	410	420	430	440	450
0.0	0.0	20.0	17.0	17.1	17.3	17.5	17.7	17.8	18.0	18.2	18.4	18.6	18.7	18.9	19.1	19.3	19.4	19.6
2.4	0.5	19.5	17.1	17.3	17.4	17.6	17.8	18.0	18.2	18.3	18.5	18.7	18.9	19.1	19.3	19.4	19.6	19.8
4.9	1.0	19.0	17.2	17.4	17.6	17.8	17.9	18.1	18.3	18.5	18.7	18.9	19.1	19.2	19.4	19.6	19.8	20.0
7.5	1.5	18.6	17.3	17.5	17.7	17.9	18.1	18.3	18.5	18.7	18.9	19.0	19.2	19.4	19.6	19.8	20.0	20.2
10.3	2.0	18.1	17.5	17.7	17.9	18.1	18.3	18.4	18.6	18.8	19.0	19.2	19.4	19.6	19.8	20.0	20.2	20.4
13.2	2.5	17.6	17.6	17.8	18.0	18.2	18.4	18.6	18.8	19.0	19.2	19.4	19.6	19.8	20.0	20.2	20.4	20.6
16.3	3.0	17.1	17.8	18.0	18.2	18.4	18.6	18.8	19.0	19.2	19.4	19.6	19.8	20.0	20.2	20.4	20.6	20.8
19.5	3.5	16.7	17.9	18.2	18.4	18.6	18.8	19.0	19.2	19.4	19.6	19.8	20.0	20.3	20.5	20.7	20.9	21.1
23.0	4.0	16.2	18.1	18.3	18.5	18.8	19.0	19.2	19.4	19.6	19.8	20.1	20.3	20.5	20.7	20.9	21.1	21.4
26.6	4.5	15.7	18.3	18.5	18.7	19.0	19.2	19.4	19.6	19.9	20.1	20.3	20.5	20.7	21.0	21.2	21.4	21.6
30.5	5.0	15.2	18.5	18.7	19.0	19.2	19.4	19.6	19.9	20.1	20.3	20.6	20.8	21.0	21.2	21.5	21.7	21.9
34.6	5.5	14.8	18.7	18.9	19.2	19.4	19.6	19.9	20.1	20.4	20.6	20.8	21.1	21.3	21.5	21.8	22.0	22.2
39.0	6.0	14.3	18.9	19.2	19.4	19.7	19.9	20.1	20.4	20.6	20.9	21.1	21.4	21.6	21.8	22.1	22.3	22.6
43.7	6.5	13.8	19.2	19.4	19.7	19.9	20.2	20.4	20.7	20.9	21.2	21.4	21.7	21.9	22.2	22.4	22.7	22.9
48.7	7.0	13.3	19.4	19.7	19.9	20.2	20.5	20.7	21.0	21.2	21.5	21.7	22.0	22.3	22.5	22.8	23.0	23.3
54.1	7.5	12.9	19.7	20.0	20.2	20.5	20.8	21.0	21.3	21.6	21.8	22.1	22.4	22.6	22.9	23.2	23.4	23.7
59.9	8.0	12.4	20.0	20.3	20.5	20.8	21.1	21.4	21.7	21.9	22.2	22.5	22.8	23.0	23.3	23.6	23.9	24.2
66.2	8.5	11.9	20.3	20.6	20.9	21.2	21.5	21.7	22.0	22.3	22.6	22.9	23.2	23.5	23.8	24.1	24.3	24.6
72.9	9.0	11.4	20.6	20.9	21.2	21.5	21.8	22.1	22.4	22.7	23.0	23.3	23.6	23.9	24.2	24.5	24.8	25.1
80.3	9.5	11.0	21.0	21.3	21.6	22.0	22.3	22.6	22.9	23.2	23.5	23.8	24.1	24.5	24.8	25.1	25.4	25.7
88.3	10.0	10.5	21.4	21.8	22.1	22.4	22.7	23.1	23.4	23.7	24.0	24.4	24.7	25.0	25.3	25.7	26.0	26.3
97.0	10.5	10.0	21.9	22.2	22.6	22.9	23.2	23.6	23.9	24.3	24.6	24.9	25.3	25.6	26.0	26.3	26.7	27.0
106.6	11.0	9.5	22.4	22.7	23.1	23.4	23.8	24.2	24.5	24.9	25.2	25.6	25.9	26.3	26.7	27.0	27.4	27.7
117.2	11.5	9.0	22.9	23.3	23.7	24.0	24.4	24.8	25.2	25.5	25.9	26.3	26.7	27.0	27.4	27.8	28.2	28.5
129.0	12.0	8.6	23.5	23.9	24.3	24.7	25.1	25.5	25.9	26.3	26.7	27.1	27.5	27.9	28.3	28.7	29.1	29.5
142.1	12.5	8.1	24.2	24.6	25.0	25.4	25.9	26.3	26.7	27.1	27.5	28.0	28.4	28.8	29.2	29.6	30.0	30.5
156.8	13.0	7.6	24.9	25.4	25.8	26.3	26.7	27.2	27.6	28.1	28.5	28.9	29.4	29.8	30.3	30.7	31.2	31.6
173.4	13.5	7.1	25.8	26.3	26.8	27.2	27.7	28.2	28.6	29.1	29.6	30.1	30.5	31.0	31.5	31.9	32.4	32.9
192.4	14.0	6.7	26.8	27.3	27.8	28.3	28.8	29.3	29.8	30.3	30.8	31.3	31.8	32.3	32.9	33.4	33.9	34.4
214.2	14.5	6.2	27.9	28.5	29.0	29.6	30.1	30.6	31.2	31.7	32.3	32.8	33.4	33.9	34.4	35.0	35.5	36.1
239.4	15.0	5.7	29.2	29.8	30.4	31.0	31.6	32.2	32.8	33.4	33.9	34.5	35.1	35.7	36.3	36.9	37.5	38.1
269.2	15.5	5.2	30.8	31.5	32.1	32.7	33.4	34.0	34.6	35.3	35.9	36.6	37.2	37.8	38.5	39.1	39.8	40.4
304.7	16.0	4.8	32.7	33.4	34.1	34.8	35.5	36.2	36.9	37.6	38.3	39.0	39.7	40.4	41.1	41.8	42.5	43.2

Table 11-12.

FUEL OIL #5—60°F to 750°F

EXIT GAS HEAT LOSSES

% EXCESS AIR	% OXYGEN	% CO2	NET STACK TEMPERATURE DEG F EXIT FLUE GAS TEMPERATURE - COMBUSTION AIR TEMPERATURE															
			600	610	620	630	640	650	660	670	680	690	700	710	720	730	740	750
0.0	0.0	16.7	15.6	15.7	15.9	16.1	16.3	16.5	16.6	16.8	17.0	17.2	17.4	17.5	17.7	17.9	18.1	18.3
2.3	0.5	16.3	15.8	16.0	16.2	16.4	16.5	16.7	16.9	17.1	17.3	17.5	17.6	17.8	18.0	18.2	18.4	18.6
4.7	1.0	15.9	16.1	16.3	16.4	16.6	16.8	17.0	17.2	17.4	17.6	17.8	17.9	18.1	18.3	18.5	18.7	18.9
7.2	1.5	15.5	16.3	16.5	16.7	16.9	17.1	17.3	17.5	17.7	17.9	18.1	18.3	18.5	18.6	18.8	19.0	19.2
9.9	2.0	15.1	16.6	16.8	17.0	17.2	17.4	17.6	17.8	18.0	18.2	18.4	18.6	18.8	19.0	19.2	19.4	19.6
12.7	2.5	14.7	16.9	17.1	17.3	17.5	17.7	17.9	18.1	18.3	18.5	18.7	19.0	19.2	19.4	19.6	19.8	20.0
15.6	3.0	14.3	17.2	17.5	17.7	17.9	18.1	18.3	18.5	18.7	18.9	19.1	19.3	19.5	19.7	19.9	20.2	20.4
18.8	3.5	13.9	17.6	17.8	18.0	18.2	18.4	18.7	18.9	19.1	19.3	19.5	19.7	19.9	20.1	20.4	20.6	20.8
22.1	4.0	13.5	17.9	18.2	18.4	18.6	18.8	19.0	19.3	19.5	19.7	19.9	20.1	20.4	20.6	20.8	21.0	21.2
25.6	4.5	13.1	18.3	18.5	18.8	19.0	19.2	19.4	19.7	19.9	20.1	20.4	20.6	20.8	21.0	21.3	21.5	21.7
29.3	5.0	12.7	18.7	19.0	19.2	19.4	19.7	19.9	20.1	20.3	20.6	20.8	21.0	21.3	21.5	21.7	22.0	22.2
33.3	5.5	12.3	19.2	19.4	19.6	19.9	20.1	20.3	20.6	20.8	21.1	21.3	21.5	21.8	22.0	22.3	22.5	22.7
37.5	6.0	11.9	19.6	19.9	20.1	20.4	20.6	20.8	21.1	21.3	21.6	21.8	22.1	22.3	22.6	22.8	23.1	23.3
42.0	6.5	11.5	20.1	20.4	20.6	20.9	21.1	21.4	21.6	21.9	22.1	22.4	22.6	22.9	23.2	23.4	23.7	23.9
46.8	7.0	11.1	20.6	20.9	21.1	21.4	21.7	21.9	22.2	22.5	22.7	23.0	23.3	23.5	23.8	24.1	24.3	24.6
52.0	7.5	10.7	21.2	21.5	21.7	22.0	22.3	22.5	22.8	23.1	23.4	23.6	23.9	24.2	24.5	24.7	25.0	25.3
57.5	8.0	10.3	21.8	22.1	22.4	22.6	22.9	23.2	23.5	23.8	24.1	24.3	24.6	24.9	25.2	25.5	25.8	26.0
63.6	8.5	9.9	22.4	22.7	23.0	23.3	23.6	23.9	24.2	24.5	24.8	25.1	25.4	25.7	26.0	26.3	26.6	26.9
70.1	9.0	9.5	23.2	23.5	23.8	24.1	24.4	24.7	25.0	25.3	25.6	25.9	26.2	26.5	26.8	27.1	27.4	27.7
77.1	9.5	9.1	23.9	24.2	24.6	24.9	25.2	25.5	25.8	26.2	26.5	26.8	27.1	27.4	27.7	28.1	28.4	28.7
84.8	10.0	8.7	24.8	25.1	25.4	25.8	26.1	26.4	26.8	27.1	27.4	27.8	28.1	28.4	28.8	29.1	29.4	29.8
93.2	10.5	8.4	25.7	26.0	26.4	26.7	27.1	27.4	27.8	28.1	28.5	28.8	29.2	29.5	29.9	30.2	30.6	30.9
102.4	11.0	8.0	26.7	27.1	27.4	27.8	28.2	28.5	28.9	29.2	29.6	30.0	30.3	30.7	31.1	31.4	31.8	32.2
112.6	11.5	7.6	27.8	28.2	28.6	29.0	29.3	29.7	30.1	30.5	30.9	31.3	31.6	32.0	32.4	32.8	33.2	33.6
123.9	12.0	7.2	29.1	29.5	29.9	30.3	30.7	31.1	31.5	31.9	32.3	32.7	33.1	33.5	33.9	34.3	34.7	35.1
136.5	12.5	6.8	30.4	30.9	31.3	31.7	32.2	32.6	33.0	33.4	33.9	34.3	34.7	35.1	35.6	36.0	36.4	36.9
150.7	13.0	6.4	32.0	32.5	32.9	33.4	33.8	34.3	34.7	35.2	35.6	36.1	36.5	37.0	37.4	37.9	38.4	38.8
166.7	13.5	6.0	33.8	34.3	34.7	35.2	35.7	36.2	36.7	37.2	37.6	38.1	38.6	39.1	39.6	40.1	40.5	41.0
184.9	14.0	5.6	35.8	36.3	36.8	37.3	37.9	38.4	38.9	39.4	39.9	40.4	41.0	41.5	42.0	42.5	43.0	43.5
205.8	14.5	5.2	38.1	38.7	39.2	39.8	40.4	40.9	41.5	42.0	42.6	43.1	43.7	44.2	44.8	45.4	45.9	46.5
230.2	15.0	4.8	40.9	41.5	42.1	42.7	43.3	43.9	44.5	45.1	45.7	46.3	46.9	47.5	48.1	48.7	49.3	49.9
258.8	15.5	4.4	44.1	44.7	45.4	46.0	46.7	47.4	48.0	48.7	49.3	50.0	50.6	51.3	51.9	52.6	53.2	53.9
292.9	16.0	4.0	47.9	48.7	49.4	50.1	50.8	51.5	52.3	53.0	53.7	54.4	55.1	55.9	56.6	57.3	58.0	58.7

Table 11-13.

FUEL OIL #2—450°F to 600°F

EXIT GAS HEAT LOSSES

% EXCESS AIR	% OXYGEN	% CO2	NET STACK TEMPERATURE DEG F EXIT FLUE GAS TEMPERATURE - COMBUSTION AIR TEMPERATURE															
			450	460	470	480	490	500	510	520	530	540	550	560	570	580	590	600
0.0	0.0	15.7	14.2	14.4	14.6	14.7	14.9	15.1	15.3	15.5	15.6	15.8	16.0	16.2	16.3	16.5	16.7	16.9
2.3	0.5	15.3	14.4	14.6	14.8	14.9	15.1	15.3	15.5	15.7	15.8	16.0	16.2	16.4	16.6	16.8	16.9	17.1
4.6	1.0	15.0	14.6	14.8	15.0	15.1	15.3	15.5	15.7	15.9	16.1	16.3	16.5	16.6	16.8	17.0	17.2	17.4
7.1	1.5	14.6	14.8	15.0	15.2	15.4	15.6	15.7	15.9	16.1	16.3	16.5	16.7	16.9	17.1	17.3	17.5	17.7
9.8	2.0	14.2	15.0	15.2	15.4	15.6	15.8	16.0	16.2	16.4	16.6	16.8	17.0	17.2	17.4	17.6	17.7	17.9
12.5	2.5	13.8	15.2	15.4	15.6	15.8	16.0	16.2	16.4	16.6	16.8	17.0	17.2	17.4	17.6	17.8	18.0	18.2
15.5	3.0	13.5	15.5	15.7	15.9	16.1	16.3	16.5	16.7	16.9	17.1	17.3	17.5	17.7	17.9	18.2	18.4	18.6
18.5	3.5	13.1	15.7	15.9	16.1	16.4	16.6	16.8	17.0	17.2	17.4	17.6	17.8	18.1	18.3	18.5	18.7	18.9
21.8	4.0	12.7	16.0	16.2	16.4	16.6	16.9	17.1	17.3	17.5	17.7	18.0	18.2	18.4	18.6	18.8	19.0	19.3
25.3	4.5	12.3	16.3	16.5	16.7	16.9	17.2	17.4	17.6	17.8	18.1	18.3	18.5	18.7	19.0	19.2	19.4	19.6
29.0	5.0	12.0	16.6	16.8	17.0	17.3	17.5	17.7	18.0	18.2	18.4	18.7	18.9	19.1	19.4	19.6	19.8	20.0
32.9	5.5	11.6	16.9	17.1	17.4	17.6	17.9	18.1	18.3	18.6	18.8	19.0	19.3	19.5	19.8	20.0	20.2	20.5
37.0	6.0	11.2	17.2	17.5	17.7	18.0	18.2	18.5	18.7	19.0	19.2	19.5	19.7	19.9	20.2	20.4	20.7	20.9
41.5	6.5	10.8	17.6	17.9	18.1	18.4	18.6	18.9	19.1	19.4	19.6	19.9	20.1	20.4	20.7	20.9	21.2	21.4
46.2	7.0	10.5	18.0	18.3	18.5	18.8	19.1	19.3	19.6	19.8	20.1	20.4	20.6	20.9	21.2	21.4	21.7	21.9
51.4	7.5	10.1	18.4	18.7	19.0	19.2	19.5	19.8	20.1	20.3	20.6	20.9	21.1	21.4	21.7	22.0	22.2	22.5
56.9	8.0	9.7	18.9	19.2	19.4	19.7	20.0	20.3	20.6	20.9	21.1	21.4	21.7	22.0	22.3	22.5	22.8	23.1
62.8	8.5	9.3	19.4	19.7	20.0	20.2	20.5	20.8	21.1	21.4	21.7	22.0	22.3	22.6	22.9	23.2	23.5	23.8
69.2	9.0	9.0	19.9	20.2	20.5	20.8	21.1	21.4	21.7	22.0	22.3	22.6	23.0	23.3	23.6	23.9	24.2	24.5
76.2	9.5	8.6	20.5	20.8	21.1	21.4	21.7	22.1	22.4	22.7	23.0	23.3	23.7	24.0	24.3	24.6	24.9	25.2
83.8	10.0	8.2	21.1	21.4	21.8	22.1	22.4	22.8	23.1	23.4	23.8	24.1	24.4	24.8	25.1	25.4	25.8	26.1
92.1	10.5	7.9	21.8	22.1	22.5	22.8	23.2	23.5	23.9	24.2	24.6	24.9	25.3	25.6	26.0	26.3	26.7	27.0
101.2	11.0	7.5	22.6	22.9	23.3	23.7	24.0	24.4	24.7	25.1	25.5	25.8	26.2	26.6	26.9	27.3	27.7	28.0
111.3	11.5	7.1	23.4	23.8	24.2	24.5	24.9	25.3	25.7	26.1	26.5	26.8	27.2	27.6	28.0	28.4	28.8	29.1
122.5	12.0	6.7	24.3	24.7	25.1	25.5	25.9	26.4	26.8	27.2	27.6	28.0	28.4	28.8	29.2	29.6	30.0	30.4
134.9	12.5	6.4	25.4	25.8	26.2	26.7	27.1	27.5	27.9	28.4	28.8	29.2	29.6	30.1	30.5	30.9	31.4	31.8
148.9	13.0	6.0	26.5	27.0	27.5	27.9	28.4	28.8	29.3	29.7	30.2	30.6	31.1	31.5	32.0	32.4	32.9	33.3
164.7	13.5	5.6	27.9	28.4	28.8	29.3	29.8	30.3	30.8	31.3	31.7	32.2	32.7	33.2	33.7	34.2	34.6	35.1
182.7	14.0	5.2	29.4	29.9	30.4	30.9	31.5	32.0	32.5	33.0	33.5	34.0	34.6	35.1	35.6	36.1	36.6	37.1
203.4	14.5	4.9	31.2	31.7	32.3	32.8	33.4	33.9	34.5	35.0	35.6	36.1	36.7	37.3	37.8	38.4	38.9	39.5
227.5	15.0	4.5	33.2	33.8	34.4	35.0	35.6	36.2	36.8	37.4	38.0	38.6	39.2	39.8	40.4	41.0	41.6	42.2
255.8	15.5	4.1	35.6	36.3	36.9	37.6	38.2	38.9	39.5	40.2	40.9	41.5	42.2	42.8	43.5	44.1	44.8	45.4
289.5	16.0	3.7	38.5	39.2	40.0	40.7	41.4	42.1	42.8	43.5	44.3	45.0	45.7	46.4	47.1	47.9	48.6	49.3

Table 11-14.

FUEL OIL #6—300°F to 450°F

EXIT GAS HEAT LOSSES

% EXCESS AIR	% OXYGEN	% CO2	300	310	320	330	340	350	360	370	380	390	400	410	420	430	440	450
0.0	0.0	16.7	10.2	10.4	10.5	10.7	10.9	11.1	11.3	11.4	11.6	11.8	12.0	12.2	12.3	12.5	12.7	12.9
2.3	0.5	16.3	10.3	10.5	10.7	10.9	11.0	11.2	11.4	11.6	11.8	12.0	12.1	12.3	12.5	12.7	12.9	13.1
4.7	1.0	15.9	10.4	10.6	10.8	11.0	11.2	11.4	11.6	11.8	11.9	12.1	12.3	12.5	12.7	12.9	13.1	13.3
7.2	1.5	15.5	10.6	10.8	11.0	11.1	11.3	11.5	11.7	11.9	12.1	12.3	12.5	12.7	12.9	13.1	13.3	13.5
9.9	2.0	15.1	10.7	10.9	11.1	11.3	11.5	11.7	11.9	12.1	12.3	12.5	12.7	12.9	13.1	13.3	13.5	13.7
12.7	2.5	14.7	10.9	11.1	11.3	11.5	11.7	11.9	12.1	12.3	12.5	12.7	12.9	13.1	13.3	13.5	13.7	13.9
15.6	3.0	14.3	11.0	11.2	11.4	11.6	11.9	12.1	12.3	12.5	12.7	12.9	13.1	13.3	13.5	13.7	13.9	14.1
18.8	3.5	13.9	11.2	11.4	11.6	11.8	12.0	12.3	12.5	12.7	12.9	13.1	13.3	13.5	13.8	14.0	14.2	14.4
22.1	4.0	13.5	11.4	11.6	11.8	12.0	12.2	12.5	12.7	12.9	13.1	13.3	13.6	13.8	14.0	14.2	14.4	14.7
25.6	4.5	13.1	11.6	11.8	12.0	12.2	12.5	12.7	12.9	13.1	13.4	13.6	13.8	14.0	14.3	14.5	14.7	14.9
29.3	5.0	12.7	11.8	12.0	12.2	12.5	12.7	12.9	13.2	13.4	13.6	13.9	14.1	14.3	14.5	14.8	15.0	15.2
33.3	5.5	12.3	12.0	12.2	12.5	12.7	12.9	13.2	13.4	13.7	13.9	14.1	14.4	14.6	14.8	15.1	15.3	15.6
37.5	6.0	11.9	12.2	12.5	12.7	12.9	13.2	13.4	13.7	13.9	14.2	14.4	14.7	14.9	15.2	15.4	15.7	15.9
42.0	6.5	11.5	12.5	12.7	13.0	13.2	13.5	13.7	14.0	14.2	14.5	14.7	15.0	15.3	15.5	15.8	16.0	16.3
46.8	7.0	11.1	12.7	13.0	13.2	13.5	13.8	14.0	14.3	14.6	14.8	15.1	15.3	15.6	15.9	16.1	16.4	16.7
52.0	7.5	10.7	13.0	13.3	13.5	13.8	14.1	14.4	14.6	14.9	15.2	15.5	15.7	16.0	16.3	16.5	16.8	17.1
57.5	8.0	10.3	13.3	13.6	13.9	14.1	14.4	14.7	15.0	15.3	15.6	15.8	16.1	16.4	16.7	17.0	17.3	17.5
63.6	8.5	9.9	13.6	13.9	14.2	14.5	14.8	15.1	15.4	15.7	16.0	16.3	16.6	16.9	17.2	17.4	17.7	18.0
70.1	9.0	9.5	14.0	14.3	14.6	14.9	15.2	15.5	15.8	16.1	16.4	16.7	17.0	17.3	17.6	18.0	18.3	18.6
77.1	9.5	9.1	14.4	14.7	15.0	15.3	15.6	16.0	16.3	16.6	16.9	17.2	17.5	17.9	18.2	18.5	18.8	19.1
84.8	10.0	8.7	14.8	15.1	15.4	15.8	16.1	16.4	16.8	17.1	17.4	17.8	18.1	18.4	18.8	19.1	19.4	19.8
93.2	10.5	8.4	15.2	15.6	15.9	16.3	16.6	17.0	17.3	17.7	18.0	18.4	18.7	19.1	19.4	19.8	20.1	20.5
102.4	11.0	8.0	15.7	16.1	16.5	16.8	17.2	17.6	17.9	18.3	18.7	19.0	19.4	19.8	20.1	20.5	20.9	21.2
112.6	11.5	7.6	16.3	16.7	17.1	17.5	17.8	18.2	18.6	19.0	19.4	19.8	20.1	20.5	20.9	21.3	21.7	22.1
123.9	12.0	7.2	16.9	17.3	17.7	18.1	18.5	18.9	19.4	19.8	20.2	20.6	21.0	21.4	21.8	22.2	22.6	23.0
136.5	12.5	6.8	17.6	18.0	18.5	18.9	19.3	19.8	20.2	20.6	21.0	21.5	21.9	22.3	22.8	23.2	23.6	24.0
150.7	13.0	6.4	18.4	18.9	19.3	19.8	20.2	20.7	21.1	21.6	22.0	22.5	22.9	23.4	23.8	24.3	24.8	25.2
166.7	13.5	6.0	19.3	19.8	20.3	20.7	21.2	21.7	22.2	22.7	23.2	23.6	24.1	24.6	25.1	25.6	26.0	26.5
184.9	14.0	5.6	20.3	20.8	21.3	21.9	22.4	22.9	23.4	23.9	24.4	25.0	25.5	26.0	26.5	27.0	27.5	28.1
205.8	14.5	5.2	21.5	22.0	22.6	23.1	23.7	24.2	24.8	25.4	25.9	26.5	27.0	27.6	28.1	28.7	29.2	29.8
230.2	15.0	4.8	22.8	23.4	24.0	24.6	25.2	25.8	26.4	27.0	27.6	28.2	28.8	29.4	30.0	30.6	31.2	31.8
258.8	15.5	4.4	24.4	25.1	25.7	26.4	27.1	27.7	28.4	29.0	29.7	30.3	31.0	31.6	32.3	32.9	33.6	34.3
292.9	16.0	4.0	26.4	27.1	27.8	28.5	29.2	30.0	30.7	31.4	32.1	32.8	33.6	34.3	35.0	35.7	36.4	37.2

COMBUSTIBLE LOSSES (CARBON MONOXIDE)

When combustion is incomplete, unburned fuel can escape with the exit flue gases. One measurement of this loss is the Carbon Monoxide (CO) level. Carbon Monoxide also has a toxic effect on humans so its deadly presence could mean much more than an energy loss.

Explanation

See Tables 11-15 and 11-16. Using the excess air level in the left column and the measured Carbon Monoxide (CO) in parts per million (PPM) along the top, the intersection of the two measurements will provide a value for the losses involved.

1. Burner repairs have lowered the carbon monoxide level in the combustion gasses of

Table 11-15

Carbon, Monoxide Energy Loss Table for Natural Gas

NATURAL GAS

CARBON MONOXIDE HEAT LOSSES (%)

CARBON MONOXIDE PARTS PER MILLION (PPM)

% EXCESS AIR	% OXYGEN	% CO2	100	200	300	400	500	600	700	800	900	1000	1100	1200	1300	1400	1500	1600
0.0	0.0	11.7	0.03	0.06	0.08	0.11	0.14	0.17	0.20	0.23	0.25	0.28	0.31	0.34	0.37	0.39	0.42	0.45
2.2	0.5	11.4	0.03	0.06	0.09	0.12	0.14	0.17	0.20	0.23	0.26	0.29	0.32	0.35	0.37	0.40	0.43	0.46
4.4	1.0	11.1	0.03	0.06	0.09	0.12	0.15	0.18	0.21	0.24	0.27	0.30	0.33	0.35	0.38	0.41	0.44	0.47
6.8	1.5	10.9	0.03	0.06	0.09	0.12	0.15	0.18	0.21	0.24	0.27	0.30	0.33	0.36	0.39	0.42	0.45	0.48
9.3	2.0	10.6	0.03	0.06	0.09	0.12	0.16	0.19	0.22	0.25	0.28	0.31	0.34	0.37	0.40	0.44	0.47	0.50
12.0	2.5	10.3	0.03	0.06	0.10	0.13	0.16	0.19	0.22	0.26	0.29	0.32	0.35	0.38	0.42	0.45	0.48	0.51
14.8	3.0	10.0	0.03	0.07	0.10	0.13	0.16	0.20	0.23	0.26	0.30	0.33	0.36	0.39	0.43	0.46	0.49	0.53
17.7	3.5	9.8	0.03	0.07	0.10	0.14	0.17	0.20	0.24	0.27	0.30	0.34	0.37	0.41	0.44	0.47	0.51	0.54
20.8	4.0	9.5	0.03	0.07	0.10	0.14	0.17	0.21	0.24	0.28	0.31	0.35	0.38	0.42	0.45	0.49	0.52	0.56
24.1	4.5	9.2	0.04	0.07	0.11	0.14	0.18	0.21	0.25	0.29	0.32	0.36	0.39	0.43	0.47	0.50	0.54	0.57
27.6	5.0	8.9	0.04	0.07	0.11	0.15	0.18	0.22	0.26	0.30	0.33	0.37	0.41	0.44	0.48	0.52	0.55	0.59
31.4	5.5	8.6	0.04	0.08	0.11	0.15	0.19	0.23	0.27	0.31	0.34	0.38	0.42	0.46	0.50	0.53	0.57	0.61
35.4	6.0	8.4	0.04	0.08	0.12	0.16	0.20	0.24	0.28	0.32	0.35	0.39	0.43	0.47	0.51	0.55	0.59	0.63
39.6	6.5	8.1	0.04	0.08	0.12	0.16	0.20	0.24	0.29	0.33	0.37	0.41	0.45	0.49	0.53	0.57	0.61	0.65
44.2	7.0	7.8	0.04	0.08	0.13	0.17	0.21	0.25	0.30	0.34	0.38	0.42	0.46	0.51	0.55	0.59	0.63	0.68
49.0	7.5	7.5	0.04	0.09	0.13	0.18	0.22	0.26	0.31	0.35	0.39	0.44	0.48	0.53	0.57	0.61	0.66	0.70
54.3	8.0	7.2	0.05	0.09	0.14	0.18	0.23	0.27	0.32	0.36	0.41	0.45	0.50	0.55	0.59	0.64	0.68	0.73
60.0	8.5	7.0	0.05	0.09	0.14	0.19	0.24	0.28	0.33	0.38	0.43	0.47	0.52	0.57	0.61	0.66	0.71	0.76
66.1	9.0	6.7	0.05	0.10	0.15	0.20	0.25	0.30	0.34	0.39	0.44	0.49	0.54	0.59	0.64	0.69	0.74	0.79
72.8	9.5	6.4	0.05	0.10	0.15	0.21	0.26	0.31	0.36	0.41	0.46	0.51	0.57	0.62	0.67	0.72	0.77	0.82
80.0	10.0	6.1	0.05	0.11	0.16	0.21	0.27	0.32	0.38	0.43	0.48	0.54	0.59	0.64	0.70	0.75	0.81	0.86
88.0	10.5	5.9	0.06	0.11	0.17	0.23	0.28	0.34	0.39	0.45	0.51	0.56	0.62	0.68	0.73	0.79	0.84	0.90
96.7	11.0	5.6	0.06	0.12	0.18	0.24	0.30	0.35	0.41	0.47	0.53	0.59	0.65	0.71	0.77	0.83	0.89	0.95
106.3	11.5	5.3	0.06	0.12	0.19	0.25	0.31	0.37	0.44	0.50	0.56	0.62	0.68	0.75	0.81	0.87	0.93	1.00
117.0	12.0	5.0	0.07	0.13	0.20	0.26	0.33	0.39	0.46	0.53	0.59	0.66	0.72	0.79	0.85	0.92	0.99	1.05
128.9	12.5	4.7	0.07	0.14	0.21	0.28	0.35	0.42	0.49	0.56	0.63	0.70	0.76	0.83	0.90	0.97	1.04	1.11
142.3	13.0	4.5	0.07	0.15	0.22	0.30	0.37	0.44	0.52	0.59	0.66	0.74	0.81	0.89	0.96	1.03	1.11	1.18
157.4	13.5	4.2	0.08	0.16	0.24	0.32	0.39	0.47	0.55	0.63	0.71	0.79	0.87	0.95	1.02	1.10	1.18	1.26
174.6	14.0	3.9	0.08	0.17	0.25	0.34	0.42	0.51	0.59	0.68	0.76	0.84	0.93	1.01	1.10	1.18	1.27	1.35
194.4	14.5	3.6	0.09	0.18	0.27	0.36	0.45	0.55	0.64	0.73	0.82	0.91	1.00	1.09	1.18	1.28	1.38	1.45
217.4	15.0	3.3	0.10	0.20	0.30	0.39	0.49	0.59	0.69	0.79	0.89	0.99	1.08	1.18	1.28	1.38	1.48	1.58
244.4	15.5	3.1	0.11	0.21	0.32	0.43	0.54	0.64	0.75	0.86	0.97	1.07	1.18	1.29	1.40	1.50	1.61	1.72
276.7	16.0	2.8	0.12	0.24	0.35	0.47	0.59	0.71	0.83	0.95	1.06	1.18	1.30	1.42	1.54	1.65	1.77	1.89

Table 11-16

Carbon Monoxide Energy Loss Table for Fuel Oil

FUEL OIL

CARBON MONOXIDE HEAT LOSSES (%)

CARBON MONOXIDE PARTS PER MILLION (PPM)

% EXCESS AIR	% OXYGEN	% CO2	100	200	300	400	500	600	700	800	900	1000	1100	1200	1300	1400	1500	1600
0.0	0.0	11.7	0.03	0.06	0.09	0.12	0.15	0.18	0.21	0.24	0.27	0.30	0.33	0.36	0.39	0.42	0.45	0.47
2.2	0.5	11.4	0.03	0.06	0.09	0.12	0.15	0.18	0.21	0.24	0.27	0.30	0.33	0.36	0.40	0.43	0.46	0.49
4.4	1.0	11.1	0.03	0.06	0.09	0.12	0.16	0.19	0.22	0.25	0.28	0.31	0.34	0.37	0.40	0.44	0.47	0.50
6.8	1.5	10.9	0.03	0.06	0.10	0.13	0.16	0.19	0.22	0.26	0.29	0.32	0.35	0.38	0.42	0.45	0.48	0.51
9.3	2.0	10.6	0.03	0.07	0.10	0.13	0.16	0.20	0.23	0.26	0.30	0.33	0.36	0.39	0.43	0.46	0.49	0.52
12.0	2.5	10.3	0.03	0.07	0.10	0.13	0.17	0.20	0.24	0.27	0.30	0.34	0.37	0.40	0.44	0.47	0.51	0.54
14.8	3.0	10.0	0.03	0.07	0.10	0.14	0.17	0.21	0.24	0.28	0.31	0.35	0.38	0.42	0.45	0.48	0.52	0.55
17.7	3.5	9.8	0.04	0.07	0.11	0.14	0.18	0.21	0.25	0.28	0.32	0.36	0.39	0.43	0.46	0.50	0.53	0.57
20.8	4.0	9.5	0.04	0.07	0.11	0.15	0.18	0.22	0.26	0.29	0.33	0.37	0.40	0.44	0.48	0.51	0.55	0.59
24.1	4.5	9.2	0.04	0.08	0.11	0.15	0.19	0.23	0.26	0.30	0.34	0.38	0.42	0.45	0.49	0.53	0.57	0.60
27.6	5.0	8.9	0.04	0.08	0.12	0.16	0.19	0.23	0.27	0.31	0.35	0.39	0.43	0.47	0.51	0.55	0.58	0.62
31.4	5.5	8.6	0.04	0.08	0.12	0.16	0.20	0.24	0.28	0.32	0.36	0.40	0.44	0.48	0.52	0.56	0.60	0.64
35.4	6.0	8.4	0.04	0.08	0.12	0.17	0.21	0.25	0.29	0.33	0.37	0.42	0.46	0.50	0.54	0.58	0.62	0.66
39.6	6.5	8.1	0.04	0.09	0.13	0.17	0.21	0.26	0.30	0.34	0.39	0.43	0.47	0.52	0.56	0.60	0.64	0.69
44.2	7.0	7.8	0.04	0.09	0.13	0.18	0.22	0.27	0.31	0.36	0.40	0.45	0.49	0.53	0.58	0.62	0.67	0.71
49.0	7.5	7.5	0.05	0.09	0.14	0.18	0.23	0.28	0.32	0.37	0.42	0.46	0.51	0.55	0.60	0.65	0.69	0.74
54.3	8.0	7.2	0.05	0.10	0.14	0.19	0.24	0.29	0.34	0.38	0.43	0.48	0.53	0.58	0.62	0.67	0.72	0.77
60.0	8.5	7.0	0.05	0.10	0.15	0.20	0.25	0.30	0.35	0.40	0.45	0.50	0.55	0.60	0.65	0.70	0.75	0.80
66.1	9.0	6.7	0.05	0.10	0.16	0.21	0.26	0.31	0.36	0.42	0.47	0.52	0.57	0.62	0.67	0.73	0.78	0.83
72.8	9.5	6.4	0.05	0.11	0.16	0.22	0.27	0.33	0.38	0.43	0.49	0.54	0.60	0.65	0.70	0.76	0.81	0.87
80.0	10.0	6.1	0.06	0.11	0.17	0.23	0.28	0.34	0.40	0.45	0.51	0.57	0.62	0.68	0.74	0.79	0.85	0.91
88.0	10.5	5.9	0.06	0.12	0.18	0.24	0.30	0.36	0.42	0.47	0.53	0.59	0.65	0.71	0.77	0.83	0.89	0.95
96.7	11.0	5.6	0.06	0.12	0.19	0.25	0.31	0.37	0.44	0.50	0.56	0.62	0.69	0.75	0.81	0.87	0.93	1.00
106.3	11.5	5.3	0.07	0.13	0.20	0.26	0.33	0.39	0.46	0.52	0.59	0.66	0.72	0.79	0.85	0.92	0.98	1.05
117.0	12.0	5.0	0.07	0.14	0.21	0.28	0.35	0.42	0.48	0.55	0.62	0.69	0.76	0.83	0.90	0.97	1.04	1.11
128.9	12.5	4.7	0.07	0.15	0.22	0.29	0.37	0.44	0.51	0.59	0.66	0.73	0.81	0.88	0.95	1.03	1.10	1.17
142.3	13.0	4.5	0.08	0.16	0.23	0.31	0.39	0.47	0.55	0.62	0.70	0.78	0.86	0.93	1.01	1.09	1.17	1.25
157.4	13.5	4.2	0.08	0.17	0.25	0.33	0.42	0.50	0.58	0.66	0.75	0.83	0.91	1.00	1.08	1.16	1.25	1.33
174.6	14.0	3.9	0.09	0.18	0.27	0.36	0.45	0.53	0.62	0.71	0.80	0.89	0.98	1.07	1.16	1.25	1.34	1.42
194.4	14.5	3.6	0.10	0.19	0.29	0.38	0.48	0.58	0.67	0.77	0.86	0.96	1.05	1.15	1.25	1.34	1.44	1.53
217.4	15.0	3.3	0.10	0.21	0.31	0.42	0.52	0.62	0.73	0.83	0.93	1.04	1.14	1.25	1.35	1.45	1.56	1.66
244.4	15.5	3.1	0.11	0.23	0.34	0.45	0.57	0.68	0.79	0.91	1.02	1.13	1.25	1.36	1.47	1.59	1.70	1.81
276.7	16.0	2.8	0.12	0.25	0.37	0.50	0.62	0.75	0.87	1.00	1.12	1.25	1.37	1.50	1.62	1.74	1.87	1.99

an oil fired boiler from 1400 PPM to 200 PPM The excess oxygen level is steady at 6%. How has this changed the efficiency? _____
(*Ans.: combustible losses have dropped from 0.58% to 0.08% with an efficiency increase of 0.5%*)

2. In a natural gas fired boiler, the Carbon Monoxide level has been lowered from 1000 PPM to 100 PPM with a constant 10% oxygen in the flue gases. How has this affected efficiency? _____
(*Ans.: losses have dropped from 0.54% to 0.05% for an efficiency improvement of 0.49%*)

ANALYZING COSTS AND BENEFITS OF AN
ECONOMIZER AND AN OXYGEN
TRIM SYSTEM

Part 1: Economic Analysis, Estimating Percent Savings and Dollar Savings

As found conditions:

Fuel   Natural Gas
Annual Fuel Bill $100,000
Oxygen O_2 11.0%
Net stack temperature   600F [316C]
Efficiency   70.3%

Option 1. Install an Oxygen Trim system with an installed cost of $10,000 to reduce excess air to 2.0%.

New oxygen O_2 level   2%
Net stack temperature   600F [316C]
New efficiency   79.6%

1. Efficiency Improvement   _____%
    (Ans.: 9.3%)

2. Percent Fuel Savings   _____%
    (Ans.: 11.7%)

3. Fuel Savings Dollars $_____/YR
    (Ans.: $11,683/YR)

Option 2. Install economizer with an installed cost of $30,000 to reduce stack temperature to 200F.

Oxygen O_2 level   11.0%
Net stack temperature   200F [93C]
New efficiency   83.7%

4. Efficiency Improvement _____%
    (Ans.: 13.4%)

5. Percent Fuel Savings   _____%
    (Ans.: 16%)

6. Fuel Savings Dollars $_____/YR
    (Ans.: $16,000/YR)

Part II: Economic analysis, Payback Period

$$\text{Payback Period} \quad PP = \frac{FC}{S - C}$$

PP = Payback period
PC = First cost
S= Annual fuel savings
C= Annual maintenance costs

Option 1. Calculate payback period for the oxygen trim system if the maintenance cost is $1,000/yr.

$$PP = \frac{10,000}{11,700 - 1,000} = \text{———}YR.$$

(Ans.: .93 YR.)

Option 2. Calculate payback period for an economizer if the maintenance cost is $1,000/yr.

$$PP = \frac{30,000}{16,000 - 1,000} = \text{———}YR.$$

(Ans.: 2.0 YR.)

Part III: Economic Analysis, Return on Investment

Return on investment

$$ROI\,(\%) = \frac{S - DC}{PC} \times 100$$

ROI = Return on Investment
S= Annual fuel savings $/Yr
DC = Depreciation charge
      First cost/Estimated Lifetime
PC = First Cost

Option 1. Calculate the ROI for the oxygen trim system if the estimated lifetime is 15 years:

$$\text{ROI (\%):} \quad \frac{11{,}700 - 667}{10{,}000} = \underline{\quad\quad}\%$$

(Ans.: 110%)

Option 2. Calculate the ROI for an economizer if the estimated lifetime is 15 years:

$$\text{ROI (\%):} \quad \frac{16{,}000 - 2000}{30{,}000} = \underline{\quad\quad}\%$$

(Ans.: 46.7%)

Part IV: Economic Analysis, Installing Oxygen Trim and Economizer Separately and Together

Option 3. Assuming the oxygen trim system has been installed, what are the benefits from installing the economizer under the new operating conditions?

Efficiency with oxygen trim system	79.6%
Efficiency with economizer	86.8%
Annual fuel savings(%)	8.3%
Annual fuel savings ($) $8,295/Yr	

$$\text{Payback Period} = \frac{30{,}000}{8{,}295 - 1{,}000} = \underline{\quad\quad}$$

(Ans.: 4.1 Yr.)

$$\begin{array}{c} \text{Return} \\ \text{on} \\ \text{Investment} \end{array} = \frac{8{,}295 - 2{,}000}{30{,}000} = \underline{\quad\quad}$$

(Ans.: 21%)

Option 4. What would the payback period and return on investment be if the oxygen trim and economizer were both installed as the same project?

As found efficiency	70.3%
Optimized efficiency	86.8%
Optimized fuel savings (%)	19.0%
Optimized fuel savings ($)	$19,000/Yr

$$\text{Payback Period} \quad \frac{30{,}000 + 10{,}000}{19{,}000 - 2{,}000} = \underline{\quad\quad}$$

(Ans.: 2.35 Yr.)

$$\begin{array}{c} \text{Return} \\ \text{on} \\ \text{Investment} \end{array} = \frac{19{,}000 - 2{,}667}{40{,}000} = \underline{\quad\quad}$$

(Ans.: 40%)

COST OF MONEY

The cost of money may affect energy saving projects. The purpose of this section is to illustrate how the longer payback periods are affected by the cost of money and also how interest rates change the time needed to recoup investments.

Your company must borrow money at various interest rates, how long will it take to recoup the investment required for energy conservation projects with different payback periods and for various interest rates? (See page ___)

(1) Find the discount rate column.
(2) Locate simple payback period under the assumed interest rate.
(3) The approximate time to recoup investment will be found in the left column under the "lifetime" heading.

(**Note.** These values are approximate and are given only to illustrate relative values)

	Time to Payback Period	Discount Rate	Recoup Investment	
Option 1.	1.5 Yr.	20%	————	(Ans.: <2 yr.)
Option 2.	2.6 Yr.	20%	————	(Ans.: <3 yr.)
Option 3.	5.0 Yr.	20%	————	(Ans.: >25 yr.)
Option 4.	5.0 Yr.	15%	————	(Ans.: 10 yr.)
Option 5.	5.0 Yr.	10%	————	(Ans.: 8 yr.)
Option 6.	5.0 Yr.	5%	————	(Ans.: 6 yr.)

Replacing an Old Inefficient Boiler
Size: 20 Million Btu/hr (20,000 PPH) (10 Ton/hr)
Fuel: No. 2 Distillate
Cost: $1.00/Gallon
250,000 Gallons Fuel per year
137,000 Btu/Gallon
Net Stack Temperature 750F, *The economizer needs replacement.*
Oxygen 9.5%, Excess Air 76%
Stack losses 30%
Surface losses 2.5% of full rating—(500,000 Btu/hr)

Table 11-17 is calculated from the following equation:

$$\text{PWF} \quad \frac{1 - (1 + D)^{-EL}}{D}$$

where D is discount rate expressed as a fraction and EL is the expected lifetime of the project in years.

1. The boiler is fired for 6,000 hours a year, what is its utilization rate?

$$\text{Theoretical fuel usage} = \frac{20 \text{ MBtu/hr} \times 6{,}000 \text{ hr/YR}}{137{,}000 \text{ Btu/gallon}} = 896{,}000 \text{ gallons/YR}$$

$$\frac{\text{Actual Fuel Usage}}{\text{Theoretical Fuel Usage}} = \frac{250{,}000 \text{ gallons}}{896{,}000 \text{ gallons}} = \text{about 30\% (Utilization)}$$

2. Corrected surface losses $\frac{2.5\%}{30\%} = 8.3\%$

3. How much fuel is being wasted each year with this old boiler?

(30% Stack Loss + 8.3% Surface Loss) × 250,000 gallons/yr = $95,750 gallons or ($95,750/yr)

TABLE 11-17

Present Worth Factors (PWF)

Lifetime (EL)	Discount Rate (D)				
	5%	10%	15%	20%	25%
1	0.952	0.909	0.870	0.833	0.800
2	1.859	1.736	1.626	1.528	1.440
3	2.723	2.487	2.283	2.106	1.952
4	3.546	3.170	2.855	2.589	2.362
5	4.329	3.791	3.352	2.991	2.689
6	5.076	4.355	3.784	3.326	2.951
7	5.786	4.868	4.160	3.605	3.161
8	6.463	5.335	4.487	3.837	3.329
9	7.108	5.759	4.772	4.031	3.463
10	7.722	6.145	5.019	4.192	3.571
11	8.306	6.495	5.234	4.327	3.656
12	8.863	6.814	5.421	4.439	3.725
13	9.394	7.103	5.583	4.533	3.780
14	9.899	7.367	5.724	4.611	3.824
15	10.380	7.606	5.847	4.675	3.859
16	10.838	7.824	5.954	4.730	3.887
17	11.274	8.022	6.047	4.775	3.910
18	11.690	8.201	6.128	4.812	3.928
19	12.085	8.365	6.198	4.843	3.942
20	12.462	8.514	6.259	4.870	3.954
21	12.821	8.649	6.312	4.891	3.963
22	13.163	8.772	6.359	4.909	3.970
23	13.489	8.883	6.399	4.925	3.976
24	13.799	8.985	6.434	4.937	3.981
25	14.094	9.077	6.464	4.948	3.985

The above table is calculated from the following equation:

$$PWF = \frac{1 - (1 + D)^{-EL}}{D}$$

where D is discount rate expressed as a fraction and EL is the expected lifetime of the project in years.

4. If a smaller boiler with an efficiency of 83% (17% stack loss) and with a corrected standby loss of 2% were installed, how much fuel would be saved?

(17% Stack Loss+ 2% Surface Loss) x 250,000 gallons/yr = $47,500 gallons or
($47,500/yr)

As Found Losses – New Losses   = Savings/yr
  $95,750 – $47,500   = $48,250/yr

5.   If the estimate to install a new boiler is $72,000, what is the payback period and return on investment if the new boilers life is 20 years.

$$\text{Payback Period} = \frac{\text{First Cost}}{\text{Savings}} = \frac{72,000}{\$48,250} = 1.5 \text{ yr.}$$

$$\text{Return on Investment} = \frac{\text{Savings - Cost/life yrs}}{\text{First Cost}}$$

$$\text{ROI} = \frac{48,250 - 72,000/20}{72,000} = 62\%$$

PROGRAMMING BOILER OPERATIONS

The purpose of this example is to show the wisdom of conducting efficiency tests of boilers then setting up a program for the most economical operation possible. For this example three 100,000 Pounds Per Hour (PPH), (50 Ton) (100 million Btus/hr) units are used.

1.   Review the efficiency profile of Boiler #1. (Figure 11-1)

2.   Review the efficiency profile of Boiler #2. (Figure 11-2)

3.   Review the efficiency profile of Boiler #3. (Figure 11-3)

4.   Review the combined efficiency profile of boilers #1, #2 and #3. How would you program these boilers to insure maximum economy in the boiler plant? (Figure 11-4)

5.   Review the graphic of combined boiler operation between 20 and 300 thousand PPH. (Figure 11-5)

INTRODUCTION TO THE STEAM TABLES
FOR ACCOUNTING FOR ENERGY

The purpose of the following examples is to illustrate how to account for energy in steam systems. (See Tables 11-18 and 11-19.)

1.   At 100 psig what is the total heat in the steam?
     (Ans.: 1189.6 Btu/lb)

2.   What is the latent heat of 100 psig steam?
     (Ans. 880.7 Btu/lb)

3.   What is the sensible heat in boiler water at 100 psig?
     (Ans.: 308.9 Btu/lb)

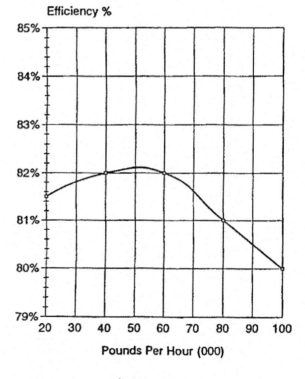

Figure 11-1

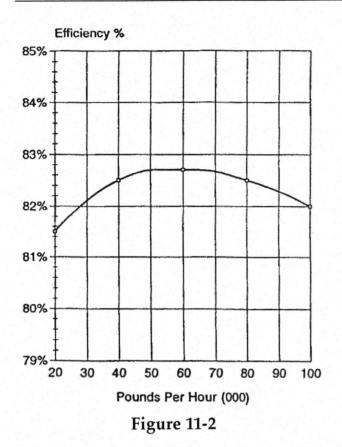

Figure 11-2

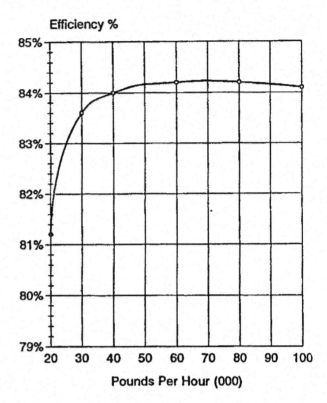

Figure 11-4

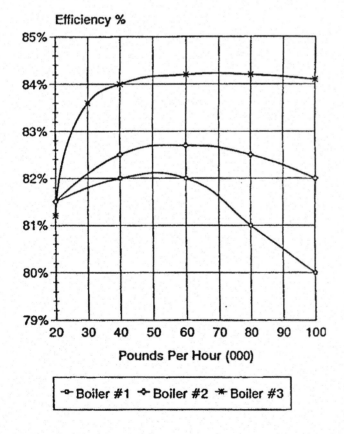

─○─Boiler #1 ─◇─Boiler #2 ─*─Boiler #3

Figure 11-3

4. What is the temperature of the steam and
    water at 100 psig?
    (Ans.: above 338 F)

5. Condensate is returning to the boiler plant at
    120F [80C] instead of 185 F [85C], how many
    Btu's are being lost in each pound of water?

185 F water [85C]	153 Btu/lb
120 F water [49C]	88 Btu/lb
Difference	65 Btu/lb

    *Comment: Starting at 32F, adding 1 Btu raises the
    temperature of water 1F. (185F-32F = 153 Btus). More
    exact tables exist.*

6. If the boiler's efficiency is 80% how much
    fuel energy (Btus) will it take to replace the
    65 Btu/lb in the example above?

    $$\frac{65 \text{ Btu/lb}}{80\%} = 81.25 \text{ Btu/lb}$$

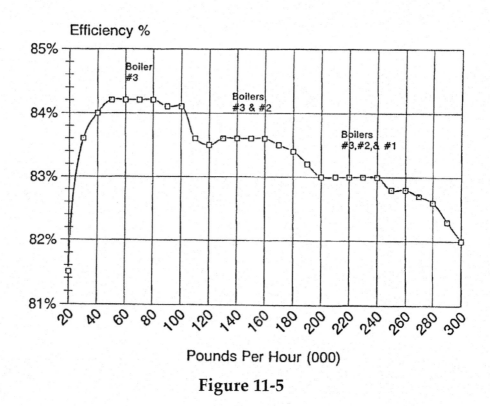

Figure 11-5

7. If a boiler is operating at a pressure of 100 psig and the total heat in the steam is 1190 Btu/lb, what percentage of the total heat in the steam is being lost because of the cold condensate in the example above?
(*Ans.: 81.25 Btu/lb 11190 Btu/lb = 6.8%*)

8. How much heat is in boiler water at:

	(*Ans.*)
a. 300 psig	(*398.7 Btu/lb*)
b. 5 psig	(*195.5 Btu/lb*)

9. What is the latent heat of 5 psig steam? _____
(*Ans. 960.5 Btu/lb*)

10. In a steam system, all the steam should condense back into water and must go through a steam trap which lowers the pressure from system pressure to a lower pressure. Blowdown water from the boiler also undergoes this same pressure reduction. As can be seen from the above data, the water at the lower pressure contains less heat. This heat difference is given off as flash steam.

Calculate the percentage of flash steam formed at 5 psig when dropping pressure from 300 psig. Use values from 8 and 9 above.

$$\frac{398.7 - 195.5}{960.5} = 21\%$$

11. What is the temperature of steam at different pressures? See Table 11-19.

Pressure (PSIG)	Temperature	(Ans.)	
15	_____	249.8 F	[121C]
75	_____	320.1 F	[160C]
100	_____	337.9 F	[170C]
150	_____	369.9F	[187C]
200	_____	387.8 F	[197C]
600	_____	488.8 F	[253C]

Table 11-18

Gage Pressure (psig)	Saturation or Boiling Temperature (Degrees F)	Specific Volume (Cu. Ft./Lb.)	Heat Content Above 32 Degrees F		
			Sensible Heat or Heat of Liquid (Btu/lb.)	Latent Heat or Heat of Evaporation (Btu/lb.)	Total Heat (Btu/lb.)
0	212.0	26.80	180.1	970.3	1150.4
1	215.5	25.13	183.6	968.1	1151.7
2	218.7	23.72	186.8	966.0	1152.8
3	221.7	22.47	189.8	964.1	1153.9
4	224.5	21.35	192.7	962.3	1155.0
5	227.3	20.34	195.5	960.5	1156.1
6	229.9	19.42	198.2	958.8	1157.0
7	232.4	18.58	200.7	957.2	1157.9
8	234.9	17.81	203.2	955.6	1158.8
9	237.2	17.11	205.6	954.1	1159.7
10	239.5	16.46	207.9	952.5	1160.4
11	241.7	15.86	210.1	951.1	1161.2
12	243.8	15.31	212.2	949.7	1161.9
13	245.9	14.79	214.3	948.3	1162.6
14	247.9	14.31	216.4	946.9	1163.2
15	249.8	13.86	218.3	945.6	1163.9
16	251.7	13.43	220.3	944.3	1164.6
17	253.6	13.03	222.2	943.0	1165.2
18	255.4	12.66	224.0	941.8	1165.8
19	257.1	12.31	225.7	940.6	1166.3
20	258.8	11.98	227.5	939.5	1167.0
21	260.5	11.67	229.2	938.3	1167.5
22	262.2	11.37	230.9	937.2	1168.1
23	263.8	11.08	232.5	936.1	1168.6
24	265.4	10.82	234.1	935.0	1169.1
25	266.9	10.56	235.6	934.0	1169.6
30	274.1	9.45	243.0	928.9	1171.9
35	280.7	8.56	249.8	924.2	1174.0
40	286.8	7.82	256.0	919.8	1175.8
45	292.4	7.20	261.8	915.7	1177.5
50	297.7	6.68	267.2	911.8	1179.0
55	302.7	6.23	272.4	908.1	1180.5
60	307.3	5.83	277.2	904.6	1181.8
65	311.8	5.49	281.8	901.3	1183.1
70	316.4	5.18	286.2	898.0	1184.2
75	320.1	4.91	290.4	894.8	1185.2
80	323.9	4.66	294.4	891.9	1186.3
85	327.6	4.44	298.2	899.0	1187.2
90	331.2	4.24	301.9	886.1	1188.0
95	334.6	4.06	305.5	883.3	1188.8
100	337.9	3.89	308.9	880.7	1189.6
105	341.1	3.74	312.3	878.1	1190.4
110	344.2	3.59	315.5	875.5	1191.0
115	347.2	3.46	318.7	873.0	1191.7
120	350.1	3.34	321.7	870.7	1192.4

Table 11-19

Boiler Plant Calculations

Gage Pressure (psig)	Saturation or Boiling Temperature (Degrees F)	Specific Volume (Cu. Ft./Lb.)	Heat Content Above 32 Degrees F		
			Sensible Heat or Heat of Liquid (Btu/lb.)	Latent Heat or Heat of Evaporation (Btu/lb.)	Total Heat (Btu/lb.)
125	352.9	3.23	324.7	868.3	1193.0
130	355.6	3.12	327.6	865.9	1193.5
135	358.3	3.02	330.4	863.7	1194.1
140	360.9	2.93	333.1	861.5	1194.6
145	363.4	2.84	335.8	859.3	1195.1
150	365.9	2.76	338.4	857.2	1195.6
155	368.3	2.68	340.9	855.0	1195.9
160	370.6	2.61	343.4	853.0	1196.4
165	372.9	2.54	345.9	850.9	1196.8
170	375.2	2.47	348.3	848.9	1197.2
175	377.4	2.41	350.7	846.9	1197.6
180	379.5	2.35	353.0	845.0	1198.0
185	381.6	2.30	355.2	843.1	1198.3
190	383.7	2.24	357.4	841.2	1198.6
195	385.8	2.19	359.6	839.2	1198.8
200	387.8	2.13	361.9	837.4	1199.3
210	391.7	2.04	366.0	833.8	1199.9
220	395.5	1.95	370.1	830.3	1200.4
230	399.1	1.88	374.1	826.8	1200.9
240	402.7	1.81	377.8	823.4	1201.3
250	406.1	1.74	381.6	820.1	1201.7
260	409.4	1.68	385.2	816.9	1202.1
270	412.6	1.62	388.7	813.7	1202.4
280	415.7	1.56	392.1	810.5	1202.7
290	418.8	1.52	395.5	807.5	1202.9
300	421.8	1.47	398.7	804.5	1203.2
400	448.2	1.12	428.1	776.4	1204.6
500	470.0	0.90	452.9	751.3	1204.3
600	488.8	0.75	474.6	728.3	1202.9

HEAT LOSSES IN THE CONDENSATE RETURN SYSTEM

The purpose of this section is to illustrate how energy is lost in the condensate recovery system serving a 100 psig steam system.

Example 1. Assuming that condensate should return to the receiver at 194F, what losses are involved for each pound of cold make-up water used to replace steam and condensate lost from the system?

City Water	50F	18 Btu/lb
Hot condensate	194 F	162 Btu/lb
Water heating (50F to 194F)		144 Btu/lb
Fuel energy used to heat city water to 194 F with 80% efficient boiler		180 Btu/lb
Water heating (194F-338F)		144 Btu/lb
Fuel energy used to bring water to saturation 338F with boiler		180 Btu/lb
Latent Heat of 100 psig steam		880 Btu/lb

Fuel energy used to form steam with 80% efficient boiler	1100 Btu/lb

Fuel Energy invested in making steam
(180 + 180 + 1100 = 1460 Btu/lb) 1460 Btu/lb

Fuel energy used to make original steam with 80% efficient boiler
(1189 Btu/lb – 162 Btu/lb)/80% = 1283 Btu/lb

$$\textbf{Fuel Loss} = \frac{\mathbf{1460 - 1283}}{\mathbf{1283}} = \mathbf{13.7\%}$$

Comment: A 144F temperature loss (194 – 50F) cost a 13.7% fuel loss, about 1% per each 10.5F.

Example 2. What is the total loss involved in steam leaks?

Replacing Lost Steam	
City Water SOP	18 Btu/lb
Heat in 100 psig steam	1190 Btu/lb
Net Heat	1172 Btu/lb
Fuel energy used to make replacement steam with 80% efficient boiler	1465 Btu/lb

Fuel energy used to make original steam with 80% efficient boiler using condensate @

194F [162 Btu/lb] $\dfrac{1190 - 162}{.8}$ 1285 Btu/lb

$$\text{Fuel Loss} \quad \frac{1465\ \text{Btu/lb} =}{1285\ \text{Btu/lb}} = 114\%$$

Example 3. Assume the condensate should be returning to the plant at 194F, it has been returning at 120F instead. What losses are involved?

Heat in 100 psig steam		1190 Btu/lb
Hot condensate	194F	162 Btu/lb

Actual condensate	120F	88 Btu/lb
Energy loss		74 Btu/lb
Fuel energy loss including boiler (80% Eff)		93 Btu/lb

Fuel energy used to make original steam with 80% efficient boiler using condensate @

194F [162 Btu/lb] $\dfrac{1190 - 162}{.8}$ 1285 Btu/lb

$$\text{Fuel Loss} = \frac{93\ \text{Bu/lb}}{7.2\%} = 1285\ \text{Btu/lb}$$

Comment: a 74 F Temperature drop of the condensate cost a 7.2% fuel loss, about 1% per each 10.3 degrees.

Example 4. Derive the "Rule of Thumb" relationship about how boiler feedwater temperature influences boiler efficiency. How many degrees must the feedwater or condensate change to cause approximately a one percent change in boiler system efficiency?

Given for water	
1 Degree F = 1 Btu/lb	
100 PSI steam	1190 Btu/lb
150 F Condensate	118 Btu/lb
	1070 Btu/lb

1% of 1070 Btu/lb = 10.7 F

Rule of Thumb: "A 1% increase in efficiency occurs with each 11°F rise in condensate return or feedwater input temperature.

Example 5. What is the efficiency of recovering 5 psig flash steam per pound?

100 psi steam	1190 Btu/lb
Fuel energy invested to make 100 psig steam with 80% boiler	1487 Btu/lb
Heat in 5 psig steam	1156 Btu/lb
Fuel energy saved by recovering steam in a 80% boiler system	1445 Btu/lb

$$\text{Savings } \frac{1445 \text{ Btu/lb}}{1487 \text{ Btu/lb}} = 97\%$$

The recovered steam has 97% of the energy of the original 100 psig steam.

DISTRIBUTION SYSTEM LOSSES

The purpose of this section is to illustrate the major losses in steam distribution systems.

Distribution system pressure	125	PSIG
Steam system average load	20,000	lb/hr
Net energy in steam (1193 – 88)	1105	Btu/lb

Steam system average input	22.1 MBtu/hr
Condensate system pressure	Atmospheric
Condensate return temperature	120F
Make up water temperature	60F
Boiler Efficiency	60F

1. What is the flash loss?

Heat in the condensate on the high pressure side of the steam trap	325 Btu/hr
Heat in condensate atmospheric pressure	180 Btu/lb
Latent heat of flash steam @ atmospheric pressure	970 Btu/lb

$$\text{Flash} = \frac{325-180}{970} = 15\%$$

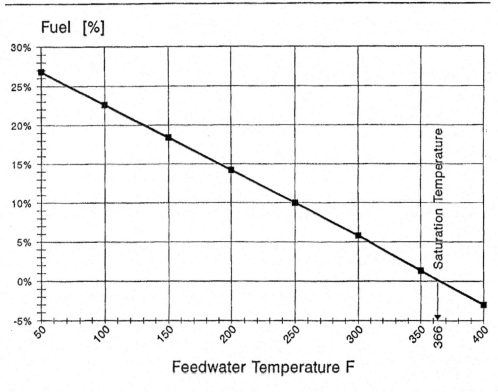

Fuel [%] to Heat Feed Water
Compared to 366F Feed Water & 150 PSI Boiler

Figure 11-6

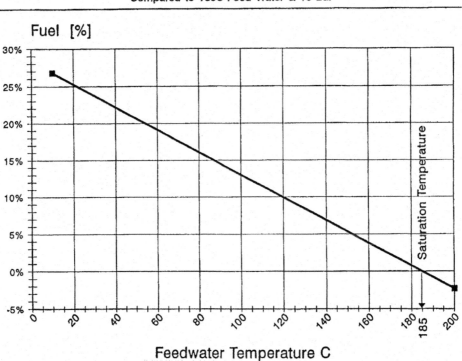

Figure 11-7

Table 11-20. Flash Steam Table

INITIAL PRESSURE PSIG	TEMP OF LIQUID	% FLASH AT ATMOSPHERIC PRESSURE	PERCENT OF STEAM FLASHED AT REDUCED PRE				
			5 PSI	10 PSI	15 PSI	20 PSI	25 PSI
100	338	13.0	11.5	10.3	9.3	8.4	7.6
125	353	14.5	13.3	11.8	10.9	10.0	9.2
150	366	16.0	14.6	13.2	12.3	11.4	10.6
175	377	17.0	15.8	14.4	13.4	12.5	11.6
200	388	18.0	16.9	15.5	14.6	13.7	12.9
225	397	19.0	17.8	16.5	15.5	14.7	13.9
250	406	20.0	18.8	17.4	16.5	15.6	14.9
300	421	21.5	20.3	19.0	18.0	17.2	16.5
350	435	23.0	21.8	20.5	19.5	18.7	18.0
400	448	24.0	23.0	21.8	21.0	20.0	19.3
450	459	25.0	24.3	23.0	22.0	21.3	20.0
500	470	26.5	25.4	24.1	23.2	22.4	21.7
550	480	27.5	26.5	25.2	24.3	23.5	22.8
600	488	28.0	27.3	26.0	25.0	24.3	23.6
BTU PER POUND OF FLASH STEAM		1150	1155	1160	1164	1167	1169
DEG F OF FLASH STEAM & LIQUID		212	225	240	250	259	267
STEAM VOLUME CUFT/LB		26.8	21.0	16.3	13.7	11.9	10.5

Energy lost with flash steam:

$$15\% \times 20,000 \text{ lb/hr} \times 970 \text{ Btu/lb} \quad = \quad 2.91 \text{ million Btu/hr}$$

Steam Energy loss $\quad = 2.91/22.1 \quad = \quad 13.2\%$

Fuel Energy Loss $\quad = 13.2\%/80\% \quad = \quad 16.5\%$

Note: The flash steam mingles with condensate in the piping and gives up heat to the cooling condensate before it vents to atmosphere so the loss is less and depends on the actual situation. This is a theoretical illustration.

2. Average trap leakage is 10% live steam in the system.

Energy loss = .1 × 20,000 PPH × 1105 Btu/lb $\quad = \quad$ 2.2 Million Btu/Hr

$2.2/22.1 \quad = \quad 10\%$

3. Because of steam and condensate leaks, 35% make-up feedwater at 60°F is required.

Energy loss = 35% × 20, 000 lb/hr × 60 Btu/lb = $\quad$ 0.42 Million Btu/hr

Steam Energy Loss $\quad 0.42/22.1 \quad = \quad 1.9\%$

Fuel Energy Loss $\quad 1.9\%/80\% \quad = \quad 2.4\%$

4. Condensate system losses. Condensate reaches the receiver at 120F, having cooled from 212F losing 92 Btu/lb. (35% makeup water)

.65 × 20,000 lb/HR × 92 Btu/lb $\quad = \quad$ 1.2 Million Btu/hr

Energy loss $\quad 1.2/22.1 \quad = \quad 5.40\%$

Fuel Energy Loss $\quad 5.4/.8 \quad = \quad 6.75\%$

	Steam System Loss	Fuel Energy Loss
Flash steam loss	13.2%	16.5%
Trap losses	10.0%	12.5%
Makeup water heat loss	1.9%	2.4%
Cold condensate return	5,4%	6.15%
Total system losses	30.5%	38.15%
Efficiency	**69.5%**	**62%**

BOILER BLOWDOWN HEAT RECOVERY

The purpose of this section is to demonstrate the potential savings available from managing the boiler water system.

1. Flash Steam Recovery:

What percentage of blowdown will flash to steam at 5 psig from the blowdown of a boiler operating at 150 psig?

$$\% \text{ Steam Flashed} = \frac{h_1 - h_2}{V_1}$$

h_1 Heat in water leaving boiler.
h_2 Heat in water at flash pressure.
V_1 Latent heat of steam at flash pressure.

STEAM TABLE INFORMATION

Boiler pressure 150 psig $h_1 =$ ____
    [Answer: 338.4 Btu/lb]

Flash tank pressure 5 psig $h_2 =$ ____
    [Answer: 195.5 Btu/lb]

Latent heat of steam at 5 psig $V_1 =$ ____
    [Answer: 960.5 Btu/lb]

% Steam Flashed= _____ [Answer: 15%]

2   How many Btu/day will be recovered from the flash tank if the daily blowdown is 80,000 lb/day.

Btu/day = % steam flashed x blowdown lb/day x (Btu/lb @ 5 psi Steam) = (At 5 psig the energy in the flash steam is 1156 Btu/lb)
Btu/day=...
[Answer: Btu/Day = .15 x 80,000 x 1156 = 13.9 Million Btu/day]

3. If you are firing a boiler with no. 2 oil with 140,000 Btu/gal which costs $1.00 per gallon, how much money can be saved in a day?

$$\$ \text{ saved/day} = \frac{\text{Btu/Day x cost/gal}}{\text{Btu/gal x E}} =$$

(E = Boiler efficiency = 75%)

Dollar savings =..
[Answer: 13.9 Million Btu/Day x $1.00/140,000 x.75 = $132.90/Day]

4. Heat exchanger recovery
The heat exchanger now reduces the blowdown water temperature to 90F, how many Btus a day can be recovered from this water?

Btu/DAY recovered = Blowdown (lb/day) x 1- % flashed x h_2-h_4 =

Heat in water at flash pressure (5 psig)...........h_2 =
_____
[Answer: 195.5 Btu/lb]

Heat in water at heat exchanger outlet (90F)....h_4 =
_____
[Answer: 58 Btu/lb]

Btu/day recovered = ... = ...
[Answer: 9.35 Million

Dollar savings/= day $$\frac{\text{Btu/Day] x \$1.00}}{140,000 \text{ Btu/GAL x .75}} =$$ _____

[Answer: $89.05/Day]

5. Total value of blowdown heat recovered:

Flash steam $————/day
[Answer: $132.90/Day]

Heat Exchanger $————/day
[Answer: $89.05/Day}

Total $————/day
[Answer: $221.95/Day]

Annual Savings $————/Yr
[Answer: $81,011.75/Yr]

SAVING ENERGY BY IMPROVING CONDENSATE RETURN

In this case there is no blowdown heat recovery system and the challenge is to conserve energy by reducing the blow-down as much as possible. There are several ways to do this such as repairing system leaks, improve feedwater quality and increasing cycles of concentration.

Calculate the savings possible in a plant exporting 2 million pounds of 100 psig steam a day.

Existing Blowdown	280,000 lb/day
Target Blowdown	160,000 lb/day

$$\$ \text{ Savings} = \frac{BR \times H \times C}{V \times \%E}$$

BR = blowdown reduction lb (280,000 -16,000 = 120,000)

H = Heat content of blowdown (100 psi)

$H = h_1 - h_5$

h_1-heat in war leaving boiler as blowdown.
(100 psig – 309 Btu/lb)

h_5-heat in makeup feedwater *(60F - 28 Btu/lb)*

$= h_1 h_5$ $(309 – 28) = 281$ Btu/lb

C = Cost of fuel $1.00/Gal

V = Heating Value of Fuel (No.2 fuel – 140,000 Btu/lb)

E = Boiler Efficiency - 80%

$$\$Savings = \frac{120,000 \text{ 1h/day} \times 281 \text{ Btu/lb} \times \$100/\text{Gal}}{140,000 \text{ Btu/Gal} \times .80}$$

$Savings	=	$301/day
$Savings	=	$109,865/Yr

REDUCING BLOWDOWN LOSSES BY INCREASING CYCLES OF CONCENTRATION

Example 1. A 300 psi boiler rated at 250,000 pounds of steam per hour uses 50% make-up water averaging 1,875,000 pounds a day. It is anticipated that the cycles of concentration can be raised from 7 to 14. What will the savings be?

$$\text{Formula A} = \frac{C2-C1}{(C1 \times C2)- C1} = 0.08$$

C_1 =Present cycles of concentration $= 7$
C_2 =Anticipated cycles of concentration $= 14$

Figure 11-8 indicates a savings of 3.00 million Btu per 100,000 lb of makeup reduction. With a cost of steam of $8.00 per million Btus, estimate the dollar savings.

Savings = 18.75 x $8.00 x 3	= $450/day
	= $164,250/Yr

CONDENSATE SYSTEM WATER LEAKS

Situation: Leaks are developing in the steam and condensate systems and the make-up water usage has risen from 50,000 gallons to 750,000 gallons per month. How much fuel is it costing to heat make-up water?

Fuel is natural gas costing $5.00 per million Btus.

Condensate weighs 8.34 pounds per gallon.
Boiler efficiency 78%

Makeup water 50F	18 Btu/lb
Normal condensate return 180F	148 Btu/lb
Btu loss per pound of make-up	130 Btu/lb
Fuel energy to heat water with boiler 80% eff	162.5 Btu/lb

How much fuel are the system leaks costing?

700,000 gallons x 8.34lb/gallon
= 5.81 million lb

5.81 x 162.5 Btu/lb
= 944 million Btus/month

944 MBtu/mo x 5.00
= $4,720/mo

$56,640/yr (plus chemicals & water costs)

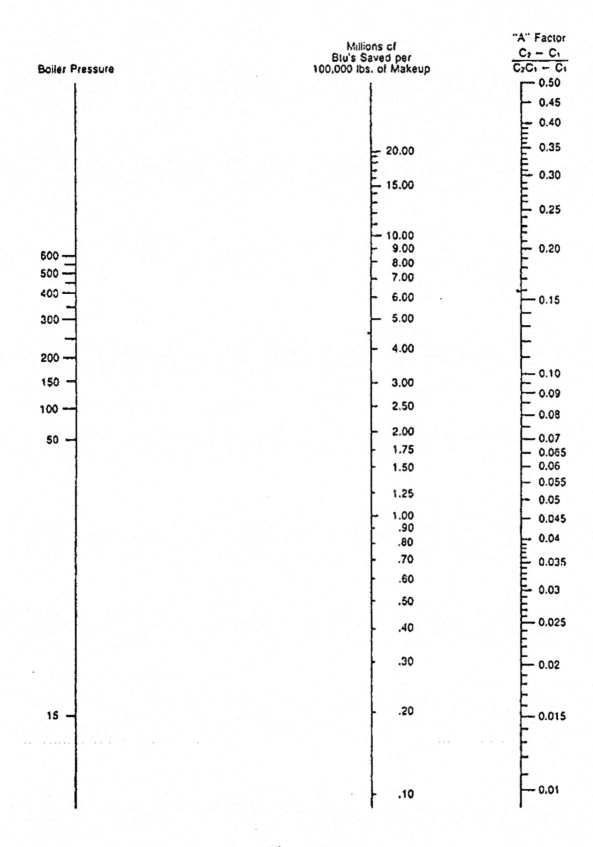

Figure 11-8
Btus Saved by Increasing Cycles of Concentration

STEAM LEAK RELIEF VALVE

A relief valve on a low pressure, 30 psig, header has started to blow steam. An examination of the make-up water readings indicates an increase of 28,776 gallons per day starting when the relief valve began blowing steam.

Fuel is distillate no. 2 costing
$4.50 per million Btus
City water 50F
18 Btu/lb
Fuel Btus needed to make 30 psig steam with a boiler efficiency of 80%
$(1172 – 18)/.80 = 1443$ Btu/lb)

1. How many pounds of steam is escaping from the steam system a day?
   28,776 gallons/day x 8.34lb/gal = 240,000 lb/day

2. How much is this costing in fuel dollars?
   240,000 lb. x 1443 Btu/lb =
   346 MBtus/day
   346 x $4.50 = $1,557/day

DISTRIBUTION SYSTEM INSULATION LOSSES

Situation: You are investigating putting a new boiler in a building 2,500 ft from the boiler plant to cut losses. You need to know how much money you can save by eliminating the 2,500 ft of6," 250 psi steam piping and 2" condensate piping.

1. How would you find out how much heat is being lost in the steam piping?
   (*Ans.: You weigh the water formed in its steam traps and collect 1,165 pounds an hour.*)

2. How much energy was given up by the steam to form this water?
   (*Ans.: The latent heat of 250 psi steam is 837 Btu/lb.*)

3. Steam costs $8.00/MBtus, how much is the steam piping insulation losses costing?

1,165 lb/hr x 837 Btu/lb = 975,000 Btu/hr
(*Ans.: $8.00 x 0.975 MBtu/hr = $7.80/h, ($187.20/day) ($5,616.00/mo) ($67,392.00/yr)*)

4. The condensate temperature drops from 200F to 155F returning to the boiler plant, how much is this costing?

Average condensate return	100,000 lb/hr
Heat loss	45 Btu/hr
Replace heat with 80% efficient boiler	56 Btu/lb

(*Ans.: 56 Btu/lb x $8.00 MBtu x
100,000 lb/hr = $44.80/hl
($1,075.20/day) ($32,256/mo) ($387,072/yr)*)

Condensate Losses	$387,072/yr
Steam Pipe losses	$ 67,392/yr
	$454,464/yr

ELECTRICAL LOSSES

The purpose of this section is to show how electrical losses play a pmt in the overall energy consumption of a boiler plant. Operating electrical equipment, especially if it is oversized for the load on the plant can add to unnecessary plant losses. The examples in this section are based on nameplate data, field measurements of electrical loads should be taken to confirm actual conditions.

Example 1. You find a grossly oversized 40 hp blower in your plant. A 5 hp could do the same job how much money could be saved by changing out the 40 hp unit?

With a cost for electrical power of 10 cents per kilowatt hour, what will the difference in cost be to run with the smaller blower?

A. How much does it cost to operate the 40 hp blower per hour?   (*Ans.: $3.50/hr*)

B. How much does it cost to operate the 5 hp blower per hour?
    (Ans.: $0.45/hr)

C. Assuming 8,000 hours of operation a year, how much more will it cost to operate the 40 hp blower than the 5 hp blower?
    (Ans.: $3.05 x 8,000 HR = $24,400/YR)

Example 2. A plant keeps a 100 hp feed pump on the line continuously where a 30 hp pump could be used, how much can be saved in a 8,000 hr year by installing a 30 hp pump?
A. How much does it cost to run the 100 hp pump per year?
    (Ans.: $10.00/hr x 8,000 hr = $80,000/yr)
B. How much does it cost to run the 30 hp pump per Year?
    (Ans.: $2.75 x 8,000 hr = $22,000/yr)
C. What will the annual savings be if the 100 hp pump is replaced by a 30 hp pump?
    (Ans.: $58,000/Yr (approx.))

As a practical matter you will never get anyone in a boiler plant to downsize equipment. The boiler feed pumps *must* be able to maintain proper water levels and even increase boiler water level at maximum firing rates and during emergency conditions. Blowers must be able to provide enough backup capacity to prevent fuel-rich firing under all conditions. An adjustable speed drive (ASD) controllers will probably be the most practical and safest approach.

Example 3. Pumps and fans follow certain laws known as the "Affinity Laws":
    1. Flow varies with RPM
    2. Head Varies with RPM- [Squared]
    3. Horsepower varies with RPM- [Cubed*]

*Experience has shown that actual horsepower requirements may with the square function rather than the cube junction when static back pressure is present.

What savings are possible by controlling a boiler feed pump with ASD rather than modulating the Feed Control Valve using the "square" relationship?

ENERGY SAVINGS WITH ADJUSTABLE SPEED DRIVE [ASD]

RPM	100%	90%	85%	80%	75%
Energy Required	100%	81%	72%	64%	56%
Savings	100%	19%	28%	36%	44%

SETTING THE FUEL OIL HEATER FOR PROPER FUEL OIL VISCOSITY AT THE BURNER

When the fuel oil temperature is not set to deliver oil at the proper viscosity, the burner can smoke and foul heat exchange surfaces. Also, higher levels of excess air will be needed to compensate for this problem. If the fuel oil heater temperature is too high the flame is affected also, becoming ragged with sparklers.

Situation:
As an economy measure, you have been given some heavy oil to burn which has a viscosity rating of 4,000 Sabolt Universal Seconds (SUS) at 100 degrees F. You notice that you have to use a lot more excess air with this fuel and you have to use your soot blowers more often.
Your burner technical manual indicates that your burner requires a viscosity of about 150 SUS for proper atomization.
What *new* temperature is required for the fuel oil heater?

Solution:
Use the **Viscosity-Temperature** Chart.

1. Find 4,000 SUS at 100 degrees F.
2. Move down a line parallel to the No. 6 fuel oil line to 150 SUS and read the corresponding temperature for the fuel oil.
    Answer: 220°F

Note: If there is a significant piping run between the heater and the burner the actual temperature at the burner may be lower and the heater may have to be set for a higher temperature.

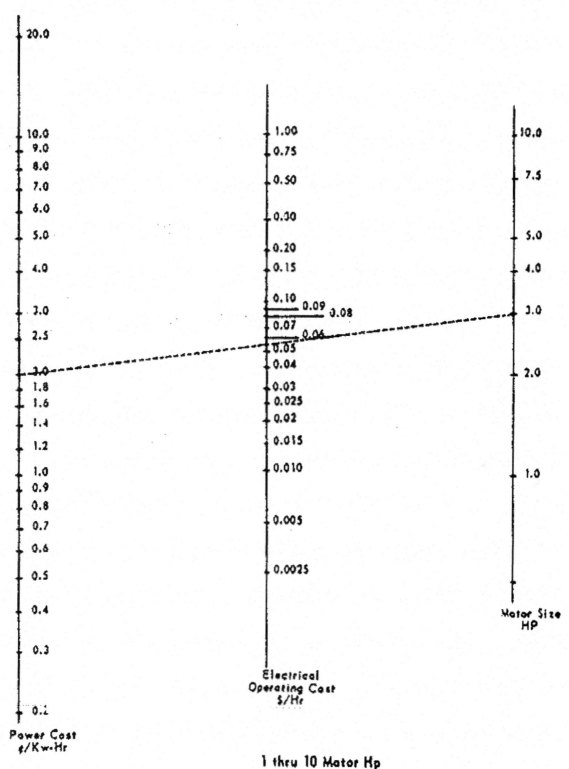

1 thru 10 Motor Hp

Figure 11-9
Hourly Electrical Operating Cost

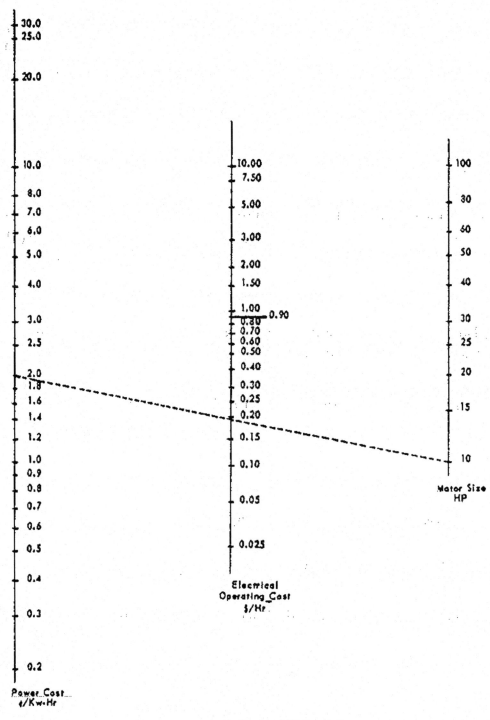

10 thru 100 Motor Hp

Figure 11-10
Hourly Electrical Operating Cost

Figure 11-11

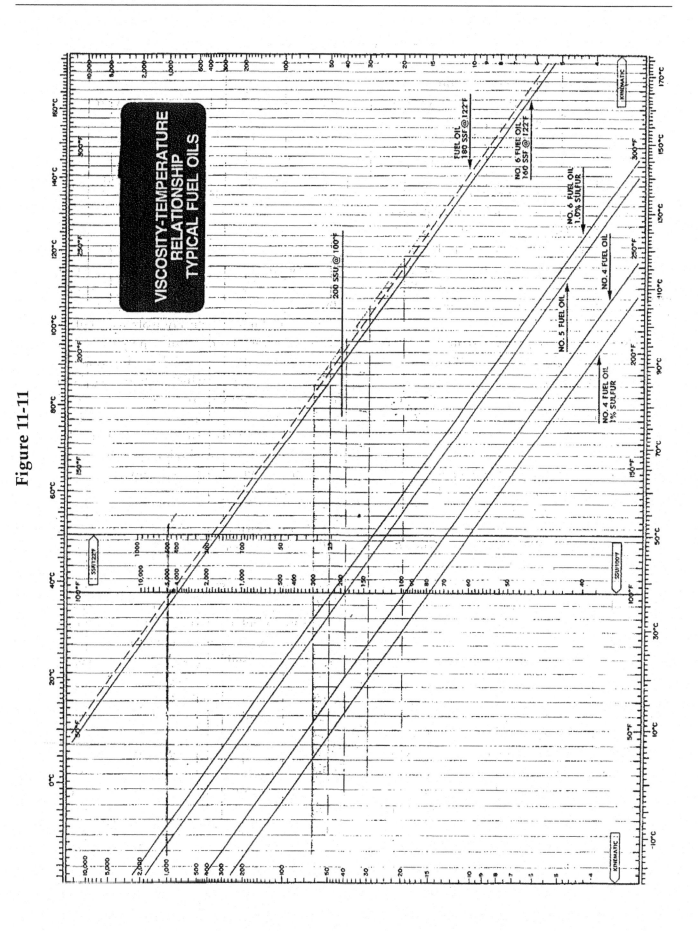

VISCOSITY-TEMPERATURE
RELATIONSHIP
TYPICAL FUEL OILS

CONVERSION FACTORS

S.I. Units [Metric Units}

S.I. units are the International system Units that were adopted by the General conference on weights and Measures in 1960 and subsequently endorsed by the International Organization for Standardization.

The Base units of the- S.I. system

	S.I Units	Symbol
(1) Length	meter	m
(2) Mass kilogram	Kg	
(3) Time	Second	s
(4) Electric Current	Ampere	A
(5) Thermodynamic Temperature	Kelvin	K
[°C + 273 = °K]		
(6) Luminous Intensity	Candela	cd

The following frequently used units are derived from the base ones.

	S.I Units	Symbol
(1) Force	Newton	$N = Kg\text{-}m/s^2$
(2) Pressure	Pascals	$P = N/m^2$
(3) Work, Energy, Heat	Joule	$J = N\text{-}m$
(4) Power	Watt	$W = J/s$

Other Important Units

	Symbol
(1) Enthalpy	$H = j/Kg$
(2) Entropy, Heat Capacity	$s = J/Kg\text{-}°k$
(3) Dynamic Viscosity	$\mu = N\text{-}s/m^2$
(4) Thermal Conductivity	$K = W/m\text{-}°k$
(5) Heat Transfer Coefficient	$h = W/M^2\text{-}°k$

Prefixes

10^{-6}	micro
10^{-3}	milli
10^3	kilo
10^6	mega
10^9	Giga

CONVERSION FACTORS

Temperature	$°F = (1.8)C+32$	$°C = (°F-32)/1.8$
Length	1 in = 25.4 mm 1 ft = 0.3048 m	1 mm = 0.03937 in 1 m = 3.2808 ft
Area	1 in^2 = 64.5 mm^2 1 ft^2 = 0.092903 m^2	1 mm^2 = 0.00155 in^2 1 m^2 = 10.76391 ft^2
Volume	1 U.S. gal = 3.7854 L 1 in^3 = 16387 mm^3 1 ft^3 = 0.02832 m^3	1 L = 0.26417 gal 1 mm^3 = 0.000061 in^3 1 m^3 = 35.3107 ft^3
Mass	1 oz = 28.352 g 1 lb = 0.4535923 Kg	1 g = 0.0352709 oz 1 Kg = 2.2939794 lb
Mass Flow	1 lb/hr = 125.9978·10^{-6} Kg/s 1 Ton/Hr = 0.27777 Kg/s	1 Kg/s = 7936.6465 lb/hr 1Kg/s = 3.6 Ton/Hr
Force	1 ft-lb = 4.4482 N	1 N = 0.22481 ft-lb
Energy	1 Btu = 1.05506 KJ 1 Million Btus = 1.05506 Giga J [1 x 10^6 Btus = 1.05506 x 10^9 J]	1 KJ = 0.947813 Btu 1 Giga J = 0.947813 million Btus [1 x 10^9 J = 0.947813 x 10^6 Btus]
Calorific Value	1 Btu/ft^2 = 37.259 KJ/m^3	1 KJ/m^3 = 0.02684 Btu/ft^2
Volumetric Flow	1 ft^2/s = 0.028316 m^2/s	1 m^2/sec = 35.3232 ft^2/s

Density
  Specific volume
	1 lb/ft^3 = 16.019 Kg/m^3
	1 ft^3/lb = 0.062426 m^3/Kg
		1 Kg/m^3 = 0.062426 lb/ft^3
		  m^3/Kg = 16.019 ft^3/lb

Presure	1 lb/in^2 = 6.8948 KN/M^2 1 lb/in^2 = 0.068948 Bar 1 Bar = 10^5 Nm^3 1 ft-lb/ft^2 = 47.880 N/m^2	1 KN/M2 = 0.145037 lb/in^2 1 Bar = 14.5037 lb/in^2 1 Nm^3 = 10^{-5} Bar 1 N/m^2 = 0.0208855 ft-lb/ft^2
Dyn viscosity	1 lb/ft-s = 1.4882 N-s/m^2	1 N-s/m^2 = 0.671953 lb/ft-s

CONVERSION FACTORS

Power	1 ft-lb/s = 1.3558 W 1 Btu/hr = 0.29307 W	1 W = 0.737572 ft-lb/s 1 W = 3.4121541 Btu/hr
Heat Transfer	1 Btu/hr-ft²-°F = 5.6783 W/m²-°K	1 W/m²-°K = 0.17611 Btu/hr-ft²-°F
Fouling Factor	1 hr-ft²-°F/Btu = 0.17611 m²-°K/W	1 m²-°K/W = 5.67831 hr-ft²-°F/Btu
Heat flux Density	1 Btu/hr-ft² = 3.1546 W/m²	1 W/m² = 0.316997 Btu/hr-ft²
Thermal Conductivity	1 Btu/hr-m²-°C = 1.7307 W/m-°K	1 W/m-°K = 0.57781 Btu/hr-m²-°C
Volumetric Heat Release	1 Btu//ft³-hr = 10.34975 W/m³	1 W/m³ = 0.0966211 Btu//ft³-hr
Heat Storage Capacity	1 Btu-in²/lb = 15.3023 KJ/Bar	1 KJ/Bar = 0.06535 Btu-in²/lb
Specific Heat	1 Btu/lb-°F = 4.1868 KJ/Kg-°K	1 KJ/Kg-°K = 0.238846 Btu/lb-°F

Chapter 12

Waste Heat Recovery

The value of waste heat comes from the fact that it supplants additional input energy, reducing overall energy costs.

WASTE HEAT RECOVERY OPPORTUNITIES

Flue gasses from a boiler represent a 17% to 30% (plus) opportunity for savings investments. This chapter will cover the technology for waste heat recovery including practical approaches to boiler efficiency improvement and other concepts for utilizing the recovered energy from flue gasses.

On the average the temperature of flue gasses leaving boilers is about 400°F ranging between 350°F and 650°F. Flue gas leaves the boiler at a temperature higher than the steam temperature for heat transfer to take place.

The boiler exhaust approach temperature, in general, will rise from 400F to 150°F from low load to maximum (Figure 12.1). Many tests have shown these numbers to vary widely, so each boiler should be tested for its characteristic exhaust temperature, preferably after a cleaning and tuneup, to establish the ideal temperatures for that particular boiler.

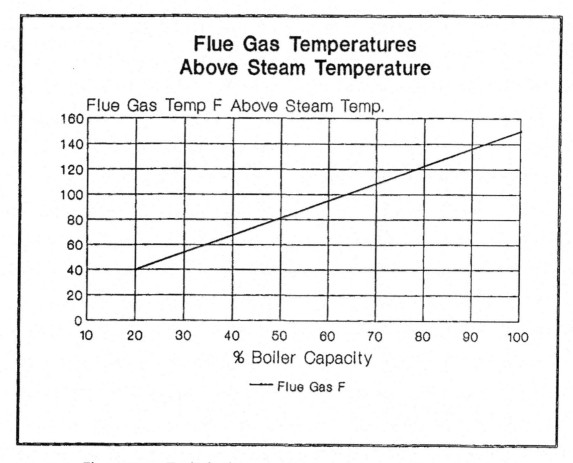

Figure 12.1— Typical exit gas temperature above steam temperature.

WHERE WASTE HEAT CAN BE USED

A suitable use for waste energy is critical to any waste heat recovery project because it doesn't matter how much energy you can recover, the only thing that is going to save you money is to actually use the energy in your facility, and by doing so, decrease the amount of outside energy you have to purchase.

Typical uses for waste heat energy are:

a. Boiler feedwater heating
b. Makeup water heating
c. Combustion air preheating
d. Process heating
e. Domestic hot water
f. Generating electricity

If waste heat can be utilized in the boiler itself, a considerable advantage is gained by the fact that it is a self-controlling process requiring simple or no controls to regulate its application. If this same energy were to be used in a plant or building, it would be supplying a demand which would vary from the typical boiler operation and need additional controls. There might be periods when the energy wouldn't be needed thereby wasting it. Using the boiler for waste heat recovery provides an uninterruptable use of this energy.

ACID FORMATION A LIMITING FACTOR IN WASTE HEAT RECOVERY

One of the most important factors influencing stack gas heat recovery is the corrosion problem accompanying the cooling of the gas. Because the sulfuric acid dew point is higher than the water vapor dew point, heat recovery efforts must eventually contend with the acid dew point problem (Figure 12.2).

The acid dew point is that temperature at which acid begins to form (Figure 12.3). This temperature varies with the sulfur content of the fuel (Figure 12.4). To avoid the corrosive effects of acids the traditional practice has been to limit the heat recovery to a minimum stack gas exit temperature of 350°F. Within the past decade, however, corrosion-resistant materials have been developed and applied to heat recovery systems, many of which operate below the acid dew point.

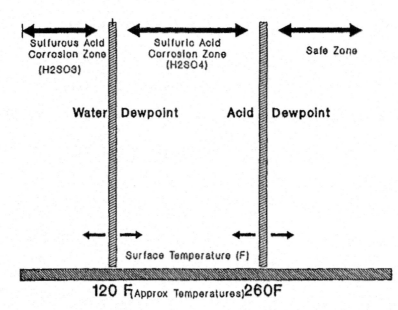

Figure 12.2. —The relationship of acid dew point and water dew point to the formation of acids from the sulfur in fuel.

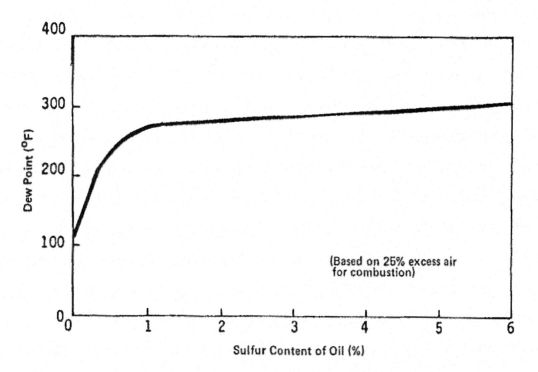

Figure 12.3. —The relationship of acid dew point to the sulfur content of fuel.

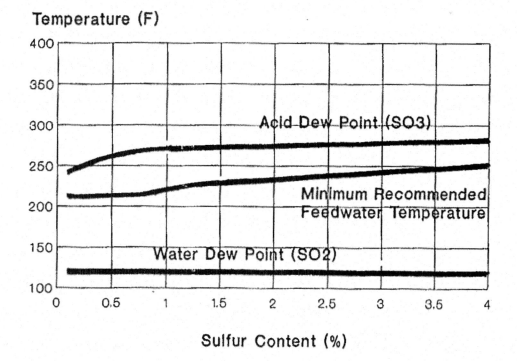

Figure 12.4. — Minimum recommended feedwater temperature to avoid economizer tube corrosion.

WHAT IS FLUE GAS?

Before we go on, let's explore some basics in the field of waste heat recovery. The first applies to the nature of flue gasses which contains both dry gas and moisture. The amount of moisture in flue gasses varies with the amount of hydrogen in the fuel, each pound of hydrogen in the fuel combines with oxygen to form approximately 9 pounds of water (Table 12.1). This water is in the superheated state containing more than half the energy in the flue gas stream. As each fuel has a different carbon to hydrogen ratio the energy in the moisture in the flue gas stream will vary from fuel to fuel.

CARBON AS FUEL

The dry gas in the flue gas is formed from the combustion of carbon to carbon dioxide plus nitrogen which does not take part in the combustion process and any excess air above and beyond the amount needed for combustion.

DRY GAS AND MOISTURE LOSSES

The carbon and hydrogen composition of various fuels determines the dry gas and moisture losses. Table 12.2 shows how this affects flue gas losses and the minimum stack losses that can be expected.

SENSIBLE AND LATENT HEAT

Sensible heat is that heat which can be sensed with a thermometer or other temperature sensing instrument.

The formula for sensible heat transfer is:

$$Q = M\,C_p\,(T_2 - T_1)$$

Q = Heat content Btu/hr

M = Flow rate LB/HR

C_p = Specific heat BTU/LB/F
T_1 = Stream temperature
T_2 = Reference temperature

Latent heat is the heat required for phase change, i.e. to change a liquid to a vapor (water to steam) or vice versa, without a change in temperature.

The formula for latent heat transfer is:

$$Q = M\,H$$

Q_1 = Heat content Btu/hr
M = Flow rate lb/hr
H = Heat of vaporization Btu/hr

FUEL

Combustion products from burning fuels with higher hydrogen content contain more water vapor and larger amounts of latent heat loss potential. Gas fired boilers are inherently less efficient than heavy oil fired units and represent better candidates for heat recovery. The type of fuel will also affect the maintainability and service life of a heat recovery system.

Natural gas is a clean-burning fuel and causes minimal corrosion problems in heat recovery hardware.

Fuel oil contains varying amounts of sulfur, which leads to acid corrosion problems.

REGENERATORS AND RECOUPERATORS

Different terminologies have developed over the years in different industries referring to heat recovery process. The term regenerators has come to refer to the alternate heating and cooling of a media, such as plates or brickwork or other heat absorbing material with hot exhaust gasses, and then recapturing the heat by warming combustion or process air over the same media by manipulating gas and air streams.

Recuperators refers to the continuous operating (static) type or heat recovery unit using

Table 12.1— percentage weight by species and higher heating values for fuel types.

Species	Percentage Weight by Species for Fuel Types		
	Natural Gas	No. 2 Oil	Coal
Carbon	74.7	87.0	75.1
Hydrogen	23.3	12.5	4.8
Higher Heating (Btu/lb)	22,904	19,520	13,380

Table 12.2—flue gas losses due to dry gas and moisture.

Fuel Min. Dry	Min. Gas loss Loss (%)	Min. Moisture Loss (%)	Stack Gas Loss (%)
Natural Gas	2.9	10.1	13.0
No.2 Oil	5.1	6.4	11.5
No.6 Oil	6.6	6.2	12.8
Coal 5.5	10.0	15.5	

an intermediate wall between the hot and cold streams.

In boiler plants it is more common to hear the heat recovery apparatus referred to by name, (i.e. air preheater, economizer, etc.) than regenerator or recuperator.

CONVENTIONAL ECONOMIZER

An economizer performs two functions, it reduces stack temperature and also heats boiler feed water (Figure 12.5).

A practical rule of thumb is that for every 40°F the stack temperature is reduced there is a corresponding 1% efficiency increase. On the water side, an increase of approximately 1% in efficiency is expected for each 11'F rise in feedwater temperature.

Economizers have been in use for a long time and it has been found that boilers operating at pressures of 75 psig or greater are excellent applications. One of the strongest points for installing an economizer is its compact size compared to other options.

Some general guidelines for economizer installations are:

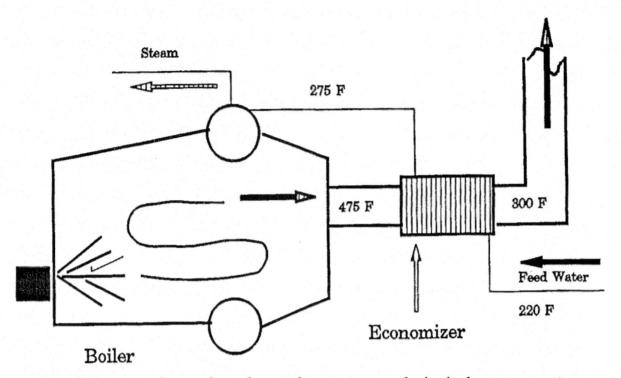

Figure 12.5—Economizer used to reduce stack temperature and raise feedwater temperature.

a.   Average stack gas temperature of 450°F
b.   More than 2,500 operating hours a year.
c.   Stack flow rate more than 15,000 lb/hr

The maximum benefit that can be safely achieved is governed by a number of technical and physical limitations.

a.   Economic considerations. Many economizers have paid for themselves in two years or less, the economizer then continues to return dividends from reduced energy costs long after it has been paid for.

b.   When limited fuel availability or steam production threatens plant production an economizer can increase boiler capacity from 4% to 10%. If there is a need for more steam capacity, this may be the most cost effective way to do it rather than invest in a new boiler.

c.   By the law of diminishing returns, an increase in heating surface does not provide, in equal proportion, for an increase in fuel

savings. The flue gas temperature can not be reduced below the temperature of the incoming feed water or the acid formation temperature.

d.   It may be possible that a controlling limitation may be imposed by the space available for the installation of an economizer.

e.   When a boiler operates with a stack for induced natural draft on a negative pressure furnace, a limitation is imposed on the furnace draft due to the cooler exhaust gasses caused by the economizer. In this case an induced draft fan may be required. In other cases of balanced draft or forced draft systems, the additional boiler efficiency can offset the additional draft requirement caused by the lower stack temperatures and pressure drop across the economizer.

f.   Outlet water temperature is a limitation to prevent steam formation and water hammer in the economizer. An approach temperature of 40°F is customary for variable boiler load conditions. The outlet water temperature may be the dominant limitation in selecting economizer size.

g. Acid formation and condensation on the gas side of the heating surface is determined solely by the temperature of the surface which is essentially the same as the water temperature.

h. A minimum gas temperature of 250°F in the stack is desirable. Basically, this is to assure that the flue gas will be sufficiently buoyant to escape into the atmosphere and not mushroom around the stack and cause smoke or acid rain nuisance.

COLD-END CORROSION IN ECONOMIZERS

The major portion of sulfur in fuel is burned and forms sulfur dioxide (SO_2) in the flue gas; a small portion, 3 to 5 percent, is further oxidized to sulfur trioxide (SO_3). These oxides combine with moisture to form sulfurous (H_2SO_3) and sulfuric acid (H_2SO_4) vapors (Figure 12.2). When in contact with a surface below the Acid Dew Point (ADP), condensation takes place. The ADP is directly related to the amount of sulfur in the fuel as shown in Figure 12.3.

Because of the higher heat transfer coefficient in liquid-metal than in gas-metal heat transfer, the gas side metal temperature of an economizer is closer to the water temperature than the gas side temperature.

Design parameters for a conventional economizer are shown in Table 12.3, illustrating the minimum recommended feedwater temperature and flue gas exit temperatures.

CONTROLLING ACID FORMATION

Controlling Economizer Inlet Temperature

The most efficient and effective means of controlling economizer metal temperature is with a

**Table 12.3.
Typical design parameters for a
conventional economizer.**

Fuel Type	Maximum Minimum Inlet Water Temperature	Exit Flue Temperature (°F)	Fin Density (Fins/In)
Natural Gas	210	300	5
No. 2 Fuel Oil	220	325	4
No 5 & 6 Fuel Oil	240	350	3
Coal	240	350	2

feedwater preheat system. The system is illustrated in Figure 12.6. It is essentially a feedwater preheater with sensors controlling the steam admission valves to the heater.

The steam admission valve sensors measure the water temperature entering the economizer and the temperature of the stack metal. Both of these surfaces are subject to corrosion from temperature exposure below the ADP. Using the heater insures neither surface will cool to that point. As an additional protection, the portion of the stack exposed to very cold weather conditions could be insulated to keep the metal temperature from becoming too cold, approaching the ADP.

Reduce Excess Air

Reducing excess air raises the dew point temperature in the stack gas (Figure 12.7). Research has shown a direct relationship between excess air and the formation of sulfur trioxide.

Use Corrosion-resistant Materials or Sleeves

Corrosion-resistant alloy steels can be used in preventing corrosion, however their high cost normally prohibits their actual use. Interlocking cast iron sleeves over carbon steel tubes can also be used where severe acid conditions are antici-

Waste Heat Recovery

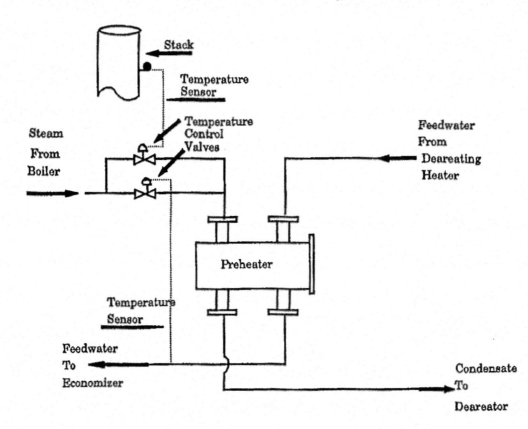

Figure 12.6—Feedwater preheater used to control cold end corrosion in the economizer and acid dew point n the stack. The steam energy used for heating is not lost as it recycles back into the boiler as heat in the feedwater.

Figure 12.7. Relationship between initial water vapor dew point of flue gasses and the excess air level.

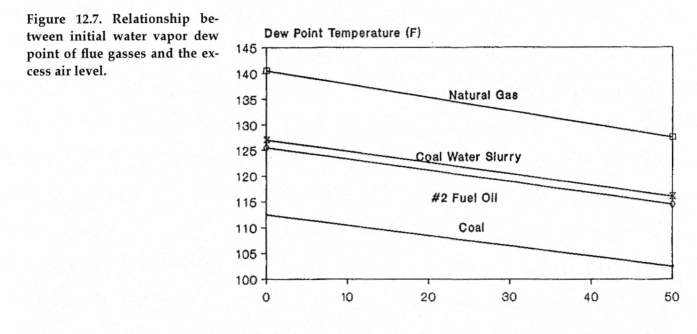

pated. The cast iron sleeve will corrode, but it can be replaced at normal maintenance intervals. The cast iron sleeves protect the carbon steel tubes from contact with corrosive gases.

Improve Fluid Flow Arrangement

Using parallel-flow tube arrangements rather than counter-flow arrangements increases economizer skin temperature. With the parallel flow arrangement, cold feedwater enters where the stack gas is the hottest, thus raising the tube surface temperature slightly.

Modulate Feedwater
Flow through the Economizer

The feedwater flow rate through the economizer economizer can be controlled by diverting feedwater flow around the economizer during periods of low flue gas temperatures (Figure 12.8).

Insulate Stack Metal

Just as with metal temperatures in the economizer, flue gas temperatures will not necessarily determine the severity of acid corrosion, the same is true for the stack. It is the temperature of the metal, or stack wall, that is in contact with the flue gas that will determine the extent of acid corrosion. Stack insulation will aid in keeping stack temperatures above the ADP.

Alternatives to insulation are high temperature corrosion-resistant stack material such as Cor-Ten or Fiberglass reinforced plastic.

CONDENSING THE MOISTURE IN FLUE GASSES

The major limitation to increased boiler efficiencies Is the amount of energy tied up in latent heat. Until recent years equipment has not been available to capture and use this large source of waste energy. Figure 12.9 shows the energy available in combustion products for natural gas at different excess air levels as the temperature drops from 600°F to the point of flue gas condensation and full recovery of latent and sensible heat.

For example, 400°F flue gasses from a natural gas fired boiler contains 18% of the total energy from the HHV of the input fuel in the form of both sensible and latent heat. Almost 62% of

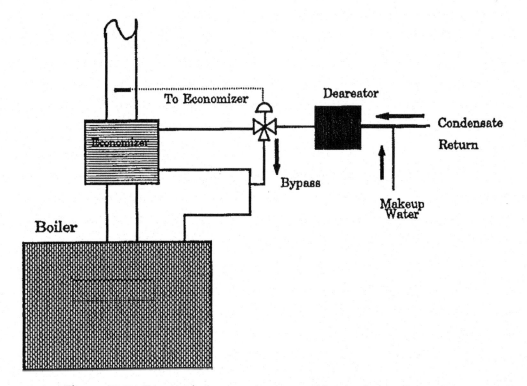

Figure 12.8—Economizer corrosion control by feedwater bypass.

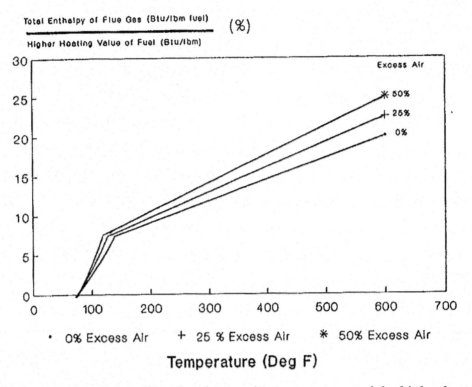

Figure 12.9—Enthalpy of combustion products as a percent of the higher heating value of natural gas as a function of temperature and excess air.

this energy which would otherwise be lost up the. stack, requires cooling the exhaust gasses below the ADP and water dew point. This temperature is a function of excess air (Figure 12.10). For natural gas with 10% excess air the initial condensation temperature is 137°F.

The condensation of flue gasses is also a function of temperature as well as excess air as shown in Figure 12.10. Note that after the initial condensation temperature is achieved, the percent of water condensed depends on lowering the temperature even further, ranging roughly from 140°F to below 80°F.

Boiler efficiencies above 95% are possible if the moisture in flue gasses can be condensed, Figure 12.11 shows how the recoverable energy increases dramatically with the condensation of flue gasses. The straight line characteristic of sensible heat recovery bulges from a 2-5% efficiency increase to 11-15% with latent heat recovery.

Various fuels, because of their hydrogen content, offer different opportunities for waste heat recovery as illustrated in Figure 12.12. It shows

at 450°F 18% for natural gas, 15% for coal-water slurry, 14% for no.2 fuel oil and 11% for coal.

Indirect-contact Condensing Heat Exchanger

The indirect-contact condensing heat exchanger is generally fabricated from corrosion-resistant materials like Teflon and glass.

Since Teflon can only be extruded over smooth surfaces and glass tubes cannot be fabricated as finned tubes, an indirect contact heat exchanger requires a greater number of tubes and will occupy a greater volume. However, the weight may not be greater since thinner tube walls can be used.

Glass tube heat exchangers are limited to applications where the flue gas temperatures do not exceed 4000F and the water pressure does not exceed 50 psig.

Teflon can be extruded over tubing as a thin film (.015 inch). It can operate with flue gas temperatures up to 500°F, but 400°F is recommended for continuous operation.

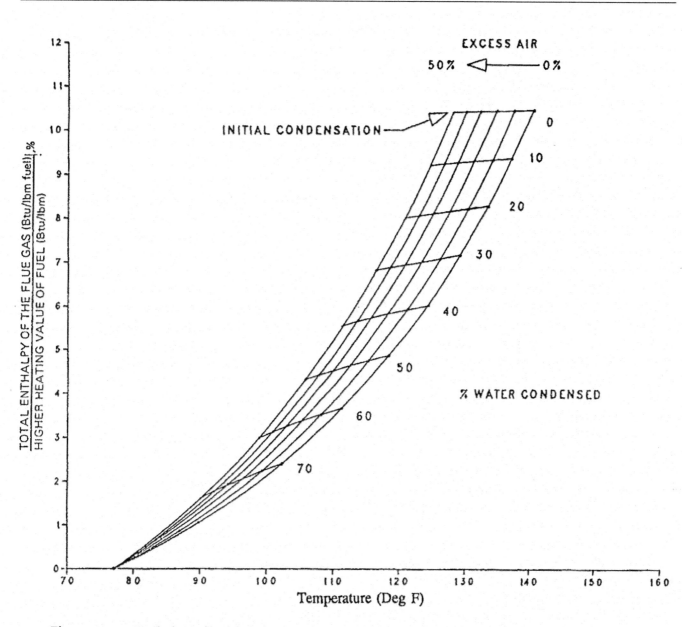

Figure 12.10—Enthalpy of combustion products during condensation, as a function of temperature and excess air for natural gas.

Teflon coated heat exchangers are capable of raising water temperature to 200 to 250°F.

Another less common heat exchanger uses stainless steel tubes. Stainless steel heat exchangers were installed in 25 hospitals with gas fired (very low sulfur fuel) boilers, and many of these heat exchangers have operated trouble free for more than 8 years. This type 304 stainless steel has been proven to be a durable material in the stack gas environment for natural gas boilers. The presence of chlorides can cause metal failure due to stress corrosion however. A common source is cleaning solvent vapors, so storage of chloride containing material near the boiler combustion air inlets could cause problems.

Metals have a very wide range of corrosion resistance. As sulfuric and sulfurous acid is the most likely attack to be encountered, material selection must take this into account. Stainless steel, for example does not stand up well under this attack. Carbon steel or "open hearth" steel is most commonly used for economizer construction, giving long and reliable service. It has a corrosion rate superior to many more expensive

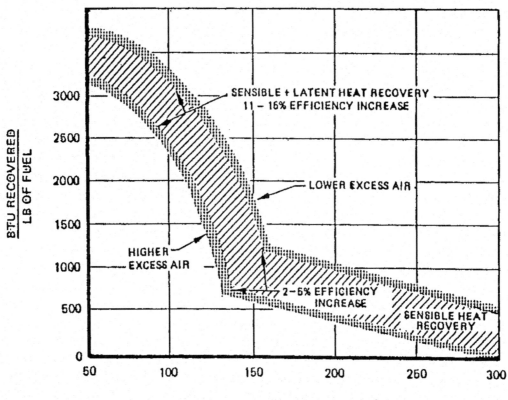

Figure 12.11—This figure demonstrates the heat recovery potential from a natural gas fired boiler by reducing exit gas temperature below the condensing point where latent heat is given up by the flue gas. The potential to recover heat depends on the hydrogen ratio of the fuel and excess air. Typically for natural gas boilers, a 11 to 15 percent efficiency increase is possible.

alloys. Cor-Ten may be another good choice for economizer material, which has a corrosion rate corresponding to titanium.

The most prevalent systems to operate continuously below the ADP to capture the large amount of latent heat usually lost with other systems include borasilicate glass tubes, ceramic coated steel or copper tubes, and coated-coated copper tubes.

Since corrosion is not a factor with these materials, stack gas can be cooled to well below the traditionally recommended safe temperatures. Lowering temperatures below the water vapor dew point promotes the recovery of latent heat from the flue gas and the recovery of large quantities of low grade energy.

The higher the hydrogen ratio in the fuel the higher the efficiency of the condensing heat re-

covery unit (Table 12.1).

The indirect-contact condensing heat exchanger also acts as a stack gas "scrubber". Researchers report that condensing heat exchangers greatly reduce stack emissions.

Direct Contact Flue Gas Condensing Heat Exchanger

In a direct contact heat exchanger, heat is transferred between the two streams, typically flue gas and water, without intervening walls. This is typical of other heat transfer equipment. It is a vertical column in which the two streams move in a counterflow direction. The flue gasses enter at the bottom and water is either sprayed or cascades over trays or travels through a packed bed from the top to the reservoir at the bottom of the column.

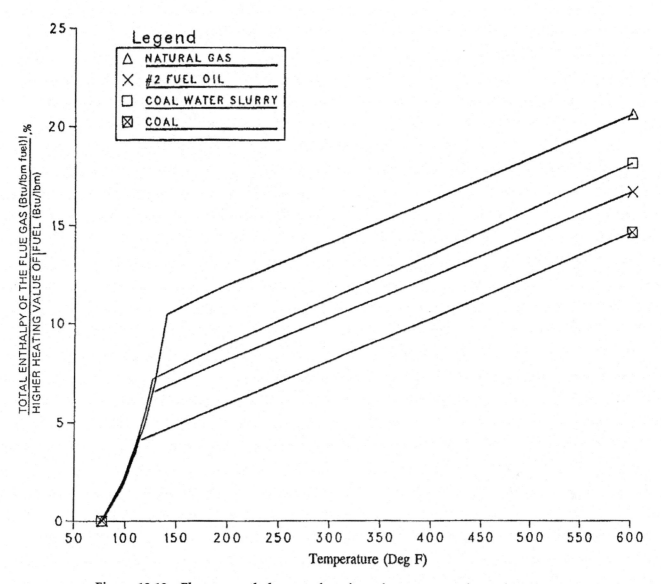

Figure 12.12—Flue gas enthalpy as a function of temperature for various fuels.

The direct contact heat exchanger is very efficient because there are no heat exchange surfaces exposed to clogging, corrosion or fouling. The elimination of an interfering wall greatly increases the heat transfer rate.

The direct contact heat exchanger is an ideal candidate for transferring latent heat from flue gas because the spraying of fine "mist like" water droplets can produce a large heat transfer surface in the presence of relatively small temperature differences between heating and cooling streams.

The packed tower design also enhances the available heat exchange surface, providing a highly effective scrubbing action which improves heat transfer efficiency and raises reservoir temperatures.

Figure 12.13 shows a direct contact flue gas condensing heat recovery system in operation. The primary water circuit includes the piping from the reservoir, a pump and the spray nozzles at the top of the column. A heat exchanger is used to transfer heat from the primary system to a secondary system which could be any low temperature water use such as domestic hot water, boiler make up feed water or heating applications.

Figure 12.14 shows the three types of columns used for direct contact heat recovery. The open spray tower presents the least obstruction to exhaust gas flow, less than one tenth inch of water column; but has the lowest reservoir temperature from 100°F to 110°F.

The tray type tower has a resistance to gas flow between 0.5 and 1.0 inches of water column and an outlet temperature from 130°F to 140°F.

The packed tower type has excellent heat transfer but also has a high pressure drop in excess of 5 to 10 inches water column, but the reservoir temperature is as high as 150°F depending on conditions.

Heat Pipe

A heat pipe is composed of a sealed pipe partially charged with water, Freon, ammonia or other suitable substance. It can be divided into three sections (Figure 12.15), the evaporator section, the adiabatic section and the condenser section.

In the evaporator section, heat is absorbed by the internal working fluid by evaporation. Pressure differential causes the vapor to flow to the cooler condensing section where the latent heat is given up to the cooler environment condensing the working fluid.

The internal circumference of the heat pipe is lined with a thin layer of wicking or mesh type material and the working fluid migrates by capillary action back to the evaporator section where

the cycle is repeated.

A thermosyphon type heat pipe does not utilize a wicking material, instead the pipe is placed in a vertical position with the condenser placed above the evaporator section. As the vapor condenses, it returns to the evaporator section by gravity. Even with wicking material it is advantageous to locate the evaporator -section-- below the condenser so condensed fluid is aided by gravity.

Heat pipes have the advantage that they can be constructed of corrosion resistant materials. They are mostly applied to transfer heat from one gas stream to another, but have been used successfully in fluid streams also. The gas streams must be adjacent to each other because of the loss of performance from routing streams of working fluid and vapor through an unusually long adiabatic section. Heat pipes can be compact, operate at low temperature differentials and have no moving parts to malfunction.

Plate Type Heat Exchanger

Plate type gas-to-gas heat exchangers can be used in counterflow or crossflow through adjacent passages separated by heat conducting walls

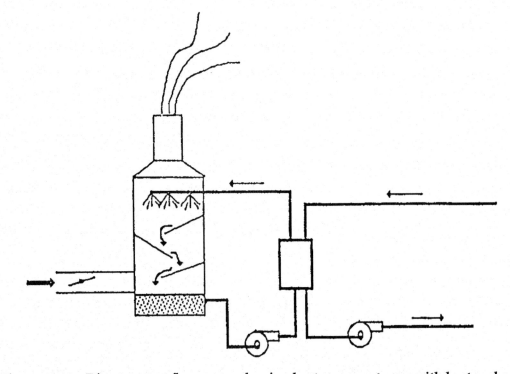

Figure 12.13—Direct contact flue gas condensing heat recovery tower with heat exchanger.

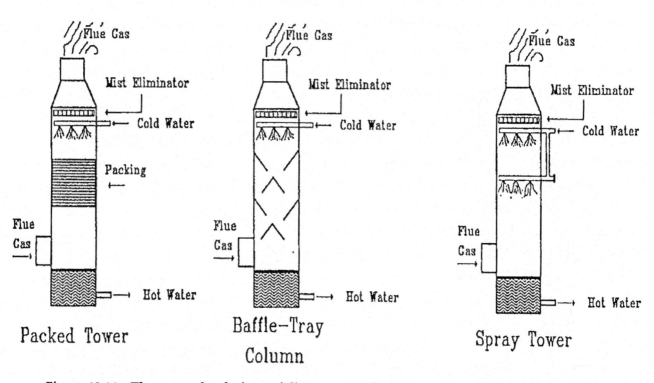

Figure 12.14—Three popular designs of direct contact flue gas condensing heat recovery tower.

(Figure 12.16). It is possible to preheat combustion air, process fluids, building make up air, etc. by this type of heat exchanger. It is a relatively simple system, but the gas streams must pass in close proximity.

One manufacturer has overcome several problems with this type of exchanger. A unit has been designed to be torqued together using resilient mountings, eliminating thermal stress and cracked welds associated with expansion and thermal shocks. Also, these plate units can be constructed of corrosion resistant materials or coated with Teflon to insure a long useful life. These units are also designed to benefit from latent heat recovery and to cope with acid corrosion and condensation problems.

The Heat Wheel

The heat wheel is a regenerative type of heat exchanger which extracts heat from one source, briefly stores it and then releases it to a cooler stream {Figure 12.17). It consists of a large rotating wheel frame, packed with a heat absorbing matrix. Its primary application is for combustion air preheating. As it rotates it passes through the

hot section where the matrix or plates are heated by the flue gasses, then on the other side of a sealing section it rotates through a cold air section giving up its heat.

Heat wheels can exceed 50 feet in diameter. Matrix materials can include aluminum, stainless steel, and ceramics for higher temperatures.

Disadvantages include cross contamination of the two streams due to sealing and purging problems, clogging of passages, gas and air flow restriction and drive motor horsepower. The large size of the units and supporting ducting is another disadvantage. These units can be used to preheat combustion air, but cannot be used to heat building air because of cross contamination of combustion products.

Tubular Heat Exchangers

Tubular heat exchangers are also used for combustion air preheating. The heater is arranged for vertical gas flow through a nest of tubes. The air passes horizontally across the tubes in sections of the air heater defined by built in baffles.

In contrast to regenerative designs, tubular or recuperative air heaters have more severe cold

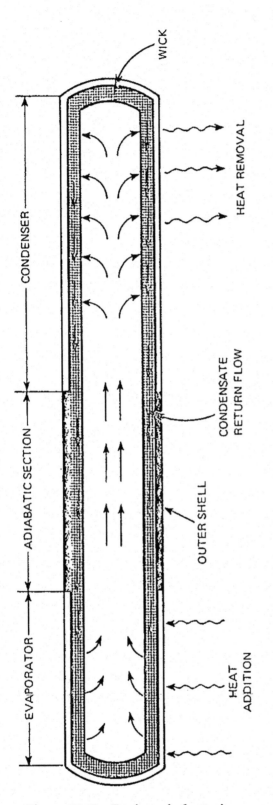

Figure 12.15—Design of a heat pipe.

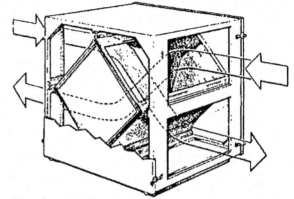

Plate heat exchanger for gases

Figure 12.16—Gas-gas plate heat exchanger.

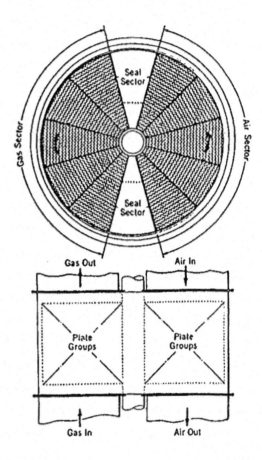

Figure 12.17—Outline of rotary air heater.

end corrosion problems.

To avoid corrosion, it is necessary to limit the air temperature entering the heat exchanger to a minimum value. This can be achieved by:

a.   recirculating some of the air from the pre-heater outlet back to the inlet.

b.   bypassing part of the inlet air around the heating surfaces.

c.   using a recuperating steam coil in the air duct upstream of the air preheater.

With high sulfur fuel, the expected life of an air preheater is short. The strategy is to extract as much energy as possible from the flue gas regardless of the corrosion potential and to specify material to keep the problem to a minimum (Table 12.4).

Combustion Air Preheating

Although preheating of combustion air is not generally recommended for smaller boilers, it could be the only way to significantly improve boiler efficiency. As a retrofit, the cost may be high because of bulky ductwork, and possibly a long distance between the stack and air inlet.

Increasing combustion-air temperature worsens emissions of NO_x, typically from 20 to 100 ppm for a 100°F rise. Reductions in SO_x, carbon monoxide and particulate emissions result from improved combustion.

In addition to making use of waste heat from flue gasses, lower excess air can be maintained owing to improved combustion at the high inlet air temperature. Higher combustion temperatures also increases heat transfer and reduces sooting.

Gas-to-air inefficient heat exchangers are very in comparison to gas-to-water exchangers. This is due thermal characteristics comparison to air.

In summary, unless corrosion resistant materials are used, combustion air entering the preheater must be suitably high to avoid cold-end corrosion. A steam coil heater, on the cold side, to raise the temperature of the incoming fresh air is recommended.

Run-around Coil

The typical run-around coil system is composed of two heat exchangers coupled together by the circulation of an intermediate fluid (Figure 12.18). The circulating fluid is heated by the hot stream and then piped to the second heat exchanger where its heat is given up to a cold stream.

Table 12.4—Cold-end air preheater temperature material selection guide.

Fuel Type	Incoming Air Temp Range	Material Specification
Oil	190-205°F	Carbon steel components.
		Corrosion resistant low alloy steel for the cold-end element.
Oil	155-190°F	Corrosion resistant low alloy steel intermediate element
		Enameled cold-end element.
		Low alloy steel for rotor and structural parts in the cold-end to the same level as the enameled elements.
Oil	Below 175°F	Enameled intermediate element.
		Enameled intermediate element.
		Corrosion resistant low allow steel rotor and supports to same level as enameled elements.
Bituminous Coal	Below 155°F	Carbon steel components.
		Corrosion resistant low allow steel cold-end element.
Pulverized Anthracite	Below 150°F	Carbon steel components
		Corrosion resistant low allow steel cold-end element.
Gas Fuel Sulfur Free		

This system can be applied to transfer heat to combustion air, process air, or building air. Since the heat exchanger requires some temperature differential to transfer heat to or from the intermediate fluid, it is inherently less efficient than a direct exchange between two primary fluids. However this system is relatively simple and more compact than a direct air/fuel gas system. A run-around system eliminates the problem of the close proximity of exhaust and inlet ducts. It is able to transfer heat from one location to another without great retrofit costs.

Figure 12.18—Run-around heat recovery system.

Organic Rankine Cycle

The Organic Rankine Cycle (ORC) is one method to convert thermal waste streams to electrical energy (Figure 12.19). It is a closed loop system filled with an organic fluid having a low boiling temperature. The working fluid is vaporized at an elevated pressure by the waste heat stream in a boiler.

The vapor expands across a turbine which is directly coupled to an electrical generator. The low pressure exhaust vapor from the turbine is condensed and pumped back to pick up more waste heat.

Both initial investment and annual maintenance can be relatively high. The application of an ORC system would depend heavily on the importance and value of electricity as a commodity.

Heat Pump

Heat pumps have the ability to raise low-temperature energy to a higher temperature level. The most common heat pump system is a closed loop filled with refrigerant (Figure 12.20).

The heat pump cycle begins with a compressor raising the temperature of the working fluid, in vapor form. This hot vapor goes to a heat exchanger (condenser) where its heat is transferred to heat air or water. The high pressure liquid refrigerant then passes through an expansion valve and its pressure is suddenly lowered causing it to cool. In the evaporator, this low temperature vapor absorbs energy from the waste heat stream raising its temperature. The refrigerant vapor

once again enters the compressor and the cycle continues.

The effectiveness of all mechanically driven vapor compression heat pumps is specified in terms of coefficient of performance (COP) defined as:

$$COP = \frac{\text{Useful thermal energy output}}{\text{Work input to the compressor}}$$

A COP of 5 means using one unit of work input (in the form of electrical energy to power a compressor) to deliver 5 units of heat output, of which four units come from the waste heat source. All five output units being at the raised output temperature.

The heat pump is becoming increasingly popular because of this ability to raise the temperature level of recovered energy.

Although heat pumps are most often driven by electrical motors, heat pumps driven by combustion engines have the advantage of being powered by fuels costing a fraction of electricity and the added advantage of using waste heat generated by the engine.

Both turbines and internal combustion engines can recover 70% to 80% of their own waste heat to add to system efficiency. Driving a compressor with a fuel-fired engine rather than an electric motor becomes attractive when the electricity costs are high and natural gas or fuel oil costs are low.

Preheating Air or Water?

Due to the superior heat transfer qualities of water, preheating feed water is more efficient and economical than preheating combustion air. However, air is always available at ambient temperature with good potential for accepting flue gas heat.

Fire Tube Boiler Air Preheating

In general, fire tube boilers are not designed to operate with preheated combustion air, and it would be impractical to attempt to improve boiler efficiency with this type of system.

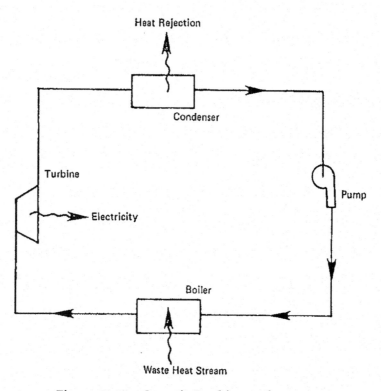

Figure 12.19—Organic Rankine cycle system.

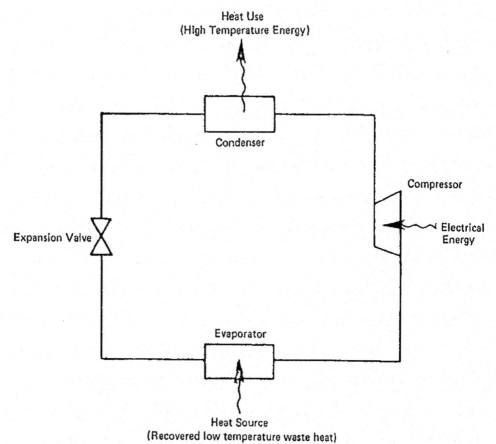

Figure 12.20—Closed-cycle vapor compression heat pump.

Which Option Is Best?

Based on a number of general assessments considering all facts involved in economic benefits, engineering criteria, installing and operating waste heat recovery systems; three primary systems have evolved as being very good prospects for waste heat recovery. The waste heat recovery systems which showed the most promise and applicability are:

a. The conventional economizer
b. The indirect-contact condensing heat exchanger
c. Direct contact flue gas condensation heat recovery

Table 12.5, summarizes the merits and limitations of various heat recovery alternatives.

Table 12.5—Merits and limitations of heat recovery equipment and approaches.

Equipment Type	Merits	Limitations
Economizer (Gas-Liquid)	Well developed and understood, easily tailor made with a variety of materials	Cold-end corrosion, water temperature approach to steam temperature 40°F
Waste Heat Boiler (Gas-Liquid /Steam)	Compact, high heat transfer easy to clean, robust, can operate in fouling environment	Not generally internal to process, must link with steam distribution system, most appropriate to high temperature exhaust streams (>600°F)
Direct Contact Flue Gas Condensing (Gas-Liquid)	Good heat transfer, simple, recovers sensible heat and latent heat, Efficiencies over 95% possible.	Lower grade heat recovery 120°F to 150°F Good for natural gas; less appropriate for oil/coal.
Indirect contact Flue gas Condensing Teflon (Gas-Gas) (Gas-Liquid)	Corrosion resistant, reliable under under thermal and mechanical shock, recovers latent heat. High efficiencies possible.	Teflon cannot be finned, operating range must be below 400 - 500°F.
Indirect Contact Flue Gas Condensing Glass, (Gas-Gas) (Gas-Liquid)	Good resistance to corrosion, easy to clean, tailor made, can recover latent heat, no cross contamination, can recover latent heat, high efficiencies	Large area of borosilicate glass needed, temperature limitation (< 400°F),sensitive to mechanical vibrations and shocks.
Plate Exchangers (Gas-Liquid) (Gas-Gas) (Liquid-Liquid)	No cross contamination, easy to install on-line cleaning, compact, temperature differentials can approach 2°F. Condensing units available.	Only small pressure differentials can be tolerated, large, inconvenient ducting. Difficult to clean in some circumstances. Subject to thermal and mechanical shock.
Heat Pipe Heat Exchangers (Gas-Gas)	No moving part , no cross-contamination, large pressures can be tolerated, compact. Various working fluids suite different temperature ranges.	Gas streams must be close.
Run-around Coil (Gas-Gas)	Covers large distances econominally, no cross contamination, easy to design and install.	Thermal efficiency less than 65%; moving parts, temperature limitations. Freeze protection may be needed. Electrical load controls needed.
Rotating Preheater Heat Wheel (Gas-Gas)	High operating efficiency 85%, low pressure drop, can recover latent heat. Very large sizes. Operating temperature to 800°F.	Cross contamination possible, moving parts, wear, large ductwork. Low differential pressure tolerance between gas streams can be a challenge.

Chapter 13

Steam System Optimization

STEAM SYSTEM OPTIMIZATION, A HUGE OPPORTUNITY FOR SAVINGS

According to the Department of Energy, nearly 160 billion dollars is spent creating steam in a single year. Steam generation accounts for fully one half of the industrial and commercial energy dollar. When you consider that steam plays a major role in generating electricity this number can be much higher. Energy conservation measures described in this chapter can play a significant role in saving over 15 billion dollars annually which is lost in steam distribution systems.

The cost of operating a steam system includes:

a. Boiler operating costs
b. Steam distribution system losses
c. Trapping system losses
d. Condensate return system losses
e. Operation and maintenance of the systems involved.

Boiler Plant Optimization

Boiler plant operating costs and losses have been covered in the earlier chapters of this book. An important fact to keep in mind is that every Btu wasted in the steam distribution system has to be replaced by the boiler and the boiler has large losses of its own. If a boiler is 80% efficient, then at least 20% of the original fuel used to generate steam for the distribution system is lost and does no useful work.

Getting the Most Energy from Steam Systems

Energy in steam consists of sensible and latent heat (Figure 13.1). It is apparent that a significant amount of heat is invested in bringing water up to boiling temperature. This sensible energy may be lost through a phenomena known as "Flashing" when hot condensate is formed at a high pressure and escapes to a lower pressure through the steam trapping system.

Figure 13.2 shows what happens to the condensate when it escapes from a higher pressure to a lower pressure. Illustrated here is a trap for a heat exchanger which is forming condensate at the rate of 100 pounds per hour. The system pressure is 100 psi and the corresponding steam temperature is 353°F. This 100 lb of condensate contains 32,500 Btu.

On the low pressure side of the trap, which is atmospheric pressure in this case, the saturation temperature is 212°F. Of the 100 pounds of condensate, 14.8 lb will form flash steam with a heat value of 17,100 Btu. The remaining 85.2 pounds of condensate formed will contain 180 Btu per pound for a total heat content of 15,400 Btu. Notice that the 32,500 Btu in the condensate on the high pressure side of the trap has formed steam and water with the same total heat content of 32,500 Btu.

Table 13.1 shows the percentage of steam that will form at the new lower pressure. This low pressure flash steam will, in fact, contain more heat than the condensate. There is a great potential for half of the energy going through the trapping system to be lost as flash steam, especially in condensate recovery systems that vent to atmosphere.

The higher the steam system pressure, the higher this flash steam loss will be, so one important factor that should be examined is how to reduce this loss.

Table 13.2 shows that the higher the steam system pressure, the less latent heat is available in the steam. This latent heat is what does the work,

247

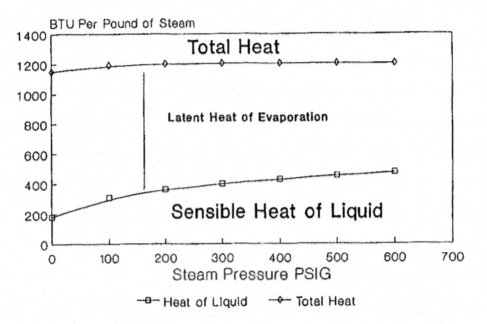

BTUs in a Pound of Steam

Figure 13.1—Heat in one pound of saturated steam at various pressures.

Table 13.1—Percent flash steam generated by condensate or boiler water when pressure is lowered.

FLASH STEAM TABLE

INITIAL PRESSURE PSIG	TEMP OF LIQUID	% FLASH AT ATMOSPHERIC PRESSURE	PERCENT OF STEAM FLASHED AT REDUCED PRESSURE				
			5 PSI	10 PSI	15 PSI	20 PSI	25 PSI
100	338	13.0	11.5	10.3	9.3	8.4	7.6
125	353	14.5	13.3	11.8	10.9	10.0	9.2
150	366	16.0	14.6	13.2	12.3	11.4	10.6
175	377	17.0	15.8	14.4	13.4	12.5	11.6
200	388	18.0	16.9	15.5	14.6	13.7	12.9
225	397	19.0	17.8	16.5	15.5	14.7	13.9
250	406	20.0	18.8	17.4	16.5	15.6	14.9
300	421	21.5	20.3	19.0	18.0	17.2	16.5
350	435	23.0	21.8	20.5	19.5	18.7	18.0
400	448	24.0	23.0	21.8	21.0	20.0	19.3
450	459	25.0	24.3	23.0	22.0	21.3	20.0
500	470	26.5	25.4	24.1	23.2	22.4	21.7
550	480	27.5	26.5	25.2	24.3	23.5	22.8
600	488	28.0	27.3	26.0	25.0	24.3	23.6
BTU PER POUND OF FLASH STEAM		1150	1155	1160	1164	1167	1169
DEG F OF FLASH STEAM & LIQUID		212	225	240	250	259	267
STEAM VOLUME CUFT/LB		26.8	21.0	16.3	13.7	11.9	10.5

Steam Trap Flash Loss

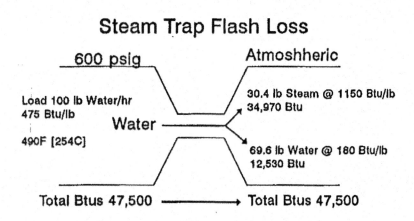

Figure 13.2—Flash Steam Energy Balance.

the rest is lost through the trap. If we were able to drop the working pressure on a piece of equipment from 125 psig to 5 psig, the latent heat would be 83.5% of the total energy instead of 73.5%.

Significant steam savings can be realized by reducing steam pressure as low as possible using pressure reducing valves in new as well as existing installations.

As shown in Table 13.3, the percent heat savings available from reducing pressure is significant. For example, there will be a 13% savings by reducing the pressure from 150 to 20 psig.

A Rule of Thumb for Optimizing Steam System Pressure

a. System steam pressure should be distributed at high enough pressure as is practical to overcome line losses and satisfy the highest pressure user.

b. Steam pressure should be reduced at the point of use to as low a pressure as is practical.

Two general points to consider in the steam distribution system is, design requirements limit steam velocities to less than 6,000 to 12,000 feet per minute and the system pressure drop should not exceed 20 percent of the total maximum pressure of the boiler.

Table 13.2 shows why steam pressure should be reduced at the point of use. It conserves the loss of energy through the steam trap.

Pressure psig	Latent Heat Btu/lb	Percent of Total Heat
125	868	73.5
50	912	77.5
5	961	83.5

Table 13.3—Heat savings available from reducing pressure at the point of use. (To calculate these numbers, flash steam reduction and the latent heat of the steam and the heat in the condensate at the old and new pressure is considered)

| Reduced Pressure | Original Pressure (psig) | | |
| | 125 | 150 | 200 |
	(Steam Savings %)		
100 psig	2.1	3.6	6.8
75 psig	4.4	5.9	9.0
50 psig	7.2	8.6	11.6
20 psig	11.5	13.0	15.8
10 psig	13.7	15.0	17.8

Heat Transfer Efficiency

Accumulations of air and noncondensible gases in the steam system can also limit steam flow, steam temperature and heat energy release.

Air is present in the system on start up and is also introduced by vacuum breakers on heat exchangers and process equipment. Noncondensible gases are liberated in the boiler by bicarbonates which forms CO_2. Oxygen is carried through the system. These noncondensible gases, when released, flow with the steam and can create heat transfer problems.

The gases cause a temperature reduction by contributing to total system pressure.

Dalton's Law of Partial Pressure states that a mixture of steam and other gases is equal to the sum of the partial pressures. This effectively reduces steam pressure, temperature and energy transfer.

Table 13.4 shows the effect of air on the temperature of a steam-air mixture.

Table 13.4—Air-steam temperatures.

Mixture Pressure	Pure Steam	5% Air	10% Air	15% Air
	Temperature °F			
2	219	216	213	210
5	227	225	222	219
10	239	237	233	230
20	259	256	252	249

Insulating Barriers

There is a second phenomena involved with noncondensible gases in the steam. When steam condenses on the heat exchange surface, the noncondensible gases also accumulate on the surface forming an insulation barrier.

The first one percent of air barrier has the most effect on reducing heat transfer (Figure 13.3). In this figure T_S is the steam temperature and T_W is the water temperature of the heat exchanger.

These gases take up volume and don't condense into a liquid as readily as steam, hence the term noncondensible. If allowed to accumulate for long periods, they take up enough volume to effectively block steam flow and energy transfer.

Bellows type thermostatic steam traps can be used as automatic air vents on heat exchange equipment. Air and noncondensibles in the system tend to be lighter than the steam and accumulate in quiet zones. If installed at these locations, the thermostatic device can sense the temperature reduction caused by the air.

Batch process cookers, large shell and tube heat exchangers and large steam coils should incorporate automatic air vents to eliminate air accumulations.

STEAM DISTRIBUTION SYSTEM LOSSES

Insulation

Steam is distributed through hot pipes which must be kept insulated to prevent excessive loss of heat and for safety. The range of surface temperature can vary from 200-500°F. The bare surface losses can vary from 300 Btu/Hr to 1700 Btu/Hr. The losses from uninsulated surfaces can be impressive, Figure 13.4 shows the losses from an uninsulated four inch gate valve for one year for steam costing $5.00, $8.00 and $11.00 per million Btus.

Steam users seem to be fully aware of the need to insulate hot surfaces. to prevent heat loss. Insulation pays for itself quite quickly, but insulating steam piping means not only the main piping, but also the unions, flanges, valve bodies, steam traps and everything else that is hot.

Removable (lace-up, foam in place and molded etc.), insulating covers are available for valve bodies and other hard to fit shapes. Mechanical steam traps such as float and thermostatic (F&T), inverted bucket (IB) should also be insulated as well as the bodies of thermodynamic disc traps (the covers should be left bare). Only Thermostatic (TS) traps and their cooling legs should be left uninsulated.

Condensate return lines are often not insulated because their heat losses are not considered

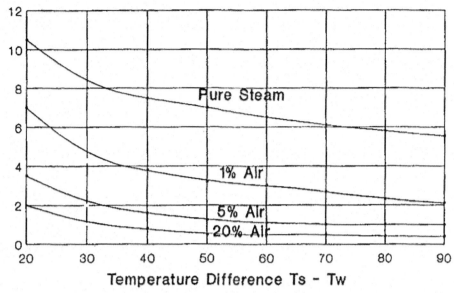

Heat Transfer Coefficient

Figure 13.3—Effect of air in steam on heat transfer.

Table 13.4—Annual dollar energy loss cost from an uninsulated 4-inch valve.

Mixture Pressure	Temperature °F			
	Pure Steam	5% Air	10% Air	15% Air
2	219	216	213	210
5	227	225	222	219
10	239	237	233	230
20	259	256	252	249

Table 13.5—Uninsulated bare metal heat loss in 80 degree still air for various steam pressures.

	Loss BTU/SQ Ft/Hr	
Steam Pressure PSIG	Temperature (F)	Heat Loss BTU/SQ Ft Per Hour
15	250	452
35	280	556
50	298	595
75	320	716
100	338	805
125	353	864
150	355	876
175	377	989
200	389	1,051
300	422	1,238
400	448	1,404
500	470	1,551
600	489	1,693

excessive, It should be a matter of concern however it is a matter of economy to insure that all of the heat possible is returned to the boiler plant for additional savings. Table 13.5 shows heat losses per hour for bare metal temperatures in 80 degree still air.

Inactive Piping

Inactive or dead piping must be positively isolated from the steam supply, this is often overlooked. Inactive steam piping wastes 100% of the steam supplied as well as being a maintenance burden.

Instead of relying on a hand operated valve, which may not shut tight or be leaking, it is better to blank off the unused part of the system or remove a section of piping to insure positively that this section is not wasting good energy.

External Steam Leaks

Steam leaks to the atmosphere are usually visible, audible or both. Because of the safety hazard they present, they are usually fixed quickly, However many leaks seem to go on being neglected for years. Their large plumes are hard to hide and can be seen for long distances (Table 13.6).

Table 13.6—Steam leak energy loss estimates based on steam plume length.

Plume Length (Ft)	Cost/Yr @ $7.00 Million Btus
4	$3,300
5	$5,900
6	$10,000
7	$18,000
8	$32,000

Internal Steam Leaks

Steam plumes are easy to spot. What about the steam leaks you can't see? Leaks inside steam systems are invisible.

Valves in bypasses around steam taps are prime candidates for this type of leakage. They are often opened to assist system start up or to vent excessive air and can easily be forgotten or only closed partially or develop leaks over time.

The need for bypasses should always be questioned. If they are needed for startup it is likely that their real purpose is the discharge of non-condensible gasses faster than a steam trap can handle them. An automatic air vent can do the job better and more economically. A trap could be installed in the bypass line as a back up in case repairs are required for the main trap or to provide more capacity during high loads. This way the liability of internal steam loss can be prevented.

Condensate Return System Losses

Condensate is certainly not waste water. It is water that has been preheated, softened, made suitable for use as boiler feedwater, and then distilled.

If it can be returned to the boiler feed system, it can be recycled with little further treatment. By reusing the condensate, you avoid the need for more raw water to be purchased and treated and you can avoid the sewage costs which can be incurred when condensate is allowed to run to the drain. Also, this condensate contains valuable heat energy that was paid for in the boiler fuel; energy that an efficient condensate recovery system will conserve.

The efficiency of a condensate return system can be improved, sometimes dramatically. For example, Table 13.7 shows annual dollar savings from improving condensate return. Savings are based on steam generation cost of $6.00 per 1,000 lbs. Make up improves from 100% to just 10%, a 90% reduction. Annual dollar savings for three boiler sizes listed include water cost, chemical treatment and energy savings.

Table 13.7—Annual cost reduction from improving condensate return.

100 BHP	250 BHP	500 BHP
$21,900	$54,750	$109,500

The Basic Challenge

The condensate system is perhaps the most ignored and least understood system in a steam using facility. An examination of the dollar savings listed in Table 13.7 suggests that it might be profitable to spend some time studying your condensate systems.

One of the most useful instruments for this job is a makeup water meter. It can be used to detect system leaks, malfunctioning system components as well provide an excellent basis for calculating wasted energy.

The rule of thumb on condensate return energy is:

For each 11°F the water temperature is cooled, it loses 1% of efficiency for the system.

The challenge is to get this heat back to the boiler plant. The flash steam formed as the condensate leaves a trap is of first concern, approximately half the energy in the trap exhaust is the energy in this flash steam. In an open system it can escape to atmosphere through a vent in the receiver tank.

If the condensate piping is not adequately insulated, additional losses will occur.

Some pumping systems cannot handle condensate above 180°F because of cavitation. There may be other operational problems that have lead to lower condensate return temperatures. Some plants have actually devised ways to cool condensate so they could get their pumps would work.

Good management of the condensate systems has additional benefits:
a. Reduction in replacement water costs.
b. Reduction in water treatment chemicals.
c. Reduction in fuel consumption used in preheating the make-up water.

One solution to getting the hot or even boiling condensate back to the boiler room is the steam pressure powered pump (Figure 13.5). It collects condensate and discharges it when the liquid level reaches a certain point. It uses system steam pressure for the pumping action, a small investment for high overall energy savings.

The condensate pressure powered pump eliminates many problems connected with handling hot condensate with electric pumps. When the condensate goes above 190°F they tend to cavitate, forming steam in the suction end of the pump. One common solution had been to cool the condensate to a temperature where the old pumps could handle it. Another common practice is to dump the condensate down a convenient drain.

This type of pump is reported to use only 3 pounds of steam for every 1,000 pounds of liquid pumped. When exhaust is vented back in a closed system, the steam is recovered and the cost of operation is negligible.

Acid Corrosion and Oxygen Attack In Piping Systems

A great part of condensate recovery system failure can be traced to a feedwater treatment problem that allows carbon dioxide to enter the distribution system piping. Oxygen also causes problems but most are confined to the boiler and economizer.

Carbon dioxide that carries over in the steam forms carbonic acid which has the capacity to combine with one and a quarter pounds of steel per pound of CO_2, forming a groove in the bottom of the piping. Over years this can eat up a lot of metal and cause countless problems. Oxygen pitting and scale formation can also destroy piping and boiler tubes as well as interfere with heat transfer and the operation of pressure reducing valves and trap mechanisms.

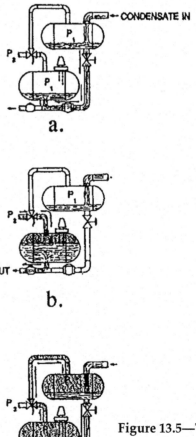

Figure 13.5—The operation of a pressure powered condensate pump: (a) is the fill cycle, (b) is the pumping cycle and (c) is the venting cycle.

Chapter 14

Steam Traps

When steam expends its energy, by giving up its heat or by doing work, it condenses back into water (Figure 14.1). This water must be removed from the pathway of the steam so it will not interfere with the function of the steam system. Once removed, the condensate, pure hot water, should be returned to the boiler where it is again heated to produce more steam.

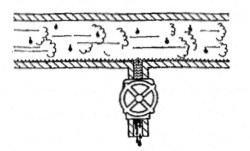

Figure 14.1—Condensate formation in steam distribution line. A valve is shown draining line instead of a steam trap.

The Purpose of Steam Traps

The job of removing condensate is handled by the steam trap (Figure 14.2). The steam trap's job is to remove condensate while preventing steam from escaping from the distribution system. It must discharge this condensate from a higher to a lower pressure. To do this job, it is designed to differentiate steam from condensate, usually by reacting to temperature, density or thermodynamic properties.

The Btus which are released by steam in heating, and process applications and by pipe radiation loss causes the steam to condense and form droplets of water that can quickly combine into larger masses. If this condensate is not effectively removed (trapped), it can reduce the efficiency of heat transfer equipment by a phenomena known as waterlogging.

Condensate accumulating at the bottom of

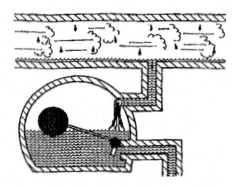

Figure 14.2—Float trap is trapping steam while allowing condensate to discharge.

a pipe is swept along by the high velocity steam (Figure 14.3). As the water moves through the pipe collecting additional droplets, forming a larger arid larger slug, it develops a high level of energy and can cause serious damage through the phenomena known as "Waterhammer." Severe waterhammer can burst the wall of a pipe, possibly causing personal injuries and damaging other equipment in the area, it is usually accompanied by a sharp metallic noise.

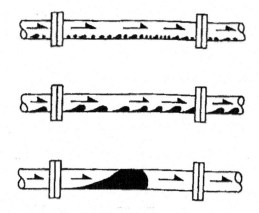

Figure 14.3—Condensate accumulates in distribution piping and forms a slug which is carried along by the steam at high velocity.

255

Severe damage can occur when the incompressible slug of water is forced to change direction by a pipe bend, fitting or valve. It produces a shock wave of tremendous momentary force which can cause great physical damage. It is a function of the mass of water and the velocity squared, so the greater part of its energy comes from the high velocities (6,000-12,000 feet per minute) achieved in the steam piping.

Some steam boilers have a tendency to produce low quality steam, which contains an appreciable amount of moisture and possibly boiler chemicals and contaminated boiler water. This introduces water into the boiler header and distribution piping. Depending on the severity, this can overload the distribution and trapping systems.

To detect this problem a Throttling Calorimeter can be used to test for steam quality (percent moisture in the steam). This instrument measures the temperature of a very small amount of steam discharging from the steam main. Dry steam will become superheated due to the pressure drop. If there is any moisture present, the temperature of the discharge will be reduced. Measuring temperature and comparing it to the maximum possible temperature will indicate steam quality.

Other methods that can be used are:

a.   Ion exchange

b.   Conductivity

c.   Sodium tracer flame photometry

d.   Specific ion electrodes.

These methods determine the solids content of the steam, including the solids carried over by water droplets.

The throttling steam calorimeter is the only direct way to determine steam quality and is most accurate below 600 psi and where moisture content is above 0.5 percent.

When low water quality is detected, it indicates a problem with the water level control, feedwater treatment, blowdown cycle, boiler sizing or steam drum internal problems. Boiler and steam

generator operating procedures may also have to be investigated and changed.

All of the steam generated by the boiler must eventually leave the system as condensate, except for leaks and steam directly used for some purpose. It is obvious that if the steam can somehow escape from the system without going through a trap as condensate, the system will be inefficient to that degree. Properly working traps insure that steam gives up its energy efficiently.

Air and Noncondensibles in the Steam System

Accumulations of air and noncondensible gases in the steam system can limit steam flow, steam temperature and heat energy release. Air is present in the system on startup and is introduced through vacuum breakers.

Noncondensible gases are liberated in the boiler. Carbon dioxide and oxygen are dissolved in boiler feedwater as carbonates and bicarbon-

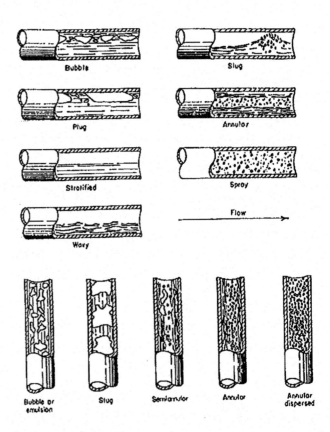

Figure 14.4—Shows several forms of steam and condensate mixing in piping.

ates. These noncondensible gases, upon release, flow with the system and can create energy problems. When steam condenses, these gases migrate to the heat exchange surface, forming a insulating film (Figure 14.5). This film can be a very effective insulator, usually only a very small percentage is needed to cause a big problem with heat transfer.

The gases also take up volume and don't condense into liquid as readily as steam, that's why they are called noncondensibles. If allowed to accumulate for long periods, they can take up enough volume to effectively block steam flow and almost all energy transfer. When condensate drainage is blocked, dangerous water hammer can occur.

Trapped air can cause heat transfer problems, especially in equipment, such as shell and tube heat exchangers, where there is room for it to accumulate at high points or in areas of low velocities. This trapped air will cause a cold spot, that's how you can find them. This would be a good indication that an air vent is needed.

When allowed to cool in the presence of condensate, carbon dioxide can combine with water to form carbonic acid. Since gas accumulation causes a temperature drop, acid formation is highly probable in any standing condensate.

The corrosion of iron forms a soluble bicarbonate which leaves no protective coating on the metal. If oxygen is also present, rust forms and CO_2 is released, which is now free to cause more corrosion.

Once gases become dissolved, they should be removed, but while in the system they usually contain some damaging acids. The corrosion they produce causes piping leaks, damaged traps and valves and formation of insulating scales.

System Protection

One application, of many that are needed in a steam system, is the removal of condensate from steam strainers (Figure 14.6). Strainers are

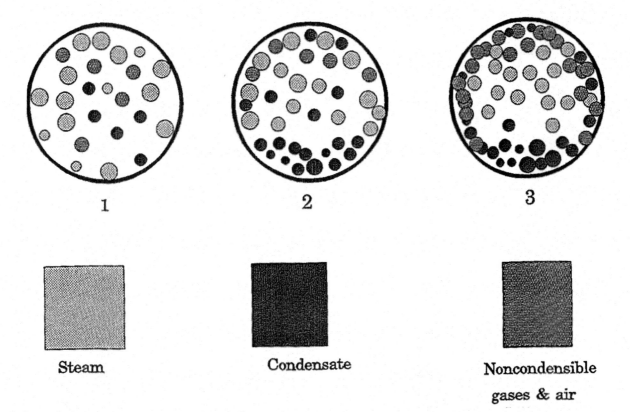

Figure 14.5—Reduction of heat transfer by noncondensible gases: (1) a steam line containing steam, condensate and gases, (2) noncondensibles are migrating to the heat exchange surface as steam condenses and (3) an insulating film is forming (darker shade) at the heat transfer surface and condensate is forming in the bottom of the line.

used to protect pressure reducing valves and traps and are generally installed to keep the system clean. The strainer body is a low point in the system which accumulates condensate naturally, reducing the effective area of strainer screen. In the case of a valve down stream that has been closed, some condensate will be picked up by the flow when it is opened, impacting the valve seat with dirty condensate.

Installing an inverted bucket steam trap on the strainer blowdown will keep the condensate drained. This will free strainer surface from blockage improving strainer performance. This may be the reason why some valves have perpetual problems.

How Steam Traps Fail

Steam traps are subject to harsh operating conditions and like all mechanical devices, their

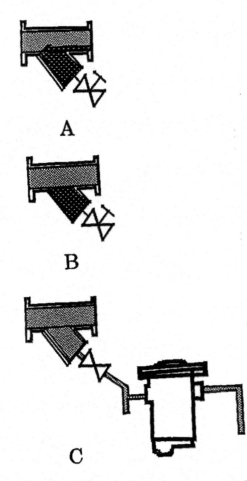

Figure 14.6—Installing a steam trap on a strainer to prevent valve damage.

moving parts are subject to wear, corrosion and eventual failure.

Most traps function intermittently. During the closed portion of their cycle, the leg of water that accumulates encourages the formation of a mildly corrosive acid and the cycling itself causes temperature fluctuations which accelerate the problem.

A typical trap may open and close several million times a year and some wear, malfunction and outright failure is inevitable. Steam losses conservatively exceed 15 billion dollars annually.

Steam trap leaks are a form of invisible steam leak. Rather than arrive at the point of use, the steam escapes to the condensate return system without accomplishing useful work.

There are four essential causes for steam trap leaks: (Figure 14.7)

a.   The trap responds too slowly, not closing fast enough to prevent the escape of some steam on the closing cycle.

b.   The trap leaks in the closed position because of either a defect in the valve closing mechanism or in the sealing surfaces allowing steam to leak through.

c.   The trap fails to close completely, because of mechanical failure.

d.   The trap fails open, allowing steam to blast through the escape orifice.

All steam traps will eventually fail. Most traps fail because mechanical parts wear out through normal operation. Others will fail because of the flashing of condensate as it passes through the trap, which can wire draw the valve seat. Still others will fail because of the stresses of steam service such as water hammer, rapid temperature fluctuations, carbonic acid corrosion or general fatigue of operating components due to millions of operational cycles each year.

Steam traps are small, relatively inexpensive and short-lived components of the steam system which represents a very large opportunity for savings in almost any plant.

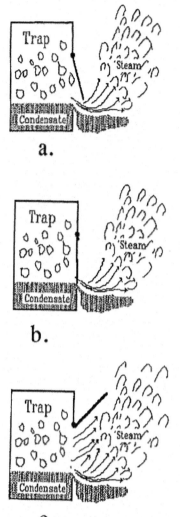

a.

b.

c.

Figure 14.7—Various ways a steam trap can waste steam: (a) is either failure to close completely or closing too slowly, (b) is leaking in the closed position and (c) trap fails open.

STEAM TRAP TYPES AND CHARACTERISTICS

Effective steam trapping practices have a direct impact upon the efficiency of any steam or condensate system.

To ensure maximum steam trap life and steam system efficiency, steam trap type and size must be properly matched to each application. It is important to emphasize there is no "UNIVERSAL" steam trap which will provide efficient operation for all applications.

Most facilities will require several types of steam traps in various sizes to provide efficient steam trapping. The user must have knowledge of the various types of traps available to properly select and specify the traps to be used.

Categories of Steam Traps

All steam traps are designed to distinguish between condensate and steam. To be effective they must stop or trap the steam and release condensate. The physical differences between steam and condensate has lead to a variety of approaches to steam trap design.

There are three basic categories of steam traps, each using a different principle to differentiate between steam, non-condensible gas and condensate. These are:

Mechanical, which are operated by differences in density between steam and condensate.

Thermodynamic, operated by kinetic energy.

Thermostatic, operated by temperature.

Mechanical Steam Traps

Mechanical steam traps are operated by the difference in density between steam and condensate. All use a float in some form to operate the valve to let condensate drain from the system.

Mechanical traps include Inverted Bucket, Open Bucket and Float and Thermostatic (F&T) types. All can operate at or near steam saturation temperature because trap operation is dependent on the density difference between steam and water which is about 700 to 1. Figure 14.8 shows the density relationship between water and steam.

All mechanical traps respond quickly to changing loads and most have large turn down ratios. However, without a separate mechanism to discharge air and non-compressible gasses, mechanical traps have a limited venting capacity.

Since all mechanical traps depend on a condensate level inside the trap for proper operation, they are position sensitive. Some will fail to operate when installed even a little off vertical.

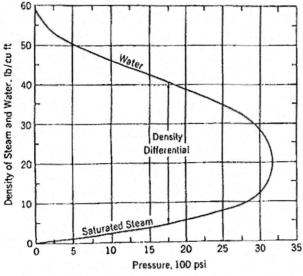

Figure 14.8—Water steam density relationship.

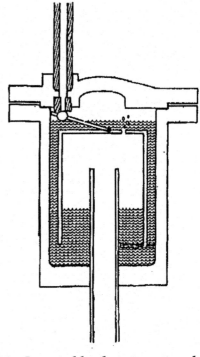

Figure 14.9—Inverted bucket steam trap has trapped steam under the bucket and floated up to close the outlet valve.

Inverted Bucket Traps

Inverted bucket traps use an open inverted bucket as a float. As with all mechanical traps, inverted bucket traps operate on the difference in density between steam and condensate.

The inverted bucket is usually attached to the valve by a lever mechanism. When the inverted bucket sinks toward the bottom of the trap body, the valve is opened. Since the linkage is designed with some "play," inverted bucket traps can operate in either the cyclic or modulation mode.

Condensate, steam and non-condensible gasses enter under the inverted bucket. The lighter steam and air lift the bucket, closing the valve (Figure 14.9). There is a bleed hole in the top of the bucket to let air and a small amount of steam bleed through so the bucket doesn't get stuck in the closed position.

During startup, the inverted bucket rests on the bottom of the trap body because of its own weight. The valve is open allowing the discharge of cold condensate, air and noncondensibles.

As the condensate enters the trap, a liquid level is formed which provides a water seal around the inverted bucket. Since the bucket is filled with air and non condensible gasses, the weight of the bucket is overcome by the buoyant force of these gasses in water, and the bucket rises closing the valve. The air and non-condensible gasses trapped under the bucket are bled through

the hole at the top of the bucket, allowing more condensate to enter. The bucket loses buoyancy, sinks and opens the valve.

If the amount of air and non-condensibles entering the trap is greater than the vent capacity of the bleed hole, the trap can air bind and fail closed. Some manufacturers have alleviated this problem by using larger bleed holes or thermostatic elements to enhance their venting capacity.

Once the valve is open, discharge continues until all of the condensate has been removed from the system and steam floats the bucket closing the valve (Figure 14.10). The trap will remain closed until enough steam has vented through the bleed hole or condensed to allow the bucket to sink, opening the valve and starting the cycle again.

The slack or play in the linkage allows the system pressure to snap the valve closed after the bucket has risen within the trap body. System steam pressure acts to keep the valve closed until the bucket has lost enough buoyancy to pull the valve off the seat. This action is affected by system pressure which requires special consideration given to the sizing of the orifice. There-

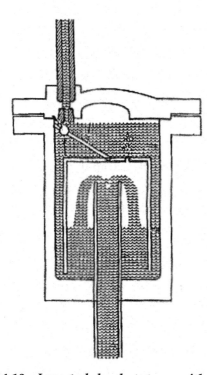

Figure 14.10—Inverted bucket trap with inverted bucket lowered and the valve open and discharging condensate.

fore, in order to maintain the proper relationships between the closing force of the pressure on the valve and the opening force of the weight of the bucket, orifice sizes must be designed to match steam system pressure.

Since the orifice must be sized to match the maximum steam system pressure, care must be taken to ensure adequate capacity is still available.

Open Bucket Traps

The open bucket trap is an older and possibly obsolete type of mechanical trap. It is simply a bucket arrangement that pivots around one end when it fills with water, opening a discharge valve. Most open bucket traps have a separate thermostatic air vent at the top.

During start up the bucket rests on the bottom of the trap body and the valve is fully open. The thermostatic element is also open, allowing the discharge of air and gases. When condensate enters the trap the enclosed trap body is filled with condensate causing the bucket to rise shutting off the valve. Condensate continues to enter

the trap as the thermostatic element allows air and gases to be dispelled from the top.

The liquid level rises above the open bucket filling it and causing it to sink, opening the valve. Condensate is discharged until steam enters the trap body and the water is forced from the open bucket, making it light enough to float up closing the discharge valve.

Without the thermostatic element or some other means to clear the trap body of air, the trap will air bind and fail closed.

Float Traps and Float and Thermostatic Traps

Float traps (Figure 14.11) are one of the oldest trap designs still in use and are not considered obsolete.

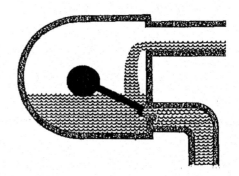

Figure 14.11—Float type steam trap.

A float trap consists of a closed float attached to a valve by mechanical linkage. During start up, the float rests on the bottom of the trap, closing the valve. Air and gases must be removed from the system, which is usually done by a thermostatic element or air vent in the top of the housing.

As condensate enters the trap, the liquid level rises lifting the float and opening the valve. As this type of trap is self regulating it will operate in the modulation mode. As the condensate level rises, the outlet valve opens further handling the greater quantity of condensate. If the condensate load diminishes, the ball lowers, partially closing the valve reducing the discharge.

The valve is located below the water level, providing a water seal preventing the loss of live steam.

Float and thermostatic (F&T) traps are a modification of the float trap in that the thermostatic air vent is an integral part of the trap (Figure 14.12).

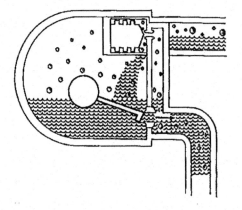

Figure 14.12—Float and Thermostatic steam trap (F&T) with thermostatic trap element in the top to bleed off air and non-condensible gases.

The thermostatic element will be open during startup, until steam enters the trap and heats the element, expanding it and closing the valve. This element will remained closed until air and gasses concentrate in the top of the trap lowering the temperature of the steam/air mixture, opening the element and discharging the mixture.

In one design of the float and thermostatic trap, there is just a "free ball" that can float up opening the orifice or sink when condensate level goes down, closing the orifice.

The failure mode for these traps is to usually have the ball float fail closed and the thermostatic element either fail open or closed.

Thermodynamic Traps
Thermodynamic traps include Disk, Piston, Lever and Orifice types.

Disc Traps
Disc traps have a single operating part, a flat disc which lifts from the valve seat to open and is forced onto the valve seat for closure.

During start up (Figure 14.13), the pressure of the cold condensate and gases pushes the disc from the valve seat. This opens the trap, allowing the discharge of condensate and gasses.

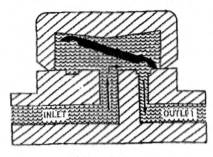

Figure 14.13—Disc trap in open position.

Discharge continues until hot condensate near steam saturation temperature enters the trap. This condensate will flash in the chamber on the outlet side of the trap, increasing the pressure on the outlet side of the disc and decreasing the pressure below the disc, caused by high velocity flow. The combined action forces the disc onto the seat, closing the trap (Figure 14.14).

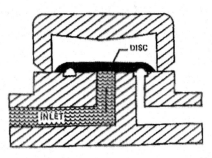

Figure 14.14—Disc trap forced closed by steam in the upper chamber.

The disc is designed to have the whole of the area of the disc exposed to pressure on the top. On the bottom, incoming side, the landing surfaces blocks pressure from acting on the whole disc area, thus unequal forces are developed. The flash steam on the top is sufficient to keep the disc seated until it cools relieving the pressure above the disc, allowing cooler condensate to flow once again.

Orifice Traps
Orifice traps consist of one or more orifices in series and have no moving parts.

Single orifice traps typically have very small orifices in the range of one tenth the diameter of thermodynamic or thermostatic trap orifices.

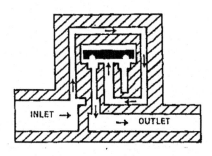

Figure 14.15—A double walled disc trap.

The operating principle is that water (condensate) flowing through the orifice is restricted. This backs up some of the condensate preventing the flow of steam through the orifice. Steam tries to flow through the orifice at a very high velocity whereas the condensate is very slow by comparison, this phenomena serves to choke the flow of steam with condensate.

The orifice is somewhat self regulating. That is, if the condensate load is less than the orifice was designed for, the condensate will be hotter, causing more flashing, which will further restrict flow. If the condensate load is greater than the orifice was designed for, the condensate will back up and cool. This cooling will reduce flashing, thus allowing a greater condensate flow.

Orifice traps have a constant discharge, not the cyclic discharge of other thermodynamic traps.

Each orifice trap must be very closely engineered to its specific load because they have the potential to either blow steam (claimed to be an insignificant amount) or back up condensate during start up or during high loads when their capacity may be exceeded.

Thermostatic Steam Traps

Thermostatic steam traps are operated by changes in temperature and include Bellows, Diaphragm, Bimetallic and Expansive Element types. Thermostatic traps respond more slowly to changing operating conditions than some other types of traps because of the heat stored within the trap materials and the condensate which collects in the trap.

Thermostatic steam trap operating principles are simple and thermostatic steam traps usually have only one moving part.

Bellows Steam Traps

Bellows traps are often called balanced pressure or thermostatic traps because the bellows contains a volatile fluid which closely parallels the temperature-pressure relationship of the steam saturation curve. The pressure created inside the bellows by this fluid as it vaporizes from the heat of the condensate closes the trap against the pressure of the steam system (Figures 14.16 and 14.17).

During start up the bellows is contracted away from the seat, allowing the discharge of condensate and gasses. Discharge continues until hot condensate enters the trap, heating the bellows and the volatile fluid. The fluid vaporizes and expands, causing the bellows to expand and close the valve.

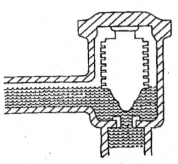

Figure 14.16—Thermostatic trap in the open position.

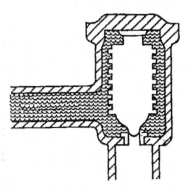

Figure 14.17—Thermostatic trap in the closed position.

The trap will remain closed until the condensate, the bellows and the volatile fluid cool. This cooling condenses some of the fluid, reducing the pressure inside the bellows. This reduced internal pressure allows the external pressure, from the steam system, to contract the bellows away from the seat. This opens the trap and it begins the cycle again.

Reduced temperature from a steam and noncondensible gas mixture will also cool the bellows and fluid, allowing the discharge of these gases.

Diaphragm Traps

Diaphragm traps are a modification of the bellows type trap. However, instead of a large bellows with many convolutions or welded elements, a single element is used.

Operation of this type steam trap closely parallels the operation of bellows traps.

Bimetallic Traps

Bimetallic traps (Figure 14.18) use the temperature of the condensate in the trap to bend bimetallic elements against the force exerted by the steam pressure on the valve. There are many different configurations of bimetallic traps which cause significant differences in operation.

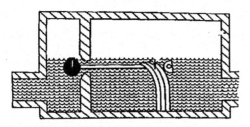

Figure 14.18—A bimetallic steam trap.

Since a bimetallic element is a nearly linear device, most manufacturers use several types of bimetal or special shapes to cause the trap to approximate the steam saturation curve over the operating range.

Because of the relatively large thermal mass of the bimetallic element, response to system changes can be slow. Additionally, these traps are sometimes affected by the back pressure which works against the opening force on the valve possibly increasing the amount of sub-cooling.

During start up the bimetallic element is relaxed, allowing the steam system pressure to open the valve discharging the cold condensate and non-condensible gases. As warm condensate

reaches the trap, the element warms and changes shape, exerting a closing force on the valve.

Expansive Element Steam Traps

Expansive element traps (Figure 14.19) are characterized by a constant discharge temperature for a given condensate load, regardless of the steam system pressure.

Various elements are used and they may be liquid or solid. In all cases, the response to changing conditions is relatively slow and dependent upon the thermal mass of the element.

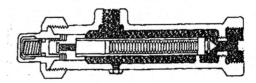

Figure 14.19—An expansive element steam trap.

During start up the element is contracted, allowing the discharge of cold condensate, air and non condensable gases. Discharge continues until warmer condensate heats and expands the element, moving the valve towards the seat. The trap will reach equilibrium condition, discharging condensate continuously at a nearly constant temperature. Only rarely will these traps cycle and then only under very light load conditions, before returning to the modulating mode.

The operation of these traps is regulated by the condensate temperature and are suitable for applications where condensate backup and a slow response to load changes is acceptable.

STEAM TRAP LOSSES

It is clear, that steam leaks are quite expensive. The high and low estimates in Table 14.1 take into account the throttling effect of condensate choking the full flow of escaping steam with variations in condensate formation load.

We can assume that the cost of steam leaks through failed traps will range from $2,000 to $50,000 per trap over the course of a year.

The economic incentive for eliminating failed steam traps clearly exists. A trap anywhere in a

Table 14.1—Annual cost of steam loss from traps for 100 psi steam at $5.00 per 1000 pounds.

Trap Orifice Diameter	High Estimate ($)	Low Estimate ($)
1/8"	3,175	2,300
1/4"	16,600	9,250
1/2"	51,000	37,000

facility may fail open at any time and the losses, if it is not found and repaired quickly, can be very large. Failure could occur at any time, even the day after the last inspection. Figure 14.20 shows how flash steam and steam from failed steam traps vents to the atmosphere.

To determine how often traps will fail has so many variables, that any conclusion should be based on actual conditions at a specific facility. Some generalizations have emerged over the years. There is very little information on this topic.

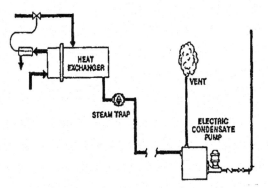

Figure 14.20—Showing heat exchanger and steam venting to atmosphere from condensate tank.

Sampling of Published Comments on Steam Trap Performance

A review leading authorities reveals that trap failures range from 5% to 50% of the trap population at any given facility.

"The failed trap percentage at any given time *should* be in the 5% to 10% range. Perfor-

mance better than that is difficult in the average plant. Failed-trap percentages as high as 50% in some plants showed up during the early days of the energy crisis, and occasionally we see such performance still."

Power Magazine
April, 1980

"Experience indicates, that in plants without planned steam trap maintenance, between 10% and 50% of the traps are malfunctioning at any given time—as a result of errors in sizing, misapplication or inadequate maintenance."

Plant Engineering
March 5, 1981

"Most plants can save 10-20% of fuel cost simply by having a formal, active steam trap program...For the first year, a return of $1 million in energy savings for each $300,000 spent upgrading the system is the rule rather than the exception."

Chemical Engineering
February 9, 1981

"Trap life should be 4-5 years average. For a plant with 2000 traps this means 400-500 replacements every year."

Power Magazine
April 1980

"The average disc trap should last six months to a year...disc traps sometimes failed 'within days of installation'..."

Energy User News
June 6, 1983

"Of 260,000 installed traps studied (in 40 industrial plants) the average performance level was found to be only 58%. In a typical plant with 2,000 traps, 840 were failed; 42% needed corrective maintenance that plant personnel were not providing... inefficiencies in the energy/management area were costing the average industrial plant over

$2000 in steam each day—even though $500 or more is spent daily on steam trap maintenance."

Industry Week
April 16, 1984

This review of some of the most prestigious and informed publications on efficient plant management are all of one voice. There is tremendous potential to waste energy in steam systems, also there is the same level of opportunity to correct the situation.

STEAM TRAP SELECTION SIZING AND INSTALLATION

Steam Trap Piping

In most plants, the removal of condensate and air from steam systems is accomplished through steam trap stations. Figure 14.21 shows the piping configuration for a steam trap station.

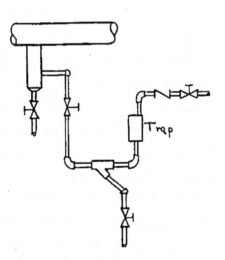

Figure 14.21—A typical steam trap station.

The first part of the steam trap station is a reservoir that connects to the lowest point of the steam line or process equipment. The reservoir is sized to collect the large quantity of condensate formed during the warmup phase when the system is brought up from the cold state. At the

bottom of the reservoir is a dirt pocket and valve to collect larger particles of scale and dirt in the system. It has a blowoff valve for clean-out.

The condensate flows from the collection point through a strainer and trap to the condensate system. The station is configured with inlet and outlet valves to isolate the trap if repairs are needed.

The test connection can be opened, coordinated with the closing of the trap outlet valve to see what is coming through the trap, steam or condensate. This is one of the best ways to test trap performance.

The "Y" strainer can be cleaned out by opening the blowoff valve.

The bypass line (not shown) enables plant personnel to bypass the entire steam trap station if necessary. Bypass valves are usually opened when there is a suspected problem with a trap or to assist with drainage during cold startup. Bypassing a steam trap may also be necessary to deal with a waterlogging problem caused by a trap malfunction.

With the bypass open, steam will be wasted and the efficiency of the plant reduced. In some cases a trap is installed in the bypass line. Then, if the main trap fails, the bypass trap can be put into service and the problem of blowing steam through the bypass is eliminated.

Selecting the Type of Trap

Properly selected, sized and installed traps are the best guarantee for efficient operation, long service life and minimum downtime.

The major considerations in selecting the type of steam trap are:

A. Type of service
   • Continuous or intermittent removal of condensate
   • Temperature of the condensate and system pressure
   • Range of load on the trap
   • Rate of change of the load

B. Operational
   • Normal steam loss during operation

- Reliability
- Failure mode most likely to occur

The above considerations were used in developing Table 14.2, the Steam Trap Selection Guide. This guide was developed by the U.S. Navy for their many and varied shore bases including hospitals, industrial facilities, air bases and training centers.

Operating characteristics of various types of traps are summarized in Table 14.3, which gives a comparison of various common trap configurations for specified service.

When installing and evaluating traps, an understanding of their limitations is essential to long, reliable and trouble free service. The limitations listed in Table 14.4 should be given careful consideration in the steam trap selection process.

Sizing Steam Traps

Factors that affect the accuracy of trap sizing are:

A. Estimating the maximum condensate load.
B. Range of operating pressure and differential pressure.
C. Selection of a safety factor.

Condensate Load

The amount of condensate generated by equipment can generally be obtained from equipment manufacturer's literature or specification sheets. Steam trap manufacturers, through years of experience, have also developed formulas, tables and graphs for estimating condensate load for most applications. Table 14.5 gives samples of simplified estimating aids Table 14.6 gives estimates of condensate loads for various sizes of steam mains and different pressures.

Pressure differential

Trap capacity is affected by the differential pressure across a trap. If a trap exhausts to atmosphere, the differential pressure will be the supply pressure. In some plants, traps are installed with the outlet connected to a pressurized return system. The trap must operate against this head plus any requirement to lift the condensate to the return system. Table 14.7 gives examples of the reduction in trap capacity caused by this back pressure which must be taken into account when sizing traps.

Safety Factor

The safety factor is a multiplier applied to the estimated condensate load since trap ratings are based on maximum discharge capacities or continuous flow ratings (Figure 14.22). Safety factors vary from 2:1 to 10:1 and are influenced by the operational characteristics of the trap, accuracy of estimated condensate load, differential pressure and expected changes in pressure and the configuration of the installation design.

Safety factors are needed to cope with the varying condensate loads experienced in the "real world " of actual use.

During start up conditions the cold piping will generate a large amount of condensate until it is up to operating temperature. Whether this occurs over several hours or a few minutes the amount of condensate will be the same, but the trap will have to handle it more quickly requiring added capacity.

To add to the challenge of condensate removal during warmup, the system pressure is lower than normal so the trap output will be lower than rated capacity. Waterhammer must be avoided so the trap must be designed for this unfavorable condition.

Other conditions affect condensate load also, cold winds and rain can reduce insulation values. Sudden surges in load on heat exchangers can also lead to waterlogging.

A boiler which has carry-over or is producing low quality steam containing unevaporated water places an unusually high load on the whole system.

Waterlogging cuts production and comfort and waterhammer can be dangerous. Insufficient trap capacity causes these problems. On the other hand, wasted energy and large dollar losses can occur if the trap is oversized and is leaking.

The answer to this situation may be an effective trap maintenance program.

Steam Trap Selection Guide

Application	Special Considerations	Primary Choice	Alternate Choice
Steam Mains and Branch Lines	• Energy conservation • Response to slugs of condensate • Ability to handle dirt • Variable load response • Ability to vent gases • Failure mode (open)	Inverted Bucket Thermostatic in locations where freezing may occur	Float and Thermostatic
Steam Separators	• Energy conservation • Variable load response • Response to slugs of condensate • Ability to vent gas • Ability to handle dirt • Failure mode (open)	Inverted bucket (large vent)	Float and Thermostatic Thermostatic (above 125 psig)
Unit Heaters and Air Handling Units	• Energy conservation • Resistance to wear • Resistance to hydraulic shock • Ability to purge system • Ability to handle dirt	Inverted Bucket (constant pressure) Float and Thermostatic (variable pressure)	Float and Thermostatic (constant pressure) Thermodynamic (variable pressure)
Finned Radiation and Pipe Coils	• Energy conservation • Resistance to wear • Resistance to hydraulic shock • Ability to purge system • Ability to handle dirt	Thermostatic (constant pressure) Float and Thermostatic (variable pressure)	Thermostatic

Table 14.2 Steam Trap Selection Guide

Steam Trap Operating Characteristics

Characteristics	Bellows Thermostatic	Bimetallic Thermostatic	Disk	F & T	Inverted Bucket
• Method of operation (discharge)	Continuous (1)	Self-modulating	Intermittent	Continuous	Intermittent
• Operates against back pressure	Excellent	Poor	Poor	Excellent	Excellent
• Venting capability	Excellent	Excellent	Good (3)	Excellent	Fair
• Load change response	Good	Fair	Poor to Good	Excellent	Good
• Freeze resistance	Excellent	Excellent	Good	Poor	Poor (5)
• Waterhammer resistance	Poor	Excellent	Excellent	Poor	Excellent
• Handles start up loads	Excellent	Fair	Poor	Excellent	Fair
• Suitable for superheat	Yes	Yes	Yes	No	No
• Condensate subcooling	5 - 30°F	50 - 100°F	Steam Temperature	Steam Temperature	Steam Temperature
• Usual failure mode	Closed (2)	Open	Open (4)	Closed	Open

1. Can be intermittent on low loads.
2. Can fail open due to wear.
3. Not recommended for very low pressure.

4. Can fail closed due to dirt.
5. May be insulated for excellent resistance.

Table 14.3—Steam Trap Operating Characteristics

Steam Trap Limitation Guide

A. Bucket trap

- Trap will not operate where a continuous water seal cannot be maintained.

- Must be protected from freezing.

- Air handling capacity not as great as other type traps.

B. Ball float trap

- Must be protected from freezing.

- Operation of some models may be affected by waterhammer.

C. Disk trap

- Not suitable for pressures below 10 psi.

- Not recommended for back pressures greater than 50 percent of inlet pressure.

- Freeze proof when installed as recommended by the manufacturer.

D. Impulse orifice trap

- Not recommended for back pressures greater than 50 percent of inlet pressure.

- Not recommended where subcooling condensate 30°F below the saturated steam pressure is not permitted.

E. Thermostatic trap

- Limited to applications in which the condensate can be held back and subcooled before being discharged.

- Operation of some models may be affected by waterhammer.

F. Combination float and thermostatic trap.

- Cannot be used on superheated steam systems.

- Must be protected from freezing.

- Operation of some models may be affected by waterhammer.

Table 14.4—Trap Limitations

Application	Lb/Hr of Condensate
Heating Water	$= \dfrac{\text{GPM}}{2} \times \text{Temperature Rise °F}$
Heating Fuel Oil	$= \dfrac{\text{GPM}}{4} \times \text{Temperature Rise °F}$
Heating air with Steam coils	$= \dfrac{\text{CFM}}{900} \times \text{Temperature Rise °F}$
Heating: pipe coils and radiation	$= \dfrac{\text{A} \times \text{U} \times (\text{Steam °F - Air °F})}{\text{L}}$

A = Area of heating surface
U = Heat transfer coefficient
    (2 for free convection)
Delta T = Steam temperature - Air temperature, °&F
L = Latent heat of steam BTU/Lb

Table 14.5—General formulas for estimating condensate loads.

Estimating Condensate Load for Insulated Steam Mains

Insulated Steam Main

Lb/Hr of condensate per 100 lineal foot
at 70°F (at 0°F, multiply by 1.5)

Steam Pressure (psig)	Size of Main Inches					
	2	4	6	8	10	12
10	6	12	16	20	24	30
30	10	18	25	32	40	46
60	13	22	32	41	51	58
125	17	30	44	55	68	80
300	25	46	64	83	103	122
600	37	68	95	124	154	182

Table 14.6—Estimating condensate load for insulated steam mains.

Table 14.7—
Percentage reduction in steam trap capacity

Inlet Pressure (psig)	Back Pressure (% of Inlet Pressure)		
	25%	50%	75%
10	5%	18%	36%
30	3%	12%	30%
100	0	10%	28%
200	0	5%	23%

Steam Trap Maintenance

The cost of steam lost through a leaking trap in one day can exceed the cost of a new trap. If a trap is backing up condensate and has become waterlogged, the reduction in capacity or efficiency of steam-using equipment can also waste more money in one day than the cost of the trap.

Good inspections that pinpoint problem traps for repair or replacement are important. While the trap is the focus of tests and inspections, the steam trap station and its piping and fittings should be checked also as these things may be contributing to trap problems. Missing strainers, open bypass lines and condensate piping pressure are a few problems that could be found.

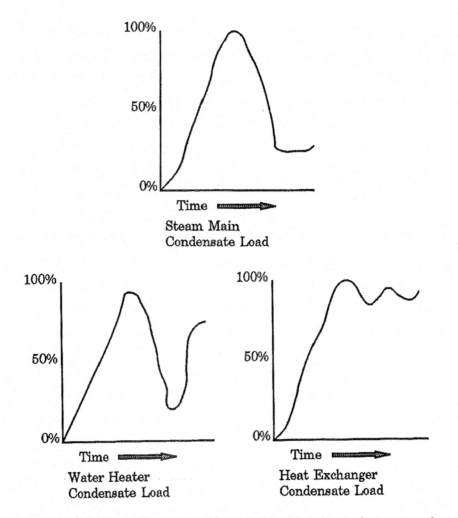

Figure 14.22—Condensate load change over time on a cold start up for a water heater, steam main and heat exchanger.

Inspection schedules

The ideal is all traps functioning properly. This must be balanced against cost effective use of time and materials. Each location offers different challenges. Dirty systems, above normal acid accumulations and problems with the trapping stations all point to judgment playing a major role in trap maintenance. Table 14.8 and 14.9 are commonly used trap inspection schedules.

Trap Failure Rate	Inspection Frequency
Over 10%	Two months
5 - 10%	Three months
Less than 5%	Six months

Table 14.8—Trap inspection schedule.

Pressure (psi)	Inspection Frequency
0 - 30	Annual
30 - 100	Semiannual
100 - 250	Quarterly or monthly
Over 250	Monthly or weekly

Table 14.9—One manufacturer recommends inspection frequencies based on system pressure.

Traps serving critical process equipment may need special inspections during very cold weather to guard against freezing damage.

The basic methods of inspecting traps are:

A. Visual observation
B. Sound detection
C. Temperature measurements.

Visual observation is the best and least costly method of checking trap operating conditions,

but none of the methods provide a cure-all for trap troubleshooting.

Table 14.10 is a steam trap troubleshooting guide, useful for trap inspections.

Visual Observation

Observing the discharge from a trap is the only positive way of checking its operation.

No special equipment is required but training and experience are necessary, particularly for recognizing the difference between flash steam and live steam.

- Flash steam is the lazy vapor formed when the hot condensate gives up 1 energy when reducing pressure through the trap.

- Live steam is a higher temperature, higher velocity discharge, and usually leaves the discharge pipe with a short clear, almost invisible, section of flow before condensation begins and a visible steam cloud develops.

Sound Detection

Some traps can be heard cycling on and off as they discharge condensate. The disk, inverted bucket and piston type of traps have this characteristic.

Some traps continually modulate like the float traps or expansive element and some thermostatic traps; they only give off flow sounds. However if they aren't working, they don't give off any sound.

An automotive stethoscope or other simple sound transmission device may be adequate. If there is a lot of noise around or a lot of traps giving off noise, then an ultrasonic listening device may be warranted. They can sense high frequency (an indication of flow) and low frequency mechanical noises (indicates operation).

Temperature Measurements

Diagnosing trap condition from temperature differences between upstream and downstream pipes is the least reliable inspection method. It can be useful in combination with visual or sound inspection as long as the potential ambiguities are recognized.

Table 14.10—Troubleshooting Steam Traps

TRAP TYPE	NORMAL OPERATION	PROBLEM INDICATION	POSSIBLE CAUSE
		VISUAL INSPECTION	
ALL TYPES	• Trap hot under operating conditions • Discharge a mixture of condensate and Flash steam. • Cycling open/closed depending on type of trap. • Relatively high inlet temperature	• Live steam discharge, little entrained liquid • Condensate cool, little flash steam • No discharge • Leaking steam at trap	Failed open Holding back condensate. Failed closed; clogged strainer: line obstruction Faulty gasket
		TEMPERATURE MEASUREMENT	
		• High temperature down stream • Low temperature upstream	Failed open Failed closed; clogged strainer; line obstruction
		SOUND INSPECTION	
FLOAT AND F & T	• Continuous discharge on normal loads may be intermittent on light load • Constant low pitched sound of continuous flow	• Noisy high pitched sound • No sound	Steam flowing through; Failed open. Failed closed
THERMOSTATIC	• Discharge continuous or intermittent depending on load, pressure, type. • Constant low pitched sound of continuous or modulating flow.	• Same as for Float, F & T above.	
INVERTED BUCKET	• Cycling sound of bucket opening and closing • Quiet steady bubbling on light load	• Steam blowing through • No sound • Discharging steadily; no bucket sound. • Discharging steadily;bucket dancing • Discharging steadily; bucket dancing after priming • Discharging steadily; no bucket sound.	Failed open Failed closed Handling air, check later Lost prime Failure of internal parts. Trap undersized
DISK	• Intermittent discharge • Opening and snap closing of disk about every 10 seconds	• Cycles faster than every 5 seconds. • Disk chattering over 60 times/min or no sound	Trap undersized or faulty. Failed open

There is a wide range of temperature measurement equipment. It ranges from infrared devices, which are handy for reading temperatures from a distance and at inaccessible locations. Standard pyrometers and surface thermocouples are also suitable. Heat sensitive color markers are also used at some locations.

Take temperature measurements immediately adjacent to, not more than two feet away, either side of the trap. Temperature readings should be in the ranges shown in Table 14.11 for the pressures in the in and out lines. For example, for a steam system with a pressure of 150 psi, if the trap inlet temperature is 340°F and the outlet side, with a pressure of 15 psi, has a temperature of 230°F this indicates expected temperature conditions.

Table 14.11—Normal pipe temperatures at 348 various operating pressures.

Steam Pressure (psig)	Pipe Surface Temperature Range (°F)
0 (Atmosphere)	212
15	225 - 238
30	245 - 260
100	305 - 320
150	330 - 350
200	350 - 370
450	415 - 435
600	435 - 465

Table 14.12 is a steam trap inspection checklist to aid personnel in their trap checking routines. They should also investigate the possibility of trap misapplication. Table 14.13 will be helpful for this purpose.

One manufacturer has developed an automatic system for detecting if steam traps are blowing-off an excessive amount of steam. The sensor chamber (Figure 14.23) has a small orifice in a division plate which is designed to pass "normal" steam flow. If this flow increases substantially, the level goes down on the upstream chamber exposing a sensor which signals failure of the trap (Figure 14.24).

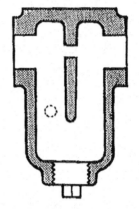

Figure 14.23—Sensor chamber for trap monitoring installation. Note the orifice and weir (partition plate).

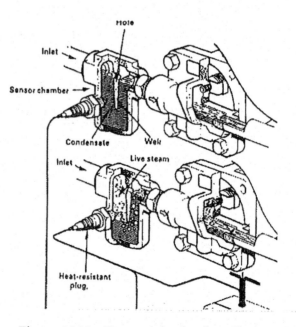

Figure 14.24—Trap monitoring installation.

Table 14.12—Steam trap inspection check off list.

FOR ALL TRAPS

- Is steam on?
- Is trap hot - at operating temperature?
- Wet test for signs of a hot trap; squirt a few drops of water on trap. Water should start to vaporize or sizzle immediately. If it does not this indicates a cold trap.
- Tag traps for maintenance check to see if this is a trap or system problem.
- Blowdown strainer.

SOUND CHECK HOT TRAP

- Listen to trap operate.
- Check for continuous flow:
    - Low pitch condensate flow
    - High pitch steam flow
- Check for intermittent flow
- Is trap cycling?
- Note mechanical sounds.

VISUAL CHECK TRAPS THAT SOUND BAD

- Close valve return line.
- Open discharge valve.
- Observe discharge for:
    - Normal condensate and flash steam
    - Live steam
    - Continuous or intermittent operation

TEMPERATURE CHECK IF NECESSARY

- Clean spots upstream and downstream of trap for measuring temperature.
- Record supply line pressure.
- Measure supply line temperature.
- Record trap discharge line pressure.
- Measure trap discharge line temperature.
- Tag failed traps for replacement or shop repair.

CHECK EXTERNAL CONDITIONS

- Supports and braces
- Insulation
- Corrosion
- leaks

Table 14.13—Inspection check list for trap misapplication.

1. Is trap installed backwards or upside down?

2. Trap located too far away from the equipment being serviced.
   Piping runs too long.

3. Traps not installed at low points or sufficiently below steam-using equipment to insure proper drainage.

4. Traps oversized for conditions. Oversized traps may allow steam to blow through.

5. More than one item of equipment served by one trap. "Group trapping" is likely to short circuit one line due to differences in pressure and other lines will not be properly drained.

6. The absence of check valves, strainers and blowdown cocks where required for efficient operation.

7. Trap vibration due to insecure mounting.

8. Bypass line with open valves. Bypass should be fitted with a standby trap.

9. Condensate line elevation higher than steam pressure through trap can lift. Inlet steam pressure must be high enough to lift condensate to drain system.

10. Inverted bucket and thermostatic traps that may be exposed to freezing. Freezing may affect operation of other traps too.

11. Thermostatic and disk traps must give off heat to operate, check to see that these types are not insulated.

12. Disk trap with excessive back pressure may not have enough differential pressure to operate properly.

Chapter 15

Boiler Water Treatment

IMPURITIES IN BOILER FEEDWATER CONCENTRATE IN THE BOILER

All boiler water contains dissolved solids. When the feed water is heated, it evaporates and goes off as distilled steam leaving impurities behind. As more and more water is distilled in the boiler, more feed water is added to replace it. As a result the amount of these solids dissolved in the boiler water gradually increases. After a while there is so much of these highly soluble solids in solution in the boiler water that it does not boil like ordinary water, it boils more like a syrup.

The bubbles of steam which rise to the surface of the water do not readily break free from the surface. Instead big bubbles form. When they break, they carry with them into the steam space some of the film that formed the bubble material. This condition is commonly referred to as carry over.

In addition some highly soluble materials are changed by the high boiler temperature to materials of low solubility such as calcium carbonate which are then precipitated. Much of this precipitation takes place where the boiler water is the hottest, where water is in contact with high heat transfer zones. The precipitate is deposited on the heating surface and forms a scale build up.

This scale is a good insulator, reducing heat transfer. As scale builds up, the steam and water is unable to keep the tube metal surfaces cool and it begins to overheat. This overheating destroys the strength of the tubing causing tube failure. This can occur as blistering which ruptures or general melting, depending on the circumstances. One of the purpose of a water treatment program is to keep certain scale forming solids in solution. Other scale forming solids are turned into a soft fluffy precipitate and carried down to the low points of the boiler. Table 15.1shows the effect of improper or inadequate water conditioning.

CYCLES OF CONCENTRATION

All of the impurities dissolved in water are usually termed Total Dissolved Solids, referred to as TDS. Modern methods utilize electronic instruments to measure the conductance, the opposite of resistance, of boiler water. These readings are called "mhos" or "micromhos" and can be mathematically converted to parts per million with respect to sodium ions by simply using a multiplier.

One part per million (ppm) is one pound in a million pounds of water. Since water weighs approximately 8.33 pounds per gallon, one ppm is one pound in 120 thousand gallons of water.

If a given water had a total dissolved solids of 500 ppm and we concentrated this water two times or two cycles, the TDS level would be 1,000 ppm. At three cycles the TDS would be 1,500 ppm and at four cycles 2,000 ppm and so on.

In the case of a boiler as small as 100 horsepower, it can evaporate more than 10,000gallons of water in 24 hours. If this water had a hardness of 340 ppm, 28 pounds of residue would be left behind in the boiler every day.

Energy loss through blowdown is minimized by maintaining the boiler water cycles of concentration as close to the recommended limit as possible. This can be best accomplished by automating the continuous blowdown. F1gure 15.1 shows how blowdown losses can be reduced by increasing cycles of concentration of boiler water.

Table 15.1—Effect of inadequate or improper water conditioning.

Effect	Problem	Remarks
Scale	Silica	Forms a hard glassy coating on internal surfaces of boiler. Vaporizes in high pressure boilers and deposits on turbine blades.
	Hardness	$CaSO_4$, $MgSO_3$, $CaCO_3$ and $MgCO_3$ form scale on boiler tubes.
Reduced heat transfer	Scale & Sludge deposits	Loss of efficiency, wasted fuel.
Corrosion	Oxygen	Causes pitting of metal in boilers and steam and condensate piping.
	Carbon Dioxide	Major cause of deterioration of condensate return lines.
	Oxygen + Carbon Dioxide	Combination is more corrosive than either by itself.
Foaming & Priming	High boiler water concentrations	Distribution system contamination, wet steam and deposits in piping, on turbine blades and valve seats.
Caustic embrittlement	High caustic concentrations	Causes intercrystalline cracking of boiler metal.
Economic losses	Repair of boilers	Repair pitted boilers and clean heavily scaled boilers.
	Outages	Reduce efficiency and capacity of plant.

Automatic control of the continuous blowdown involves the use of a continuous conductivity monitor to activate the blowdown control valve. These units can maintain boiler water conductivity within close limits and thereby minimize excessive blowdown loss, which occurs with manual control.

WATER HARDNESS

Hardness in the boiler water indicates the presence of relatively insoluble impurities. Impurities may be classed as (a) dissolved solids, (b) dissolved gasses and (c) suspended matter. Actually, in the heating and concentration of boiler water, these impurities will precipitate even more rapidly since they are even less soluble at high temperatures. Waters containing large amounts of calcium and magnesium minerals are "hard to wash with." The amount of hardness in normal water may vary from several ppm to more than 500 ppm. Because calcium and magnesium compounds are relatively insoluble in water, they tend to precipitate out, causing scale and deposit problems. The hardness of the water source is an important consideration in determining the suitability of water for steam generation.

As mentioned earlier, this process of precipitation will take place on the heat exchange surfaces and is known as scale.

CONDENSATE SYSTEM CORROSION

The most prevalent type of condensate system corrosion is caused by carbon dioxide.

Carbon dioxide enters the system with the boiler feed water in the form of bicarbonate and carbonate alkalinity. When exposed to boiler temperatures, bicarbonates and carbonates break down to form carbon dioxide. The carbon dioxide is carried away in the steam and is condensed to form carbonic acid.

THE PRIMING BOILER

Priming occurs when slugs of water enter the steam distribution system. There are a number of causes such as: high steam demand, large and sudden drops in system pressure, blowdown, quick opening valves in the distribution system and unsuitable steam nozzle and steam header size.

One way priming could happen is after an operator has brought a boiler up to operating pressure he opens the distribution header quickly. Then, within a few minutes, he notices the water bouncing up and down in the sight glass. This bouncing can become quite violent and the boiler

can shut down on low water cut off.

What has happened? The piping and heat exchange equipment serviced by the distribution header is cold and the steam is quickly condensed possibly forming a vacuum. This causes a violent boiling action in the boiler, it could even start a series of oscillations where the water surface rises or "mounds" towards the steam nozzle. The lift of the water can not be sustained and it drops starting a sloshing effect. A wave or bouncing effect can start which will allow water to enter the steam system in slugs; a dangerous condition. This condition can also upset boiler water circulation patterns.

One rule of thumb used in raising boiler water temperature is to go at a rate of 100 F per hour. Open valves slowly, allowing the system to warmup and balance out before introducing sudden surges on the system.

Deareators

Oxygen and carbon dioxide are very harmful to boiler systems. Deareators are designed to

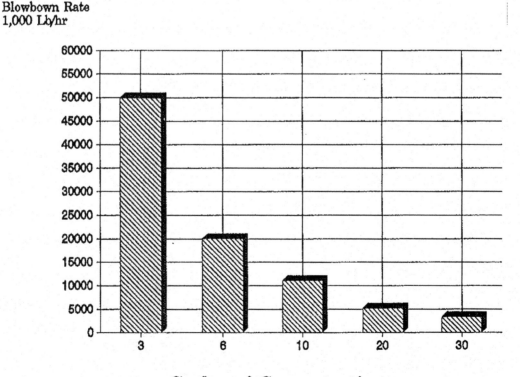

Cycles of Concentration

Figure 15.1—Blowdown is reduced dramatically when Cycles of Concentration is increased.

remove dissolved gases from the boiler feed water. They are effective and oxygen can be reduced to trace levels; about .005 ppm. While deareators are efficient, traces of oxygen can cause a significant amount of corrosion, so chemical treatment is also used.

While the deareator removes carbon dioxide from the feed water, bicarbonate and carbonate alkalinity in the boiler will produce additional carbon dioxide. This will require some water treatment in the boiler.

A deareator usually consists of a heating and deareating section. The storage section of these units are often designed to hold about 10 minutes of rated capacity of boiler feed water.

The water enters the deareator and is broken into a spray or mist and scrubbed with steam to force out the dissolved gasses.

Steam and noncondensibles into the vent condensing section where the steam is condensed. The released gases are discharged to atmosphere through the vent outlet.

Continuous Blowdown Heat Recovery

The continuous blowdown, sometimes called the surface or skimmer blowdown, is most effective in controlling the concentration in boiler water. Where continuous blowdown systems are used, the bottom blowdown is used for removal of precipitated impurities, especially those which tend to settle in the lower parts of the boiler.

Heat exchangers can be used with the continuous blowdown to recover energy from the boiler water which is being expelled from the boiler.

How pure must feedwater be?

Feedwater purity is a matter of both quantity of impurities and the nature of the impurities. Some impurities such as hardness, iron and silica for example are more concern than sodium salts. The purity requirements for any feedwater depend on how much feedwater is used as well as what the particular boiler design is. Pressure, heat transfer rate and operating equipment on the system such as turbines have a lot to do with feedwater purity.

Feedwater purity requirements can vary widely. A low pressure firetube boiler can usually tolerate higher feedwater hardness, with proper chemical treatment, while virtually all impurities must be removed from the water used with most modern high pressure watertube boilers.

WATER CARRYOVER IN STEAM

The water evaporated to produce steam should not contain any contaminating materials, however there will be water droplets carried into steam due to several processes.

MIST CARRYOVER

A fine mist is developed as water boils. This process is illustrated in Figure 15.2. A bubble of water vapor (steam) reaches the water surface and bursts, leaving a dent in the water. Liquid collapses in on the dent, with the center rising at a faster rate than the edges. This results in a small droplets being tossed free of the boiler water surface. These droplets form a fine mist. This mist is removed to a great extent in the dry portion of the boiler. However, any mist that remains entrained in the steam will have the same level of contamination as the boiler water.

Foaming Carryover

The alkalinity, TDS and suspended solids can interact to create a foam in the boiler. A light foam

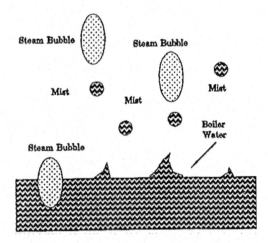

Figure 15.2—Mist formation in boiler water with high impurity levels. The volume above the water is steam, steam bubble is shown for illustration.

will reduce the problem of misting to some extent. A heavy foam layer is another source of liquid carryover into the steam. The level of foaming can normally be controlled to a reasonable level by maintaining the total alkalinity at a level less than 20 percent of the TDS and maintaining the total suspended solids at a level less than 8 percent of the TDS. Antifoam agents are added to the sodium lignosulfonate sludge dispersant to help control foaming.

PRIMING CARRYOVER

Priming carryover is caused by liquid surges into the steam drum that throw water into the steam space where it is carried into the steam header. Priming is always caused by a mechanical problem or mechanical properties such as oversensitive feedwater controls or incorrect blowdown procedures. There is no chemical control method available.

SILICA CARRYOVER

The silica in the boiler water can evaporate and enter the steam, independent of water carryover. The silica can form a deposit on turbine blades and other equipment when the steam condenses. This problem is controlled by maintaining a low silica level in the boiler water. The suggested limits are shown in Table 15.2.

Determining the Amount of Carryover

The best indication of carryover is a measurement of steam conductivity. A steam conductivity of 20 to 30 micromho indicates there is a small chance that carryover is significant.

A High Conductivity Measurement Indicates Carryover

A high conductivity measurement in the steam condensate means there is either carryover or leakage into the steam system. The hardness must also be checked in this case. If any hardness is found, then the contamination of the condensate indicated by the high conductivity is due to leakage into the condensate system rather than carryover. This is because the carryover of boiler water should be at a very low hardness due to chemical treatment or the very low makeup water requirement.

TWO KEY OPERATING CONTROLS FOR DEAREATORS

There are two key operating controls for deareators that must be watched. First, the deareator vent must be checked to see that a plume of steam is always flowing. Second, the pressure of the deareator and temperature of the outlet water must be controlled. At a given pressure the temperature should be within 2°F of the temperatures shown in Table 15.3, based on the elevation of the boiler site. If there is low or no steam flow or a low water temperature, the deareator is not operating properly.

A mixture of oxygen and water is a very corrosive combination. This corrosivity doubles with every 18°F increase in temperature.

Oxygen corrosion can be recognized by pits, typically found in the top of the steam drum or at the waterline. Oxygen can be removed from the feedwater by mechanical or chemical deareation; a combination of these methods is commonly used.

CHEMICAL REMOVAL OF OXYGEN FROM BOILER FEEDWATER

A mechanical deareator can reduce the oxygen content of feedwater to a fraction of a ppm. Complete removal requires additional chemical treatment. One process used is catalyzed sodium sulfite.

Table 15.2—Silica limits in boiler water.

Boiler Pressure (psig)	Allowable Silica (as SiO_2)
0 - 15	200
16 - 149	200
150 - 299	150
300 - 449	90
450 - 599	40
600 - 749	30
750	20

Table 15.3—Deareator water outlet temperature for boiler systems at various pressures for sea level.

Deareator Pressure (psig)	Deareator Water Outlet Temperature (°F)
0	212
1	215.3
2	218.5
3	221.5
4	224.4
5	227.1
6	229.8
7	232.2
8	234.8
9	237.1
10	239.4
11	241.6
12	244.4
13	246.4
14	248.4
15	250.3
16	252.2
17	254.1
18	255.3
19	257.0
20	258.8

Table 15.4—Boiler water sulfite levels.

Boiler (psig)	Sulfite Residual (as ppm SO_3)
0 - 15	30 - 60
16 - 149	30 - 60
150 - 299	30 - 60
300 - 449	20 - 40
450 - 599	20 - 40
600 - 749	15 - 30
750 -	15 - 30

The chemical reaction with sodium sulfite will consume 7.88 pounds of pure sodium sulfite with one pound of oxygen. In practice, about 10 pounds per pound of oxygen are added to carry a small excess of sulfite in the boiler water. The excess that should be carried is based on the boiler pressure, according to the levels shown in Table 15.4.

Higher sulfite levels can be wasteful. In addition, sulfite can break down and cause steam and condensate to become corrosive. It should be added continuously.

CONDENSATE CORROSION

Any free carbon dioxide in the feedwater should be removed by the dearator. However carbon dioxide in the combined form can enter the boiler as carbonates and bicarbonates in the feedwater. Under the influence of heat and pressure, this combined carbon dioxide will form free carbon dioxide which will leave the boiler with the steam.

Controlling CO_1 With Neutralizing Amides

Carbon dioxide in all steam boiler systems can be controlled by neutralizing with a volatile amine. This is an amine that can be added to a boiler, where it will vaporize and pass over with the steam.

Two amides usually used are morpholine and cyclohexylamine. When steam first condenses, morpholine will be present in a larger concentration. At more distant points in the distribution system, cyclohexylamine will be available in larger concentrations and can be more effective in controlling corrosion. The two chemicals are used together for best protection.

Controlling CO_1 with Filming Amides

Another way for controlling carbon dioxide corrosion is the use of filming amides such as octadecylamine. This chemical will coat the condensate pipe and prevent the carbon dioxide in the water from coming into contact with the pipe wall. These chemicals are usually used at a level from .7 to 1.0 ppm and they are difficult to handle and mix. They must be added directly to the steam header.

What is the basis for choosing between neutralizing and filming inhibitors?

The proper choice depends on the boiler system, plant layout, operating conditions and feedwater composition. In general, volatile amines are best suited to systems with low makeup, low feedwater alkalinity and good oxygen control.

Filming inhibitors usually give more economical protection in systems with high make up, air in-leakage and high feedwater alkalinity or where the system is operated intermittently. In most cases a combination of these treatments may be best to combat condensate system corrosion.

EXTERNAL WATER TREATMENT

Makeup Water

Makeup water is water added to the boiler system from an external source to replace water lost in the boiler room and in the distribution system. This includes blowdown water, steam leaks, condensate losses and steam used directly in process applications.

The usual source of makeup water is the potable water supply or what has been referred to in many cases as city water. This represents a treated water that has a predictable and uniform quality on a day-today basis. Other sources of makeup water include well water, surface water or holding ponds that are not treated to the extent that the potable water source is treated.

The uniformity of makeup water quality is important if the boiler water system is to be operated reliably.

Makeup water treatment varies on the needs of a particular installation. Various processes are used to improve makeup water quality including:

a. Lime-Soda Softening
b. Ion Exchange Process-General
c. Sodium Ion Exchange
d. Hydrogen Ion Exchange
e. Deionization
f. Dealkalization
g. Distillation
h. Reverse Osmosis
i. Electrodialysis

The makeup water is combined with the condensed steam returned from the distribution system (called condensate return) to become boiler feedwater. The feedwater is deareated to strip out noncondensible gases and treated with oxygen scavengers.

Internal Water Treatment

The removal of scale-forming materials from the boiler makeup by reducing the hardness to near zero is the best control method.

Internal treatment of boiler water refers to chemical additions required to prevent scale formation from materials not removed by makeup treatment and to prevent sludge deposits from forming due to the precipitation of these materials.

Deposit Formation

There are two basic causes of boiler deposit formation:

- Scale; the high temperatures found in boilers cause precipitation of compounds whose solubilities are inversely proportional to the solution temperature.

- Sludge; the concentration of boiler water causes certain compounds to exceed their maximum solubility at a given temperature, forcing precipitation in areas of highest concentration.

While these represent a somewhat simplified view of the mechanisms involved, they do summarize the factors essential to boiler deposits.

Scale

The build up of boiler scale constitutes a growth of boiler crystals on waterside heat transfer surfaces and is most severe in those areas of the boiler where maximum heat transfer occurs. Figure 15.3 shows the relationship between heat transfer efficiency and scale deposit thickness.

Problems caused by scaling

The steam boiler normally uses an external heat source at a much higher temperature than the boiler water. The metal tubes in the boiler are kept cool by the boiler water. When scale forms, it acts like an insulation material between the water and the metal. This results in tubes operating at higher temperatures. The greater the thickness of the scale, the greater the insulating effect, and the higher the temperature of the tubes. At sufficiently high temperatures, the tube can lose tensile strength and rupture.

Sludges

Sludges are precipitated directly in the main body of the boiler water when their solubilities are exceeded. Sludge deposition usually occurs when binders are present or when water circulation is such that it allows sludge settling on hot spots allowing it to "bake" on to hot surfaces.

TDS in a Boiler

TDS in a boiler is one of the parameters used to control the water treatment program. Dissolved solids are continually added to the boiler makeup water. These solids are not evaporated with the steam; as a result the TDS becomes more concentrated as more steam is generated.

The level to which the TDS will concentrate is determined by the amount of these salts removed in the blowdown. Control of TDS level is critical in boiler operation. The higher TDS levels result in higher boiler efficiencies, but TDS levels that are too high will interfere with boiler operation.

The Consequences of Too Little Blowdown
TDS Too High
- Corrosive to boiler metal
- Causes foaming and carryover
- Alters boiling patterns in tubes leading to deposits.

*Suspended Solids and
Sludge too High*
- Bakes onto heat transfer surfaces causing lost efficiency.

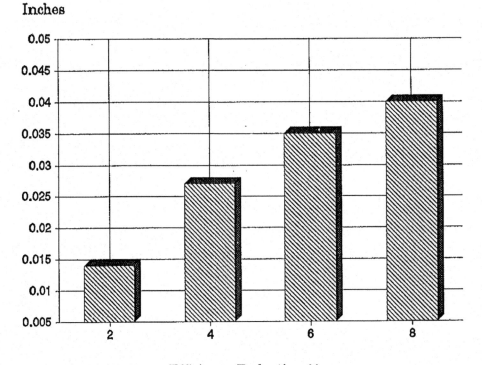

Figure 15.3—Loss of heat transfer efficiency with scale thickness.

- Alters boiling characteristic
- Dirty boilers; the high cost of cleaning and extra down time.
- Absorption of chemicals (especially P04) leading to scale and loss of efficiency.

Hardness and Salt Levels Too High
- Scale and loss of efficiency

Alkalinity Too High
- Scale and loss of efficiency

Energy Conservation through Optimum Blowdown Control

While blowdown is a key to safe, clean boiler operation, it must be remembered that blowdown water leaving the boiler carries a high level of Btus with it. Energy conservation requires maintaining the highest permissible cycles of concentration in the boiler water. To do this a margin of safety often must be sacrificed and good controls installed to insure that no damage occurs to the boiler as a result.

Many plants are able to increase their cycles of concentration and therefore reduce blowdown by reducing total solids concentration in the feedwater or altering the boiler water treatment program.

Reducing the solids content of the makeup involves a change in the plant makeup water source or altering the external water treatment program. The makeup water source can be changed or pretreatment equipment such as filters, deionizers or other equipment can be installed.

Efficient Bottom Blowdowns

Often plants will adhere to a strict schedule of bottom blowdown and still develop problems with excessive sludge buildup in the mud drum. This can be due to improper timing of blowdown periods. Experience has shown that frequent blowdowns of short duration (10-20 seconds) are more effective in removing sludge than occasional blowdowns of longer duration.

The blowdown is only effective for the first few seconds of the blowdown. Blowdowns of long duration create a great deal of turbulence in the mud drum "stirring up" the sludge level. With the sludge in suspension from this action, it can be swept up the generating tubes where it can bake onto the tube surfaces, resulting in deposits.

TDS are controlled by continuous blowdown, which is typically removed from the steam drum. Guidelines published for government plants are shown in Table 15.5.

Table 15.5—Total dissolved solids (TDS) and conductivity limits for steam boilers.

Boiler Pressure (psig)	Maximum TDS (ppm)	Maximum Conductivity ($_u$mho)
0 - 15	6,000	9,000
16 - 149	4,000	6,000
150 - 299	4,000	6,000
300 - 449	3,500	5,250
450 - 599	3,000	4,500
600 - 749	2,500	3,750
750 -	2,000	3,000

Boiler Blowdown Calculations

The rate of blowdown from a boiler is a critical operating control for TDS.

a. The water added to a boiler must equal water loss from the boiler.

$$F = E+B$$

F = feedwater lb/hr
E = Steam generated, 1b/hr
B = Blowdown, lb/hr

b. The blowdown can be related to the feedwater using cycles of concentration (COC).

$$C = FIB$$

C = Cycles of Concentration
F = Feedwater, lb/hr
B = Blowdown, lb/hr

Table 15.6 is the American Boiler Manufacturers Association (AMBA) specified limits for boiler water composition with respect to operating pressure to assure good quality steam.

Table 15.7 was developed by the American Society of Mechanical Engineers (ASME), Research Committee on Water in Thermal · Power.

It shows the need for feedwater to be extremely pure. With today's designs, heat-flux rates of 250,000 Btu/hr/sq-ft are anticipated. Combined with dimensional restrictions of modem design, this has raised the need for the limits in Table 15.7. These new guidelines are only suggested limits that will continue to be refined.

Table 15.6—American Boiler Manufacturers Association (ABMA) limits for boiler water composition.

Boiler Water Treatment

Maximum Limits for Boiler Water

Boiler Pressure (psig)	Total Solids (ppm)	Alkalinity (ppm)	Suspended Solids (ppm)	Silica (ppm)
0 - 300	3,500	700	300	125
301 - 450	3,000	600	250	90
451 - 600	2,500	500	150	50
601 - 750	2,000	400	100	35
751 - 900	1,500	300	60	20
901 - 1,000	1,250	250	40	8
101 - 1,500	1,000	200	20	2.5
1501 - 2,000	750	150	10	1.0
Over 2,000	500	100	5	0.5

Table 15.7—Boiler water limits developed by the ASME Research Committee on Water in Thermal Power Systems.

Boiler Feedwater

Drum Pressure (psig)	Iron ppm Fe	Copper ppm CU	Total Hardness ppm CaCO₃	Silica ppm SiO₂	Total Alkalinity[1] ppm CaCO₃	Specific Conductance μmho/cm
0 - 300	0.100	0.050	0.300	150	350[2]	3500
301 - 450	0.050	0.025	0.300	90	300[2]	3000
451 - 600	0.030	0.020	0.200	40	250[2]	2500
601 - 750	0.025	0.020	0.200	30	200[2]	2000
751 - 900	0.020	0.015	0.100	20	150[2]	1500
901 - 1000	0.020	0.015	0.050	8	100[2]	1000
1001 - 1500	0.010	0.010	ND[4]	2	NS[3]	150
1501 - 2000	0.010	0.010	ND[4]	1	NS[3]	100

[1]Minimum level of hydroxide alkalinity in boilers below 1000 psig must be individually specified with regard to silica solubility and other components of internal treatment.

[2]Maximum total alkalinity consistent with acceptable steam purity. If necessary, the limitation on total alkalinity should override conductance as the control parameter. If makeup is demineralized water at 600 - 1000 psig, boiler water and conductance should be as shown in the table for the 1001 - 1500 psig range.

[3]NS (Not specified) in these cases refers to free sodium - or potassium-hydroxide alkalinity. Some small variable amount of total alkalinity will be present and measurable with the assumed congruent control or volatile treatment employed at these high pressure ranges.

[4]ND is None Detectable

Index